By any standards, dogs are extraordinary animals. They have been part of human society for longer than any other domestic species. They exist in a greater variety of different shapes and sizes, and they occupy a wider ecological niche, from pampered pets and faithful servants to feral scavengers. Even our attitudes to dogs seem to oscillate between extremes – on the one hand, the dog is man's best friend, on the other, he is the despised and degraded outcast.

This unique book seeks to expose the *real* dog beneath the popular stereotypes. Its purpose is to provide a comprehensive, state-of-the-art account of the domestic dog's natural history and behaviour based on scientific and scholarly evidence rather than hearsay. Anyone with a serious interest in *Canis familiaris*, its evolution, behaviour and its place in our society, will find *The Domestic Dog* an indispensable and fascinating resource.

The domestic dog: its evolution, behaviour and interactions with people

The domestic dog

its evolution, behaviour, and interactions with people

Edited by JAMES SERPELL

Marie A. Moore Associate Professor of Humane Ethics and Animal Welfare, School of Veterinary Medicine, University of Pennsylvania

Pencil drawings by PRISCILLA BARRETT

CAMBRIDGE
UNIVERSITY PRESS

Published by the Press Syndicate of the University of Cambridge
The Pitt Building, Trumpington Street, Cambridge CB2 1RP
40 West 20th Street, New York, NY 10011-4211, USA
10 Stamford Road, Oakleigh, Melbourne 3166, Australia

First published 1995

Printed in Great Britain at the University Press, Cambridge

A catalogue record for this book is available from the British Library

Library of Congress cataloguing in publication data

The domestic dog : its evolution, behaviour, and interactions with
people / edited by James Serpell.
p. cm.
ISBN 0 521 41529 2 (hardback). – ISBN 0 521 42537 9 (paperback)
1. Dogs – Behaviour. 2. Dogs. 3. Human-animal relationships.
I. Serpell, James, 1952– .
SF433.D66 1995
599.74′442 – dc20 95–13800 CIP

ISBN 0 521 41529 2 hardback
ISBN 0 521 42537 9 paperback

WV

To Jacqui and Oscar: for persevering

Contents

Contributors

Giorgio Andreoli, *Agriconsulting SpA, Via L. Luciani 41, Roma, Italy*

Luigi Boitani, *Department of Animal & Human Biology, Universita di Roma 'La Sapienza', Viale Universita 32, 00185 Roma, Italy*

John Bradshaw, *Anthrozoology Institute, University of Southampton, Medical and Biological Sciences Building, Bassett Crescent East, Southampton SO9 3TU, UK*

Geoff Carr, *Wildlife Conservation Research Unit, Department of Zoology, University of Oxford, South Parks Road, Oxford OX1 3PS, UK*

Paolo Ciucci, *Via del Serafico 74, Roma, Italy*

Juliet Clutton-Brock, *c/o Zoology Department, The Natural History Museum, Cromwell Road, London SW7 5BD, UK*

Raymond Coppinger, *School of Cognitive Science & Cultural Studies, Hampshire College, Amherst, MA 01002, USA*

Francesco Francisci, *Via Capo di Mondo, 1, Firenze, Italy*

Benjamin Hart, *Department of Physiological Sciences, School of Veterinary Medicine, University of California, Davis, CA 95616, USA*

Lynette Hart, *Center for Animals & Society, School of Veterinary Medicine, University of California, Davis, CA 95616, USA*

Robert Hubrecht, *Universities Federation for Animal Welfare, 8 Hamilton Close, South Mimms, Potter's Bar, Herts. EN6 3QD, UK*

Andrew Jagoe, *Sambrook & Partners, Barrow Hill House, Maidstone Road, Ashford, Kent TN24 8TY, UK.*

Randall Lockwood, *Humane Society of the United States, 2100 L St. NW, Washington DC 20037, USA*

David Macdonald, *Wildlife Conservation Research Unit, Department of Zoology, University of Oxford, South Parks Road, Oxford OX1 3PS, UK*

Roger Mugford, *The Animal Behaviour Centre, P.O. Box 23, Chertsey, Surrey KT16 0PU, UK*

Helen Nott, *Waltham Centre for Pet Nutrition, Freeby Lane, Waltham-on-the-Wolds, Melton Mowbray, Leics. LE14 4RT, UK*

Valerie O'Farrell, *Royal (Dick) School of Veterinary Studies, Summerhall, Edinburgh EH9 1QH, UK*

Richard Schneider, *Department of Zoology, Box 90325, Duke University, Durham, NC 27708, USA*

James Serpell, *Department of Clinical Studies, School of Veterinary Medicine, 3850 Spruce Street, Philadelphia, PA 19104–6010, USA*

Chris Thorne, *Waltham Centre for Pet Nutrition, Freeby Lane, Waltham-on-the-Wolds, Melton Mowbray, Leics. LE14 4RT, UK*

Malcolm Willis, *Faculty of Agriculture & Biological Sciences, The University, Newcastle upon Tyne, NE1 7RU, UK*

1 Introduction

JAMES SERPELL

People's opinions about the domestic dog have a tendency to veer towards extremes. For an increasingly large sector of the population, the dog is now perceived as a dangerous and dirty animal with few redeeming qualities: a source of vicious and unprovoked assaults on children, fatal or debilitating disease risks, and unacceptable levels of organic pollution in our streets and public parks – a veritable menace to society. Recent levels of media attention given to dog attacks on people, or the dangers of dog-borne parasites and diseases, may be out of proportion to the actual risks they pose to the average individual, but they nevertheless reflect, and to some degree foster, a widespread, latent antipathy for *Canis familiaris* that should not be underestimated. Throughout the Western world, this anti-dog feeling has resulted recently in a spate of increasingly restrictive and draconian regulations to curb the activities of dogs and their owners, including legal bans on certain breeds, or even the compulsory execution of any dog of a particular type whose owner makes the mistake of allowing it to appear unmuzzled in a public place.

At the other end of the spectrum, an even larger constituency of dog lovers exists for whom this animal has become the archetype of affectionate fidelity and unconditional love. To the members of this group, dogs are more human than animal. They are given human-sounding names, like George or Mary, they are spoken to and treated like junior family members, and most of the time they are unconsciously assumed to have virtually the same thoughts, feelings and desires as people. Anthropomorphism is ubiquitous in popular 'doggy' literature, not only where one might expect it, in fictional accounts of canine exploits, but also in ostensibly factual material. Even the definitive temperament 'standards' for the different breeds are often couched in highly anthropomorphic language. The official American breed standard for the Chinese shar-pei, for instance, describes the animal as 'regal, alert, intelligent, dignified, lordly, scowling, sober and snobbish', while the ideal Rottweiler is said to be 'calm, confident and courageous . . . with a self-assured aloofness that does not lend itself to immediate and indiscriminate friendships' (American Kennel Club, 1992). Although biologically inaccurate, to say the least, this overwhelming tendency to 'personify' dogs is an inevitable and natural consequence of the kinds of relationships we have with these animals. The dog–human relationship is arguably the closest we humans can ever get to establishing a dialogue with another sentient life-form, so it is not surprising that people tend to emerge from such encounters with a special sense of affinity for 'man's best friend' (Serpell, 1986). To people of this persuasion, the idea of banning dogs, or restricting their access to public areas, is nearly equivalent to the notion of banning children or preventing youngsters from playing in parks.

To some extent, both of these polarized and misinformed attitudes continue to persist and cause problems because of a general shortage of objective and reliable scientific information about the domestic dog. When we consider how long dogs have been an integral part of human society, let alone the practical and emotional impact they have had on countless human lives, it is surprising how little we know about *C. familiaris*. The lives and loves of wolves, coyotes, jackals and most other wild canids have been studied in meticulous detail, but, with one or two notable exceptions (e.g. Lorenz, 1954; Scott & Fuller, 1965; Fox, 1978), the domestic dog has been largely ignored by scientists, except when it has become a 'problem', or when it has been used as a substitute for humans in biomedical and psychological research. This apparent lack of basic scientific interest in dogs is partly due to the 'stigma' of domestication. Despite the fact that Darwin clearly derived much of his theory of evolution by natural selection from studying the variety of *domestic* species, such as dogs (Darwin, 1868), most modern biologists and behavioural scientists seem to regard domestic animals as 'unnatural' and therefore unworthy or unsuitable as subjects for serious scientific investigation (Clutton-Brock, 1994). According to this stereotype, the domestic dog is essentially a debased and corrupted wolf, an abnormal and therefore uninteresting artifact of human design, rather than a unique biological species (or superspecies) in its own right, with its own complex and fascinating evolutionary history.

The present volume arose from a desire to rediscover the animal that lies partially hidden beneath these layers of misunderstanding, misinterpretation and mythology. Its purpose is to provide a comprehensive and up-to-date resource of objective and accurate information about the domestic dog and its place in our society, rather than the usual mixture of anecdotal observations, subjective impressions, and unsubstantiated theories that permeates so much of the existing canine literature. Each of the volume's contributors is, or was until recently, actively involved in research on some aspect of dogs and/or their behaviour, and many of the chapters contain original findings that have not been published elsewhere. This book is therefore aimed at people who wish to move beyond the conventional images of the dog embodied in popular stereotypes towards a closer understanding of this remarkable species based on the available scientific evidence. I would like to think that the material contained in this volume may also serve to stimulate further research and study, and I certainly hope that it will help to inform and moderate the ongoing, and frequently acrimonious, public debate concerning the pros and cons of dogs in our society.

For convenience, the book has been divided into three parts. Part I addresses two fundamental questions: where did the domestic dog come from? And how, in evolutionary terms, did it get to where it is today? The first of these issues is examined in Chapter 2, which provides a critical review of the archaeological and subfossil evidence linking Palaeolithic wolves and humans, and the probable processes of wolf domestication. Chapter 3 then describes and discusses the possible evolutionary mechanisms underlying the transformation of these early wolf dogs into some of the working breeds we see today, each with its own distinctive behaviour and morphology.

The theme of breed differences is resumed in Part II, which is devoted to the topic of domestic dog behaviour and behaviour problems. Chapter 4 looks at the genetics and inheritance of working ability in various breeds, while Chapter 5 explores methods of identifying and quantifying breed and gender differences in behaviour. In Chapter 6 the extensive literature on behavioural development in dogs is reviewed in the light of new evidence concerning the long-term effects of early experience, while Chapter 7 describes the specific effects of early experience and inherited predispositions on the development of canine feeding preferences. In Chapter 8 the gulf between wolves and dogs, and between the various domestic breeds, is further emphasized by focusing on important but frequently overlooked differences in social and communicatory behaviour. Chapter 9 presents a timely review of the literature on canine aggression, and the reasons why dogs sometimes bite people; Chapter 10 describes modern approaches to the diagnosis and treatment of behaviour problems in general; and Chapter 11 discusses the extent to which owner personality and attitudes can contribute to the development of canine behaviour problems.

Part III focuses on the dog's place in human society. Chapter 12 explores the remarkable physical and psychosocial benefits that humans appear to derive from canine companionship. In contrast, Chapter 13 describes the many welfare problems confronting dogs in their various relationships with human beings. One of the most tenuous of these relationships is explored in Chapters 14 and 15, both of which describe studies of the behaviour and ecology of feral dogs living as outcasts on the fringes of human society. Chapter 16 highlights the surprising degree of ambivalence that characterizes both our attitudes towards, and our relationships with, the domestic dog, despite its extraordinary contribution to human emotional and material welfare during the last 12 000 years. The final Chapter 17 concludes with a brief overview of some of the more important gaps in our knowledge of the domestic dog and its relationships with people.

This book was conceived at a conference of the same name hosted by the Companion Animal Research Group in Cambridge in 1991. I would like to thank Alan Crowden for suggesting the original idea, Donald Broom for arranging the conference facilities at St Catherine's College, Cambridge, and Elizabeth Paul and Caroline York for their invaluable organizational assistance. I also wish to express my gratitude to the Waltham Centre for Pet Nutrition and to Spillers Foods who jointly provided the funding for this meeting.

Each of the contributions to this book has been subjected to careful peer review prior to publication. In this regard, I wish to thank Marc Bekoff, Sam Berry, Luigi Boitani, Jacqueline Bowman, John Bradshaw, Donald Broom, Ray Coppinger, Ben and Lynette Hart, Robert Hubrecht, Elizabeth Jackson, Randall Lockwood, Rosie Luff, David Macdonald, Aubrey Manning, Roger Mugford, Helen Nott, Valerie O'Farrell, Elizabeth Paul, Anthony Podberscek, Harriet Ritvo and Dennis Turner for their helpful comments and suggestions on one or more manuscripts. Finally, I want to thank all of the contributors, as well as Tracey Sanderson at CUP, for their patience and forbearance during *The Domestic Dog*'s long and painful gestation.

References

American Kennel Club (1992). *The Complete Dog Book*, 18th edn. New York: Howell Book House.

Clutton-Brock, J. (1994). The unnatural world: behavioural aspects of humans and animals in the process of domestication. In *Animals and Human Society: Changing Perspectives*, ed. A. Manning & J. A. Serpell, pp. 23–35. London: Routledge.

Darwin, C. (1868). *The Variation of Animals and Plants under Domestication*, 2 vols. London: John Murray.

Fox, M. W. (1978). *The Dog: Its Domestication and Behaviour*. New York: Garland STPM Press.

Lorenz, K. (1954). *Man Meets Dog*. London: Methuen.

Scott, J. P. & Fuller, J. L. (1965). *Genetics and the Social Behaviour of the Dog*. Chicago: University of Chicago Press.

Serpell, J. A. (1986). *In the Company of Animals*. Oxford: Basil Blackwell.

I Domestication and evolution

2 Origins of the dog: domestication and early history

JULIET CLUTTON-BROCK

The dog's place in nature

The dog family or Canidae is a biologically cohesive group of carnivores that is divided into thirty-eight species, including the domestic dog (see Table 2.1). All wild canids are terrestrial, fast-running and mostly nocturnal, and they all have their young in burrows or dens. They may be solitary hunters like the fox or social like the wolf, jackal and coyote. All canids communicate with each other by means of facial expressions, body postures, tail-wagging and vocalizations such as howling and yelping (see Bradshaw & Nott, Chapter 7). With the introduction by humans of the dog and fox to Australasia, canids now inhabit almost every part of the world except Antarctica and some oceanic islands.

The dog, *Canis familiaris*, is the only member of the Canidae that can be said to be fully domesticated, although the red fox, *Vulpes vulpes*, and the raccoon dog, *Nyctereutes procyonoides*, have been bred in captivity for their fur. A description by Hamilton Smith (1839) of the domestication of the South American culpeo (*Dusicyon culpaeus*) is probably an indication that other species of canid have been tamed and even bred in captivity from time to time. It is also probable that the extinct Falkland Islands 'wolf', *Dusicyon australis*, descended from captive animals taken to the Falkland Islands by people in the early Holocene (Clutton-Brock, 1977).

Since before Darwin there has been much discussion concerning whether the domestic dog originated from the wolf, *Canis lupus*, or the golden jackal, *Canis aureus*. In 1787 John Hunter proposed that, since the dog produces fertile hybrids with both the wolf and the jackal, these three canids should be considered a single species. Linnaeus (1758), on the contrary, considered the dog to be a separate species, distinguished by its upturned tail (*cauda recurvata*), a characteristic found in no other canid. Darwin (1868) wrote as follows:

> The chief point of interest is whether the numerous domesticated varieties of the dog have descended from a single wild species or from several. Some authors believe that all have descended from the wolf, or from the jackal, or from an unknown and extinct species. Others again believe, and this of late has been the favourite tenet, that they have descended from several species, extinct and recent, more or less commingled together. We shall probably never be able to ascertain their origin with certainty.

Today, however, we are closer to this certainty. The combined results of studies of behaviour (Scott & Fuller, 1965; Fox, 1971, 1975; Hall & Sharp, 1978; Zimen, 1981), vocalizations (Lorenz, 1975; Zimen, 1981), morphology (Wayne, 1986*a*, *b*, *c*; Hemmer, 1990) and molecular biology (Wayne & O'Brien, 1987; Wayne, Nash & O'Brien, 1987) all indicate that the principal, if not the only, ancestor of the dog is the wolf, *Canis lupus*. In the 1950s, Konrad Lorenz popularized the idea that some modern breeds of dog were descended from the wolf, but that others were derived from the jackal (Lorenz, 1954). However, when he became aware of the complicated repertoire of howling found in the jackal which is quite unlike that of the dog or wolf, Lorenz rescinded this view (Lorenz, 1975).

The origin of the Australian dingo is another important question which has been hotly debated. From the results of anatomical (e.g. Newsome, Corbett & Carpenter, 1980) and molecular investigations (Clark, Ryan & Czuppon, 1975; Shaughnessy, Newsome & Corbett, 1975), it is evident that the dingo is a feral dog of ancient origin, and that it is closely related to both the pariah dogs of South-east Asia and to the wolf.

Precursors of the dog

From as early as the Middle Pleistocene period, the bones of wolves have been found in association with those of early hominids. Examples include the site of Zhoukoudian in North China, dated at 300 000 years BP (before present) (Olsen, 1985), the cave of Lazeret near Nice in the south of France, dated at 150 000 years BP (de Lumley, 1969), and the 400 000 year old site of Boxgrove in Kent, England (S. Parfitt, personal communication). As these associations demonstrate, the sites of occupation and hunting activities of humans and wolves must often have overlapped, perhaps as so poignantly envisaged by Elizabeth Marshall Thomas in her novel, *Reindeer Moon* (1987). Human hunters probably killed wolves occasionally and used their skins for clothing. Sometimes they would carry around a live pup that would often be eaten, but which now and then would

Table 2.1. *The thirty-eight species classified within the family Canidae are listed below, using the taxonomy of Clutton-Brock, Corbet & Hills (1976) and Ginsberg & Macdonald (1990)*

Species	Distribution
Canis lupus, wolf	Europe, Asia, N. America, Arctic
Canis familiaris, dog	Worldwide
Canis familiaris dingo, dingo	Australia
Canis rufus, red wolf[a]	Central N. America
Canis latrans, coyote	N. America
Canis aureus, golden jackal	SE Europe, N. Africa, S. Asia
Canis mesomelas, black-backed jackal	Africa south of the Sahara
Canis adustus, side-striped jackal	Africa south of the Sahara
Canis simensis, Simien jackal	Mountains of Ethiopia
Alopex lagopus, arctic fox	Arctic
Vulpes vulpes, red fox	Europe, N. Africa, Asia, N. America
Vulpes corsac, corsac fox	Central Asia
Vulpus ferrilata, Tibetan fox	Tibetan plateau
Vulpus bengalaensis, Bengal fox	India
Vulpus cana, Blanford's fox	SW Asia
Vulpus rueppelli, Rüppell's fox	N. Africa, SW Asia
Vulpus pallida, pale fox	Sahel
Vulpus chama, Cape fox	S. Africa
Vulpus velox, swift or kit fox	N. America
Fennecus zerda, fennec fox	N. Africa, Arabia
Urocyon cinereoargenteus, grey fox	N. America, northern S. America
Urocyon littoralis, island grey fox	Islands of California
Nyctereutes procyonoides, raccoon dog	E. Asia
Dusicyon australis, Falkland Is. 'wolf'	Falkland Islands, extinct since *c.* 1880
Dusicyon culpaeus, culpeo	S. America – Patagonian subregion
Dusicyon culpaeolus, Santa Elena zorro	Uruguay
Dusicyon gymnocercus, Azarra's zorro	E. Patagonian subregion
Dusicyon inca, Peruvian zorro	Mountains of Peru
Dusicyon griseus, grey zorro	SW Patagonian subregion
Dusicyon fulvipes, Chiloe zorro	Island of Chiloe
Dusicyon sechurae, Sechuran zorro	NW Peru, Ecuador
Dusicyon vetulus, hoary zorro	Brazil
Cerdocyon thous, crab-eating zorro	S. America – Brazil subregion
Atelocynus microtis, small-eared zorro	Central S. America – Brazil
Chrysocyon brachyurus, maned wolf	Southern Brazilian subregion
Speothos venaticus, bush dog	S. America – Brazilian subregion
Lycaon pictus, African wild dog	Africa south of the Sahara
Cuon alpinus, dhole	E. and Central Asia
Octocyon megalotis, bat-eared fox	Africa south of the Sahara

[a]Until recently many taxonomists believed that the red wolf should be classified in a separate species, *Canis rufus*, from the grey wolf, *Canis lupus*. It is possible that in the past there was a distinct species of wolf inhabiting the southern states of North America. It is evident, however, from analysis of mitochondrial DNA from blood samples and skins of red wolves, going back to 1905, that all the wolves tested were hybrids between grey wolves and coyotes (Wayne & Jenks, 1991).

become habituated to the family group and be tamed. Some wolf pups, as they matured, would have become less submissive and would no doubt have been killed or driven away. A few, however, would have remained with the humans and bred with other tamed wolves that scavenged around the settlement.

These tamed wolves were many generations away from the true domesticated dog, but they were its precursors. When the remains of such animals are found at late glacial sites (Magdalenian, *c.* 14 000 years BP), such as those recorded in Central Europe by Musil (1984), they show slight morphological differences from those of wild wolves. Musil described the finding of a canid maxilla with a shortened facial region and compacted teeth, as well as metapodial and toe bones that were more slender than those of wolves. Unfortunately his report does not include measurements.

Wolf skulls from a later period, towards the end of the last Ice Age around 10 000 years BP, have been retrieved by dredging at Fairbanks, Alaska. Because of their shortened facial regions these skulls may represent the remains of tamed wolves (Olsen, 1985). This is not improbable since humans are known to have crossed the Bering Straits into North America at least by this date.

The first dogs

Archaeological evidence indicates that the dog was the first species of animal to be domesticated and that this occurred towards the end of the last Ice Age when all human subsistence still depended on hunting, gathering and foraging. At present, the earliest find of a domesticated dog consists of a mandible from a late Paleolithic grave at Oberkassel in Germany (Nobis, 1979). It is dated at 14 000 years BP, 2000 years earlier than the sites in western Asia from where a cluster of canid remains has been identified as belonging to *Canis familiaris*. These sites belong to the cultural period known as the Epipaleolithic or Natufian and they are characterized by a dramatic change in hunting strategy. During the Paleolithic period, animals were killed by direct impact from heavy stone axes. During the Natufian, and the corresponding Mesolithic period of Europe, arrows armed with tiny stone blades called microliths came into widespread use. The success of these long-distance projectiles would have been enhanced by the new partnership with dogs who could help to track down and bring to bay wounded animals. This co-operative hunting technique would thus have resulted in greater hunting efficiency.

In western Asia the wild wolf, *Canis lupus arabs*, is the smallest of the subspecies that range over the northern hemisphere (Fig. 2.1). This makes the identification of fragments of canid bone from archaeological sites extremely difficult because there is very little reduction in size from the wolf to the dog (Harrison, 1973). For this reason the diagnosis of 'dog' based on cultural evidence, as can be done at the site of Ein Mallaha, is of special value. This Natufian site is near the Huleh lake in the upper Jordan valley in Israel, and its most important find was perhaps the skeleton of a puppy that had been buried with a human (Davis & Valla, 1978).

The site is dated at 12 000 years BP and its inhabitants were hunter–gatherers who were on the verge of becoming agriculturalists. They lived in round stone dwellings, used basalt pestles and mortars for grinding cereals, and buried their dead in stone-covered tombs. In one of these, at the entrance to a dwelling, the skeleton of an elderly human was found together with a puppy of between four and five months of age. The human skeleton lay on its right side, in a flexed position, with its hand on the thorax of the puppy (Fig. 2.2). The animal was too late to

Fig. 2.1. A male Arabian wolf, *Canis lupus arabs*. Photographed in Israel. Copyright Eyal Bartov, Oxford Scientific Films; reproduced with permission.

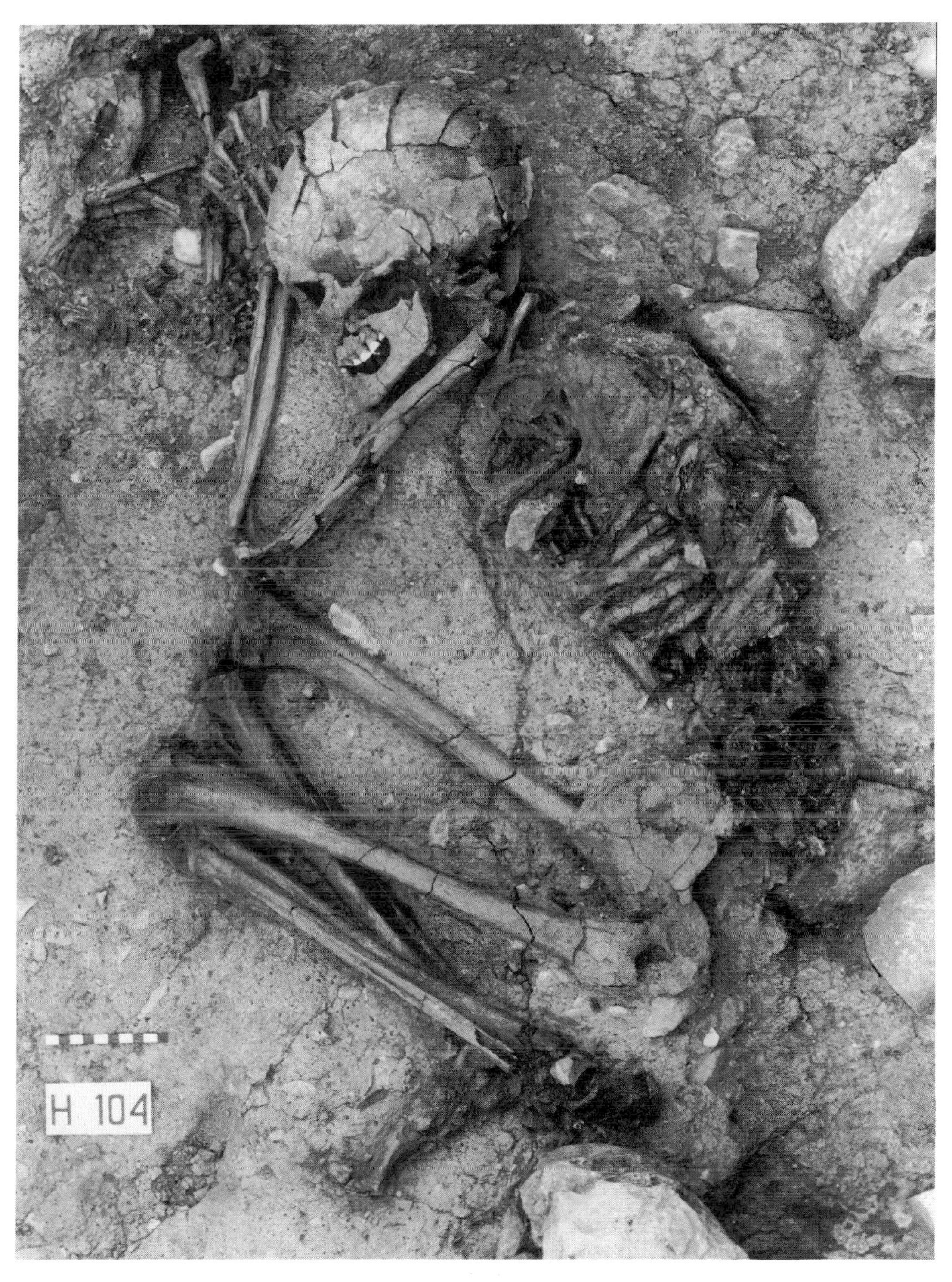

Fig. 2.2. Burial of a human with a puppy from the Natufian site of Ein Mallaha, Israel. Copyright Simon Davis; reproduced with permission.

have been a jackal and was probably a tamed wolf or 'dog'.

Since the finding of this human and canid burial at Ein Mallaha, another Natufian burial has been excavated at thc cave of Hayonim, Israel, this time of a man with two adult canids that have been identified by E. Tchernov as dogs (Valla, 1990).

During the 1930s a small number of canid skulls from Natufian sites in Palestine were identified as dog by Bate, the most notable being from the cave of Wady el-Mughara at Mount Carmel (Garrod & Bate, 1937). Thirty years later, in the 1960s, the skulls were re-examined and their domestic status was questioned. Measurements showed that they were very similar to the skulls of modern Arabian wolves, except that they were slightly smaller and rather wide in the palate (Clutton-Brock, 1962). I now believe, however, that Bate's supposition that these Natufian canids were at least tamed wolves was correct and is corroborated by the burials of canids with humans at Ein Mallaha and Hayonim. Contemporary and later sites in western Asia have also yielded canid remains, comprising jaws with compacted teeth, of the same small size as the Mount Carmel skull (see Lawrence, 1966; Clutton-Brock, 1979; Olsen, 1985). Notable amongst these is a small mandible with compacted teeth from the site of Palegawra in Iraq, dated at around 12 000 years BP (Turnbull & Reed, 1974). These finds appear to substantiate Bate's claim to have identified the very early remains of domestic dogs, especially when the small size of the auditory bulla of the skull from Mount Carmel is taken into account. (As Lawrence & Reed (1983) noted, reduction in size of the auditory bullae is a characteristic of the domestic dog.) Unfortunately, a metrical analysis of all these materials still remains to be carried out.

From the following prehistoric period, between 9000 and 7000 BP, there are a large number of dog remains from many parts of the world, and in these, the dog-like features of the skull and teeth are more developed. One hundred and thirteen fragments of skulls, teeth and skeletal bones have been identified as those of large domestic dogs from the early Neolithic site of Jarmo in Iraq (9250–7750 BP) by Lawrence & Reed (1983). In China there are dogs from 7000 years ago (Olsen, 1985) and in South America there are dogs from deposits in Fell's Cave in the southern tip of Chile, dated to 8500–6500 BP (Clutton-Brock, 1988). In this context, the status of the mysterious canid that lived wild on the Falkland Islands until the 1880s should be questioned.

The so-called Falkland Islands 'wolf', *Dusicyon australis*, bore little similarity to the mainland species of South American foxes belonging to this genus, such as the culpeo, *Dusicyon culpaeus*. It was the only wild mammal on East and West Falkland Islands until the arrival of Europeans, and it fed on birds. Unlike the mainland species of fox, which are predominantly grey in colour with long slender skulls and black tips to their tails, the Falkland Islands 'wolf' had a tan and grey pelt with white on the lower limbs and a white tip to the tail (Fig. 2.3). Anatomically, this canid bore a strong resemblance to the dingo, except in characters of the skull sutures and in the lower carnassial tooth that is remarkably trenchant. The facial region of the skull is wide and the frontal sinuses are expanded, both of which can be features of domestication (see below). It is therefore possible that this canid was taken to the Falkland Islands in the early Holocene as a domestic animal, and subsequently survived as a feral species (Clutton-Brock, Corbet & Hills, 1976; Clutton-Brock, 1977).

Regrettably, the Falkland Islands 'wolf' was exterminated by fur traders at the end of the last century, and all that is left of this intriguing canid are eleven skulls and a few skins held in museums. Recently, sequencing of mitochondrial DNA has been carried out on a sample from one of the two skins in the Natural History Museum, London, that were collected by Darwin. Preliminary results indicated that, in terms of genetic distance, the Falkland islands 'wolf' lies closest

Fig. 2.3. Engraving of the Falkland Island 'wolf', *Dusicyon australis*. From Hamilton Smith (1839).

to the coyote, *Canis latrans*, and was genetically distant from the South American 'foxes' belonging to the genus *Dusicyon* (Wayne, personal communication). This result supports the morphological evidence that the Falkland Islands 'wolf' was a species of *Canis*. Since no member of this genus is known from South America, it is hard to envisage how it reached the islands unless taken there by humans.

Until 1987, the record for the earliest dogs in North America came from Jaguar Cave in Idaho (Lawrence, 1968). The deposits in the cave were dated by radiocarbon at 10 000 BP. In 1987, however, radiocarbon accelerator dates obtained on two of the dog bones revealed that they were recent intrusions. The dates are 3220 ± 80 BP [OxA-922[1], maxilla MCZ 51776] and 940 ± 80 BP [OxA-923, mandible MCZ 52292]. These dates do not negate other archaeological evidence that humans, living in North America during the early Holocene, possessed domestic dogs. They do, however, show how necessary it is to obtain direct radiocarbon dates on key finds of subfossil animal remains, especially when they come from caves.

An increasing number of early dog-finds are being excavated from prehistoric sites in Europe. One of the first, which is also one of the earliest in date, is the skull of a five-month-old dog (Fig. 2.4) found at the famous Mesolithic site of Star Carr, about 15 km inland from the coast of Yorkshire in England (dated to 9559 ± 210 BP [Q-14] and 9490 ± 350 BP [C-35]). It was excavated in the 1950s along with the complete skull of a large wolf (Fig. 2.5; Fraser & King, 1954). More recently, there has been a remarkable sequel to this find. In 1985, during excavations at the nearby site of Seamer Carr, the neck vertebrae of a dog were retrieved that match in age and size the skull of the Star Carr dog (Clutton-Brock & Noe-Nygaard, 1990). A fragment of one of these vertebrae has provided a radiocarbon accelerator date of 9940 ± 110 BP [Oxa-1030].

It is conceivable that the skull and neck came from one individual dog, but they could also be from two dogs from the same litter or from unrelated dogs of the same size and age. At this very early period the number of dogs in Mesolithic Britain must have been extremely small and they were probably very inbred,

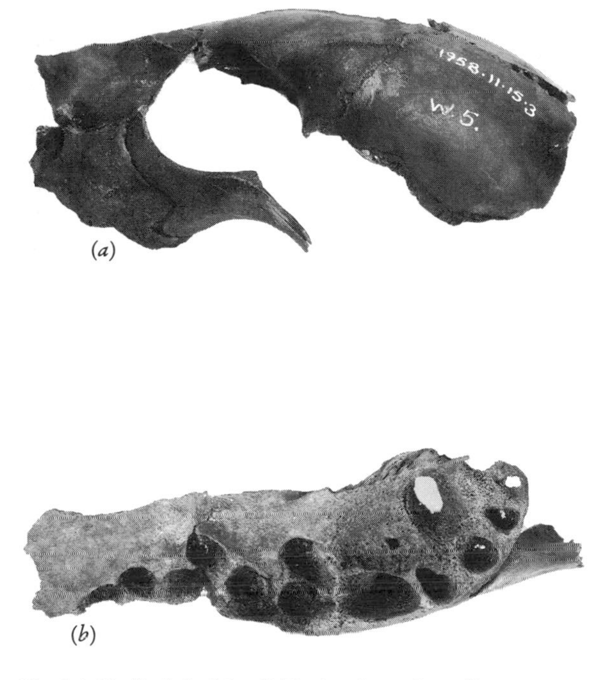

Fig. 2.4. Skull of the Mesolithic dog from Starr Carr, Yorkshire, England. (*a*) Left side of partial skull. (*b*) Palatal view of right maxilla of the same skull, showing the crowded alveoli for the permanent teeth. Natural History Museum, London.

so there would have been relatively little variation in size amongst individuals.

Another remarkable feature of the dog vertebrae from Seamer Carr is that two samples of bone yielded stable carbon isotope ratios of −14.67% and −16.97%. These ratios reveal that the dog obtained a significant part of its food from marine fish. Clutton-Brock and Noe-Nygaard (1990) have therefore postulated that the sites of Star Carr and Seamer Carr were hunting camps that were visited by people who lived for much of the year nearer to the coast and obtained most of their food by fishing.

The skull of a fully adult dog very similar in size and proportions to the Star Carr specimen has been excavated from another Mesolithic wetland site in Germany at Bedburg-Königshoven (Street, 1989). The similarity in size and osteological characteristics of most of the remains of domestic dogs found on prehistoric sites in many different parts of the world may indicate that a small population of dogs diffused from a founder group in the early prehistoric period. The skull of the dog from Bedburg-Königshoven, for example, is closer

[1] This is the reference number for the radiocarbon date, with the prefix indicating the laboratory where the determination was obtained.

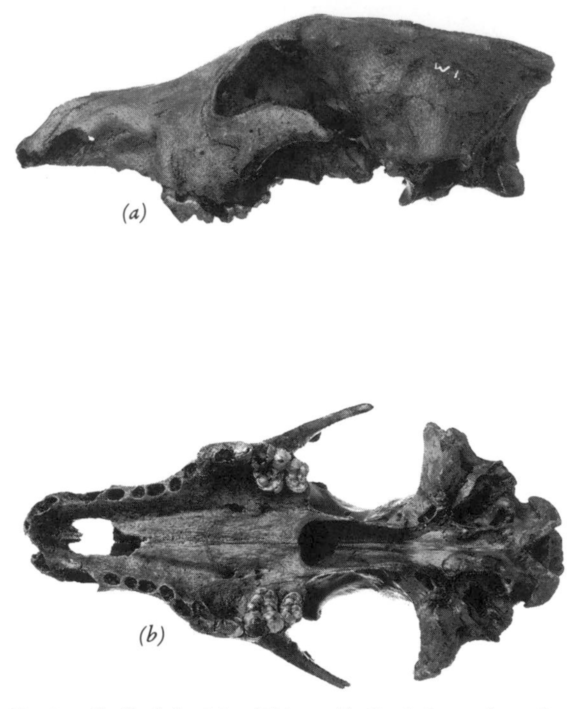

Fig. 2.5. Skull of the Mesolithic wolf, *Canis lupus*, from Star Carr, Yorkshire, England. (*a*) Left side of skull. (*b*) Palatal view. Natural History Museum, London.

in size and morphology to the remains of the earliest dogs from western Asia than it is to the very large European wolves, as exemplified by the Mesolithic wolf skull from Star Carr (Fig. 2.5). However, wolves must have been tamed and lived around human settlements in many parts of the world, and a single litter of puppies from any one of these could have provided the foundation stock for a large population of domestic dogs that subsequently became very widespread.

Remains of dogs have been found with those of the extinct Japanese wolf, *Canis lupus hodophilax*, from the rock shelter site of Tocibara in Japan, dating at around 8000 BP (Miyao *et al.*, 1984). Olsen (1977) considered that the small Chinese wolf, *Canis lupus chanco*, was the ancestor, not only of early Chinese dogs, but also of those that moved with the early human immigrants across the Bering Straits into North America. At a later period, the dogs of the Inuit and native North Americans were certainly interbred with wolves and sometimes perhaps with coyotes, while in Africa the crossing of dogs with the four species of jackal cannot be entirely ruled out (Hemmer, 1990).

Companions to more recent hunter–gatherers

The remains of dogs from archaeological sites help to provide information on the hunting practices and way of life of hunting peoples before the beginnings of agriculture. There are, however, dogs living today that may be direct descendants of these earliest hunting partners and that are still a very important element in the social and economic life of hunter gatherers. Lee (1979) discussed the use of dogs for hunting by the !Kung San of the Kalahari Desert in southern Africa. Between a third and three-quarters of the animals killed during his study period were obtained with the help of dogs that were often highly trained. Lee (1979, p. 143) described the typical !Kung dog as 'a small animal (50 cm tall at the shoulder) of undistinguished appearance. It is a short-haired breed varying in color from all black to all buff with many piebald forms'.

At present, there is no archaeological evidence of domestic dogs from sub-Saharan Africa before the first use of iron around AD 500. Therefore, it may be that the dogs of the !Kung San are a relatively recent addition to their way of life. However, the dingo in Australia and the 'singing dog' in New Guinea represent, in ever-dwindling populations, living, purebred relics of the first domestic dogs to inhabit southern and eastern Asia in the early prehistoric period.

Archaeological evidence indicates that humans first reached Australia more than 40 000 years ago. The earliest dingoes arrived less than 12 000 years ago; a deduction based on the fact that there are no remains of dogs from Tasmania, which became geographically isolated from mainland Australia by the formation of the Bass Strait at about this time. Significantly, the Australian Aborigines never acquired domestic pigs, and this would suggest that the dog was taken to the continent before the earliest domestication of the pig. In the later prehistoric period, domesticated pigs became widespread over the whole of Southeast Asia, including New Guinea and the Pacific Isles (Groves, 1981).

In skeletal anatomy, the dingo closely resembles the small wolf of India, *Canis lupus pallipes*, and the pariah dogs of Southeast Asia. It is probable, therefore, that the dingo is a direct descendant of dogs that were originally domesticated from tamed Indian wolves (Corbett, 1985). After being taken to Australia, probably in very small numbers – if not as the proverbial pregnant female – they became feral, spread rapidly, and have lived as part of the wild mammal fauna ever since. The earliest radiocarbon date obtained for dog remains from Australia is 3450 ± 95 years before present (Milham & Thompson, 1976).

For thousands of years, on the two continents of Australia and North America, the native peoples had dogs but no other domestic mammals until the arrival of Europeans. Although the dog was never used for traction in Australia, as it was by the native Americans, it was greatly valued by the Aborigines as a hunting partner, companion, bedwarmer and occasional item of food.

Early accounts by Europeans of the Aborigines and the dingo describe a relationship that was probably not dissimilar to that of hunter–gatherers and wolves all over Eurasia some 12 000 years ago, in the pre-agricultural period. The great majority of dingoes in the nineteenth century AD lived and hunted as wild carnivores and may have been as widespread as wild wolves in the early Holocene of Eurasia and North America. Aboriginal families kept some dingoes as pets, used some as hunting partners, ate them when meat was scarce and also showered them with affection. Meggitt (1965) has described the associations between the Aborigines and the dingo. These varied from hunting the adult dogs for their tails, which were worn as headdresses, to capturing young pups which, if strong, were reared as hunting partners, or, if weak, were eaten. Meggitt quoted a comment from Lumholtz (1889, p. 179), writing about the dingo in northern Queensland: 'its master never strikes, but merely threatens it. He caresses it like a child, eats the fleas off it, and then kisses it on the snout'.

These tamed dingoes were, however, very poorly fed. Apart from being given bones, they were left to scavenge for themselves, so that a tame dingo could always be distinguished from a wild one by its poor condition (Meggitt, 1965).

Today the dingo is persecuted because it kills sheep, and it is in great danger of losing its pure-bred status through interbreeding with free-ranging European dogs (Ginsberg & Macdonald, 1990). The extermination of the dingo would be a great loss because it is part of the living heritage of hunter–gatherer culture, as well as being part of Australian history.

The biology of domestication

Domestication is the result of two interwoven processes, one biological, the other cultural (Clutton-Brock, 1992). The biological process resembles natural evolution in that the parent animals become reproductively isolated from the wild population and form a small founder group, or deme, that will at first be very inbred, and which will then undergo a process of genetic drift. Over successive generations the domestic 'species' will multiply in numbers and will be genetically changed by natural selection in response to factors in the new, human environment.

With the wolf, the cultural process of domestication began when the animal was enfolded into the social structure of the human community and it became an object of ownership. In time, these tamed wolves would have become less and less like their wild forbears because inherently variable characters, such as coat colour, carriage of the ears and tail, overall size and the proportions of the limbs would have been altered by the combined effects of artificial and natural selection. In this way, the wolf became a dog; that is, it was no longer a wild carnivore but a part of human society with physical and behavioural characteristics adapted to its economic, aesthetic or ritual functions. In the latest phase of this cultural process, the individual ownership of the dog was enforced with a collar and leash, and like any other object, it could be bought, sold or exchanged at will.

Hemmer, in his book on domestication (1990), has argued that a principal factor in the process of domestication is suppression of the animal's *Merkwelt*, translated as 'perceptual world'. This means that, whereas a high degree of perception[2] combined with quick reactions to stress are essential for the survival of an animal in the wild, the opposite characteristics of docility, lack of fear and tolerance of stress are the requirements for domestication.

Alterations in the animal's perception of its environment are brought about by hormonal changes, reduction in size of the brain, less acute sight and hearing, and the retention of juvenile characteristics and behaviour into adult life. Unconscious and conscious selection for lowered perception would have begun with the earliest domestic dogs and one of the first results was probably a change in coat colour. For, as Hemmer (1990) points out, experimental evidence indicates that coat colour is closely associated with temperament in many species of domestic animals. Small dogs with a single-coloured, pale coat may have been more manageable

[2] It might also be termed 'alertness' or 'sensitivity'.

than large wolf-like individuals. That the earliest dogs were tawny-yellow in colour is supported by the survival of this coat in the dingo, which never has the wild pelage of the wolf. Over the course of time many other features have been developed that are associated with a reduction in perception and in intra-specific communication: dropped ears reduce the sense of hearing, a tightly-curled tail reduces the dog's ability to communicate, a heavy coat curtails its speed, and hair over the eyes impairs its vision (see also Bradshaw & Nott, Chapter 7).

The immediate result of taming a wolf and changing its diet would be, within a few generations, a general reduction in the size of the body and head. This trend is observed in the remains of the earliest domestic dogs. Reduction in size is a characteristic feature of the early stages of domestication and can be seen in many different species of mammal. Alterations in the proportions of the skull and a reduction in cranial capacity are also typical (Hemmer, 1990). The small overall size of early domestic mammals was partly a result of progressive stunting caused by malnutrition from the time of conception. There would also have been strong natural selection for diminution, since small animals would have survived better on little food. Stunted growth has been demonstrated in young animals kept on an inadequate diet, as in the well-known experiments on pigs by McMeekan in the 1930s (see Hammond, 1940). Tchernov & Horwitz (1991) argued that diminution of size in the early dogs was also a response to the changed ecological regime of the domestic state and they suggested that: 'The relief of selective pressures associated with domestication set in motion a cyclical reaction of accelerated maturation, increased reproductive capacity with a tendency for litter sizes to be larger, and shortened generation time. This resulted in smaller sized, younger parents with smaller sized offspring'.

During the early stages of domestication in the dog the facial region (muzzle) became shorter and wider with consequent crowding and displacement of the cheek teeth[3]. Tooth crowding occurred because

[3] Almost no whole skulls of the earliest domestic dogs have survived, so it is not possible to investigate relative changes in the different parts of the skull. The work of Wayne (1986*a*) suggests that, while the muzzle may have widened, the length of the facial region, although shortened, would have remained in proportion to the reduced cranial length.

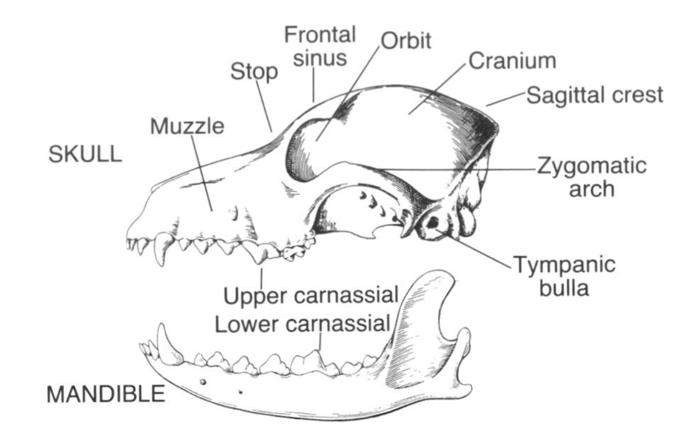

Fig. 2.6. Drawing of a dog skull and mandible to show features of domestication. After Pat Barrow, in Miller 1948.

evolutionary reduction in the size of the teeth took place at a slower rate than the shortening of the maxillary and mandibular bones. However, the teeth did become smaller in time, and modern dogs have teeth that are very much smaller than those of wolves, even in giant breeds such as the Great Dane. The shape of the mandible became more curved, and an angle developed between the facial region and the cranium, this being termed the 'stop' in modern breeds. The eyes became rounded and more forward-looking and the frontal sinuses became swollen. The tympanic bullae were reduced in size and flattened (Fig. 2.6).

In the later stages of domestication different kinds of dogs were developed as a result of artificial selection. They were selected for colour, length of coat, long or short legs, carriage of ears or tail, as well as for aspects of temperament and behaviour. In time, this selective breeding led to the development of the 400 or so different breeds of dog that inhabit the world today.

Breeds of dog

Canid bones and teeth retrieved from archaeological sites have revealed considerable diversity in size and bodily proportions within populations of dogs in the prehistoric period. There do not, however, appear to have been distinctive breeds until about 3000–4000 years ago (Harcourt, 1974; Clutton-Brock, 1984). From then onwards, dogs of the greyhound type were frequently depicted on paintings and pottery in Egypt and western Asia. The greyhound seems to be one of the most ancient of the foundation breeds, and

Fig. 2.7. Medieval hunting dogs in 'St Hubert'. A painting by Albrecht Dürer (1471–1528) in the Bridgeman Art Library. Reproduced with permission.

dogs with narrow heads, light bodies and long legs were obviously common in ancient Egypt. Dogs of the mastiff type were kept as hunting and guard dogs, and dogs with particularly short legs were also bred.

By the Roman period, most of the main breed types of dogs that exist today were well-defined, and their qualities and functions recorded. Hunting dogs, guard dogs, sheep dogs and lap dogs were all common, and the Romans were aware that selection could affect not only the appearance of a breed but also its capabilities and behaviour. They also knew that early training was very important in the rearing of a useful animal. In the first century AD, Columella recommended that a shepherd dog should be white to distinguish it from a wild wolf, and that it should not be as heavily-built as a guard dog, nor as fine-limbed as a hunting dog. A farmyard dog should be black so that it would frighten away strangers and it should be squarely-built with a large head and drooping ears (Forster & Heffner, 1968, p. 307).

Since Roman times dog breeds have been developed to serve a multitude of functions, but perhaps the main one has always been to provide companionship and to increase the personal status of the owner at home or in the hunt.

The great era for the proliferation of dog breeds in Europe was the Middle Ages, from the thirteenth to the fifteenth centuries. This was the time of feudalism and the establishment of the aristocracy to whom hunting was of supreme importance as a symbol of power and status. The laws of hunting, or 'venery' as it was known, became highly formalized and the use of different breeds of dogs for the various different hunts was an integral part of the rituals. Each beast of the chase was hunted with its own type of hound, so that there were deerhounds, wolfhounds, boarhounds and otterhounds, in addition to limers (bloodhounds) for tracking, and greyhounds for chasing game by sight (see Fig. 2.7) (Baillie-Grohman & Baillie-Grohman, 1904).

Today, enhancement of status is achieved by the dog playing the role of acolyte to its owner. In this age of the nuclear family, however, the dog is perhaps even more important as an object of affection, and small dogs are becoming even more popular. The most highly bred example is the Pekingese (Fig. 2.8), which, with its soft fur, large eyes and 'infantile' face, must represent the ideal baby substitute and the complete antithesis of the wolf. Somewhat ironically, the Pekingese was first bred – in the Imperial Palace in Beijing – to look as much as possible like the spirit-lion of Buddha, first introduced with Buddhism into China during the eastern Han Dynasty, AD 25–221 (Epstein, 1969, p. 138). The development of the skull of the Pekingese from that of the wolf must rank as one of the most extraordinary examples known of morphological variation within a single biological species.

Fig. 2.8. The Pekingese breed of dog. Copyright G. Napier.

At the present time, there are dogs in every part of the human-inhabited world. The relationship that began with commensal scavenging by tamed wolves and half-starved curs has evolved, over some 10 000 years, into a symbiosis that is indulged in by as many people as deny it. This evolution of the domestic dog was nicely summarized by the poet, John Davies, writing in the early 1590s:

Thou sayest thou art as weary as a dog,
As angry, sick, and hungry as a dog,
As dull and melancholy as a dog,
As lazy, sleepy, idle as a dog,
But why dost thou compare thee to a dog?
In that for which all men despise a dog,
I will compare thee better to a dog,
Thou art as fair and comely as a dog,
Thou art as true and honest as a dog,
Thou art as kind and liberal as a dog,
Thou art as wise and valiant as a dog.

References

Baillie-Grohman, W. A. & Baillie-Grohman, F. (eds.) (1904). *The Master of Game by Edward, Second Duke of York.* London: Ballantyne, Hanson & Co.

Clark, P., Ryan, G. E. & Czuppon, A. B. (1975).

Biochemical genetic markers in the family Canidae. *Australian Journal of Zoology*, **23**, 411–17.
Clutton-Brock, J. (1962). Near Eastern canids and the affinities of the Natufian dogs. *Zeitschrift für Tierzüchtung und Züchtungsbiologie*, **76**, 326–33.
Clutton-Brock, J. (1977). Man-made dogs. *Science*, **197**, 1340–2.
Clutton-Brock, J. (1979). The mammalian remains from the Jericho Tell. *Proceedings of the Prehistoric Society*, **45**, 135–57.
Clutton-Brock, J. (1984). Dog. In *Evolution of Domesticated Animals*, ed. I. L. Mason, pp. 198–211. London: Longman.
Clutton-Brock, J. (1988). The carnivore remains excavated at Fell's Cave in 1970. In *Travels and Archaeology in South Chile by Junius B. Bird*, ed. J. Hyslop, pp. 188–95. Iowa City: University of Iowa Press.
Clutton-Brock, J. (1992). The process of domestication. *Mammal Review*, **22**, 79–85.
Clutton-Brock, J., Corbet, G. B. & Hills, M. (1976). A review of the family Canidae with a classification by numerical methods. *Bulletin British Museum (Natural History)* (Zoology), **29**, 117–99.
Clutton-Brook, J. & Noe-Nygaard, N. (1990). New osteological and C-isotope evidence on Mesolithic dogs: companions to hunters and fishers at Star Carr, Seamer Carr and Kongemose. *Journal of Archaeological Science*, **17**, 643–53.
Corbett, L. K. (1985). Morphological comparisons of Australian and Thai dingoes: a reappraisal of dingo status, distribution and ancestry. *Proceedings of the Ecological Society of Australia*, **13**, 277–91.
Darwin, C. (1868). *The Variation of Animals and Plants under Domestication*, Vol. 1. London: John Murray.
Davis, S. J. M. & Valla, F. R. (1978). Evidence for domestication of the dog 12,000 years ago in the Natufian of Israel. *Nature*, **276**, 608–10.
de Lumley, H. (1969). Une Cabane de chasseuse acheuleenes dans la Grotte du Lazaret à Nice. *Archeologia*, **28**, 26–33.
Epstein, H. (1969). *Domestic Animals of China*. Farnham Royal, Bucks.: Commonwealth Agricultural Bureaux.
Forster, F. S. & Heffner, E. H. (transl.) (1968). *Lucius Junius Moderatus Columella on Agriculture*, vol. II. Loeb Classical Library No. 407. London: Heinemann.
Fox, M. W. (1971). *Behaviour of Wolves, Dogs and Related Canids*. London: Jonathon Cape.
Fox, M. W. (ed.) (1975). *The Wild Canids: their Systematics, Behavioural Ecology and Evolution*. New York: Van Nostrand Reinhold.
Fraser, F. C. & King, J. (1954). Faunal remains. In *Excavations at Star Carr an Early Mesolithic Site at Seamer near Scarborough, Yorkshire*, ed. J. G. D. Clark, pp. 70–95. Cambridge: Cambridge University Press.
Garrod, D. A. E. & Bate, D. M. A. (1937). *The Stone Age at Mount Carmel, Excavations at the Wady-el-Mughara*, I, pp. 175–9. Oxford: Oxford University Press.
Ginsberg, J. R. & Macdonald, D. W. (1990). *Foxes, Wolves, Jackals, and Dogs: an Action Plan for the Conservation of Canids*. Gland, Switzerland: International Union for Conservation of Nature and Natural Resources.
Groves, C. (1981). *Ancestors for the Pigs: Taxonomy and Phylogeny of the Genus* Sus. Technical Bulletin No. 3, Research School of Pacific Studies, Australian National University, Canberra.
Hall, R. L. & Sharp, H. S. (1978). *Wolf and man: Evolution in Parallel*. New York: Academic Press.
Hamilton Smith, C. (1839). *Dogs*, vol. 1. In *Naturalists Library*, ed. W. Jardine. Edinburgh: Lizars.
Hammond, J. (1940). *Farm Animals: Their Breeding, Growth, and Inheritance*. London: Edward Arnold.
Harcourt, R. A. (1974). The dog in prehistoric and early historic Britain. *Journal of Archaeological Science*, **1**, 151–75.
Harrison, D. L. (1973). Some comparative features of the skulls of wolves (*Canis lupus* Linn.) and pariah dogs (*Canis familiaris* Linn.) from the Arabian peninsula and neighbouring lands. *Bonner Zoologische Beiträge*, **24**, 185–91.
Hemmer, H. (1990). *Domestication: the Decline of Environmental Appreciation*. Cambridge: Cambridge University Press.
Hunter, J. (1787). Observations tending to show that the wolf, jackal, and dog, are all of the same species. *Philosophical Transactions of the Royal Society of London*, 77, 264–71.
Lawrence, B. (1966). Early domestic dogs. *Zeitschrift für Säugetierkunde*, **32**, 44–59.
Lawrence, B. (1968). Antiquity of large dogs in North America. *Tebiwa*, the Journal of the Idaho State University Museum, **11**(2), 43–9.
Lawrence, B. & Reed, C. A. (1983). The dogs of Jarmo. In *Prehistoric Archeology along the Zargos Flanks*, ed. L. S. Braidwood, R. J. Braidwood, B. Howe, C. A. Reed & P. J. Watson, pp. 485–94. Chicago: The Oriental Institute of the University of Chicago.
Lee, R. B. (1979). *The !Kung San: Men, Women, and Work in a Foraging Society*. Cambridge: Cambridge University Press.
Linnaeus, C. (1758). *Systema naturae*. A photographic facsimile of the first volume of the tenth edition, 1956. London: British Museum (Natural History).
Lorenz, K. (1954). *Man Meets Dog*. London: Methuen.
Lorenz, K. (1975). Foreword. In *The Wild Canids: their Systematics, Behavioural Ecology and Evolution*, ed. M. W. Fox. New York: Van Nostrand Reinhold.
Lumholtz, C. (1889). *Among Cannibals*. London: John Murray.
Meggitt, M. J. (1965). The association between Australian Aborigines and dingoes. In *Man, Culture, and Animals: the Role of Animals in Human Ecological Adjustments*, ed. A. Leeds & A. P. Vayda, pp. 7–26.

American Association for the Advancement of Science, No., 78.

Miyao, T., Nishizawa, T., Hanamura, H., & Koyasu, K. (1984). Mammalian remains of the earliest Jomon period at the rockshelter site of Tochibara, Nagano Pref., Japan. *Journal of Growth*, **23**(2), 40–56.

Milham, P. & Thompson, P. (1976). Relative antiquity of human occupation and extinct fauna at Madura Cave, south-eastern Australia. *Mankind*, **10**, 175–80.

Miller, M. E. M. (1948). *Guide to the Dissection of the Dog*. Ann Arbor, MI: Lithoprint by Edward Brothers.

Musil, R. (1984). The first known domestication of wolves in central Europe. In *Animals and Archaeology: 4. Husbandry in Europe*, ed. C. Grigson & J. Clutton-Brock, pp. 23–6. Oxford: BAR International Series 227.

Newsome, A. E., Corbett, L. K. & Carpenter, S. M. (1980). The identity of the dingo I. Morphological discriminants of dingo and dog skulls. *Australian Journal of Zoology*, **28**, 615–25.

Nobis, G. (1979). Der älteste Haushund lebte vor 14,000 Jahren. *UMSHAU*, **19**, 610.

Olsen, S. J. (1977). The Chinese wolf, ancestor of New World dogs. *Science*, **197**, 553–5.

Olsen, S. J. (1985). *Origins of the Domestic Dog: the Fossil Record*. Tucson: The University of Arizona Press.

Scott, J. P. & Fuller, J. L. (1965). *Dog Behavior: the Genetic Basis*. Chicago: The University of Chicago Press.

Shaughnessy, P. D., Newsome, A. E. & Corbett, L. K. (1975). An electrophoretic comparison of three blood proteins in dingoes and domestic dogs. *Journal of the Australian Mammal Society*, **1**, 355–60.

Street, M. (1989). *Jäger und Schamen: Bedburg-Königshoven ein Wohnplatz am Niederrhein vor 10000 Jahren*. Mainz: Römisch-Germanischen Zentralmuseums.

Tchernov, E. & Horwich, L. K. (1991). Body size diminution under domestication: unconscious selection in primeval domesticates. *Journal of Anthropological Archaeology*, **10**, 54–75.

Thomas, E. M. (1987). *Reindeer Moon*. London: Collins.

Turnbull, P. F. & Reed, C. A. (1974). The fauna from the terminal Pleistocene of Palegawra Cave, a Zarzian occupation site in northeastern Iraq. *Fieldiana Anthropology*, **63**(3), 81–146.

Valla, F. R. (1990). Le Natoufien: une autre façon de comprendre le Monde. *Mitekufat Haeven (Journal of the Israel Prehistoric Society)*, **23**, 171–5.

Wayne, R. K. (1986*a*). Cranial morphology of domestic and wild canids: the influence of development on morphological change. *Evolution*, **40**, 243–61.

Wayne, R. K. (1986*b*). Limb morphology of domestic and wild canids: the influence of development on morphologic change. *Journal of Morphology*, **187**, 301–19.

Wayne, R. K. (1986*c*). Developmental constraints on limb growth in domestic and some wild canids. *Journal of Zoology, London*, **210**, 381–99.

Wayne, R. K. & Jenks, S. M. (1991). Mitochondrial DNA analysis implying extensive hybridization of the endangered red wolf *Canis rufus*. *Nature*, **351**, 565–8.

Wayne, R. K., Nash, W. G. & O'Brien, S. J. (1987). Chromosomal evolution of the Canidae. I. Species with high diploid numbers. *Cytogenetics and Cell Genetics*, **44**, 134–41.

Wayne, R. K. & O'Brien, S. J. (1987). Allozyme divergence within the Canidae. *Systematic Zoology*, **36**, 339–55.

Zimen, E. (1981). *The Wolf: His Place in the Natural World*. London: Souvenir Press.

3 Evolution of working dogs

RAYMOND COPPINGER AND RICHARD SCHNEIDER

Introduction

When we think of dogs, we tend to think of animals that were selected for behavior performed in the service of people. Dogs pull sleds, guard property, herd sheep, guide the blind, track and retrieve game, and so on. We also think of dogs in terms of breeds, and often try to identify the breeds that make up some mongrel, as if all dogs had unadulterated, purebred backgrounds. Many see their favorite breed woven into the Bayeux tapestry or carved into the walls of some ancient tomb. Some think of breeds as if they were ancient species, separately derived from different strains of wolves, jackals or even coyotes. But breeds of dogs for the most part are modern inventions. Like other domesticated animals, dogs may have originated as scavengers, and been domesticated for use as food and fiber, or put to work, just as oxen pull ploughs, goats pull carts and pigs hunt truffles. However, unlike other domesticated animals, dogs also make excellent companions.

This chapter distils 20 years of observations and experiments at Hampshire College on the subject of breed-specific behavior. Our focus has been the working behavior of several breeds of dog, and the selective forces which resulted in dogs that are specialists in the tasks they perform. We are interested in the link between a dog's performance and its external and internal structures. We work from the premise that a dog's morphology and physiology predict and limit its behavior, and that an understanding of the development of structural differences is necessary in order to comprehend a dog's behavior.

In order to understand the evolution of working dog behavior, one must look at ways a wild animal, such as the wolf, could have been transformed into a tameable and trainable companion. It is often thought that these changes have come about through a gradual accumulation of traits, by a process similar to natural selection, but other evolutionary mechanisms were also at work. Darwin (1859) used domestic animals as an example of an unnatural selection analogous to natural selection. But discoveries in the last hundred years have suggested other evolutionary mechanisms underlying the symbiosis between humans and domestic animals.

The dogs we studied were sled dogs, livestock guarding dogs and herding dogs. Each was selected to perform different tasks. Looking at these tasks in detail provides insights into the derivation of breed-specific behavior.

Sled dogs

Modern racing sled dogs are a good illustration of how working dogs are selected and how a breed is created. Early sled dog races were recreational tests of the prowess of both hard-working, freight-pulling dogs and of the drivers who transported people, cargo and mail on the snowy frontiers of North America and Asia. These tests rapidly evolved into sprint races and dogs were selected for speed. For distances of over ten miles, sled dogs are easily the fastest land mammal. Winning teams lope (a slow gallop) at speeds averaging 20 mph for over 60 minutes. In long races of 1000 miles they must be able to trot for many hours a day, often covering over 100 miles, day after day. The goal is to complete the course in the shortest amount of time, compared with other teams which leave the starting line at regular intervals. Racing dogs must be morphologically efficient, minimizing mass and motion while maximizing speed.

The need for fast, hardy sled dogs became critical during the Alaskan Gold Rush that began in 1896. Few dogs were kept by native Alaskans before colonization by outsiders, and although these early huskies carried packs, moved cargo and hunted, they may have been just as important as sources of food and fur (Lantis, 1980). Not until the expansion of the Thule culture during the eighth and ninth centuries does evidence appear of dogs hitched to the sled in a fan arrangement (MacRury, 1991). The concept of tandem harnessing and the use of a specialized lead dog came much later, introduced to the New World by Europeans.

The few dogs belonging to fur trappers or indigenous Alaskans at the turn of the twentieth century were supplemented, during the Gold Rush, by an influx of dogs from other parts of North America and Russia. As a result, teams were composed of a variety of breeds, including bird dogs, retrievers, hounds and forms resembling Newfoundlands. By 1908, in the first All-Alaska Sweepstakes race in Nome, breed differences were still apparent, although the best teams showed a uniformity of size and conformation (Coppinger, 1977). The 1911 race was won by Scotty Allan, whose dogs (Fig. 3.1) were

Fig. 3.1. Scotty Allan's 1911 All-Alaska Sweepstakes dog team. Although the dogs are obviously mongrels, Allan has chosen them for similar size, and paired them according to type, which probably resulted in similar gaits. From Coppinger (1977); photo courtesy of Howell Book House, New York.

remarkably uniform, still looking a lot like their purebred grandparents. But several had already begun to resemble the dog we now call the Alaskan husky, a dog which dominates modern sled racing. Hybridization with the Siberian husky, a fast, hardy trail dog, imported first in 1909, gave breeders access to a diversified population of increasingly specialized dogs.

Today's best racing dogs are the product of an acquisition and culling process based not on their looks or strength but on their ability to run fast on groomed trails, and to behave themselves on a team (Fig. 3.2). Racing dogs are purchased from other racers or raised from high performance dogs that are past their racing prime. Top working animals often are not bred from because pregnancy and lactation interrupts females' racing careers, and introduces males to behavioral displays that are not appropriate on a team. Dogs from a given kennel often take on a distinct phenotype and are referred to as 'Belford dogs' or 'Lombard dogs', after mid-century champions Charles Belford or Roland Lombard. Pride in their abilities with dogs has led native Americans to pool their resources. George Attla, one of the best Alaskan drivers of the 1970s, selected dogs annually from native villages. It was an honor to have one's dog picked for his team and to have that dog distinguished as an 'Attla dog.' Dogs purchased from champion drivers become valuable as breeding animals simply because they pass through the most successful kennels. Locally superior dogs, when they become known to a wider audience, may acquire the name of their region, e.g. Alaskan husky, Quebec hound. This breed-producing process resembles that of other working breeds that bear the names of the original breeder or region, such as the Jack Russell terrier, Doberman pinscher, German shepherd, Saint Bernard, Catahoula leopard cowhog dog.

Sled dogs are also acquired and bred because of their gait. Top drivers know that gait in detail. They know, for example, that greyhounds run in a series of leaps and are said to have a 'double flight.' This means they leave the ground by pushing off with their back feet, then land on their front feet, which are used to pull/push them into the air again, and then they land on their back feet. Galloping horses have a single flight, leaving the ground with their front feet and then landing on their back feet, which are kept in place until the front feet are on the ground again. Good sled dogs lope or gallop with no flight; at least one foot always remains in contact with the ground. Dogs that have flights are known as floaters and are ineffective sled-pullers.

Speed is attained not only by rapidity of movement but also by the length of the gait. This length is determined by the dog's size and shape. The larger the dog, the longer the gait. The slope of the pelvis, the distance between the shoulder blades and the length of the back all allow for maximum extension of the front and rear legs. Good drivers say that a

Fig. 3.2. George Grove's 1975 Laconia World Championship dog team. Modern racing Alaskan huskies look like a 'breed,' showing a uniformity of breed-like characteristics, but they are strongly dependent on the diversity of hybridization. Photo by Lorna Coppinger.

dog runs with its back, and as every human racer or dancer knows, the most fatiguing activity is pulling the legs forward, much of which is done with back muscles.

As important as speed, dogs that are paired in harness need to have matched gaits, preferably mirror images. The physics of pulling a sled can be explained fairly simply. The single dog on a one-dog team pulls 100% of the load, while each dog on a two-dog team pulls about 85%. Harness vectors, plus the resulting torque on the sled, create an inefficiency that prevents any better distribution of work (ideally, 50–50 for a two-dog team). Each dog on a six-dog team actually pulls about half the load. More than 12 dogs adds no net gain. (On open-class teams, 14–16 dogs are run to provide 'spares' in case dogs have to be dropped during the race due to injuries or fatigue, since no new dogs can be added once the race has begun.) By matching the dogs' gaits, drivers can reduce the energy lost to vector forces and increase the speed, because the distance the sled and lines have to move is shortened.

Dogs are picked and trained for particular positions on a team based on observed talent. Directly in front of the sled are two 'wheel' dogs that have enough strength to steer and draw the sled away from obstacles. On a course through a forest, wheel dogs are as important as a lead dog because, on a curve, succeeding pairs of dogs are drawn into whatever obstacle defines that curve. The fastest pairs of dogs are placed at the front of the team, although there cannot be a significant difference between the fastest and slowest dogs, for that in itself will introduce inefficiencies. Any dog with a gait that differs greatly from the team average in speed, either faster or slower, will not be able to perform properly. It is a mistake to get a super-fast lead dog, for it ends up trying to tow the team.

Dogs that show agonistic behavior are disruptive; good drivers avoid situations where dogs are likely to fight. Dogs that routinely display aggressive behavior are not used. Fighting can be reduced by pairing males with females, but this also means selecting against sexual dimorphisms in order to preserve uniformity of gait. Lead dogs are often run in pairs and are often female, and chosen mainly to set the pace, since other dogs on a team know the commands. A team with 'deep' talent will have dogs that can be switched during a race to relieve fatigued leaders.

Many drivers know their dogs' characteristics without understanding the underlying biology. For example, although bigger dogs are potentially faster because of their longer stride, drivers know that dogs over 25 kg have less endurance. Running sled dogs will attain rectal temperatures of 40 °C or more; larger dogs cannot dissipate heat quickly enough and become heat-exhausted (Fig. 3.3). Bigger dogs also have to expend more energy carrying their own weight. Smaller dogs cannot conserve heat as well when not racing, nor do they have the long stride. Native dogs of the eastern Arctic typically weigh 32–36 kg, probably for reasons of heat conservation in cold environments. Such weights also indicate that they were not selected to run fast or for long distances. Siberians, malamutes or Samoyeds, which are often thought of as traditional freight-sledding breeds, were selected for qualities other than racing. They are not competitive in a modern sprint race against Alaskan huskies.

Another trait understood at one level by drivers is that the feet of some dogs 'snowball,' that is, ice and snow collect on foot hairs, and produce pain and injury as the dogs run. This condition seems to be caused by excessive sweating through the foot pad.

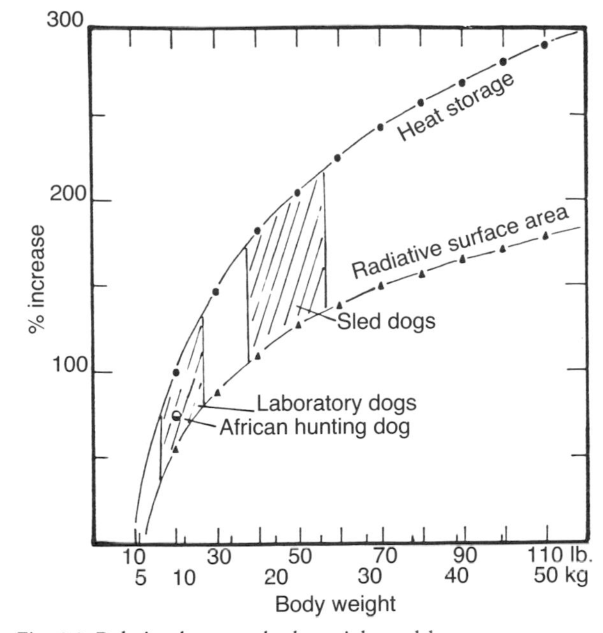

Fig. 3.3. Relation between body weight and heat storage capacity in dogs. The chart shows that large dogs cannot radiate heat fast enough to run long distances, and that small dogs have problems conserving heat. From Phillips, Coppinger & Schimel (1981).

Such sweating is universal in dogs, although arctic wolves do not have functional merocrine or apocrine glands (Sands, Coppinger & Phillips, 1977). Most drivers, unaware of the underlying physiology and anatomy, just select those animals that can go the distance without being in pain when they run.

Dogs do not pull sleds because they are forced to. Running on a team is not a typical conditioned response, for there is no obvious reward for the performance. After a race or training session the dog is given water, removed from harness and put into its travel box or kennel. Feeding is done at a given hour each day, usually in the early evening so as not to interfere with the day's activities. Dogs are not punished if they do not run well. Whipping or abusive treatment can damage a dog, decreasing its abilities and worst of all, creating a sulker. If severe, the dog is likely to quit. Cracking a whip or making novel noises may increase the speed, but to overdrive a team beyond its pace may deprive the dogs of enough energy to finish the race, and is also likely to lead to quitting, a sled dog driver's worst fear. Good drivers guard against ever having a dog quit, for if a dog learns to quit at stressful times, it cannot be relied upon and cannot be raced. Instead of punishment, good drivers keep dogs out of situations where they may misbehave. Thus, a young dog goes through a lengthy training process whereby it is placed on a team that runs at near racing speed, and throughout the training season the distance is carefully increased. The driver's job is to supervise the process. The method is one of establishing a routine; there are few commands. There is, however, considerable social facilitation between individuals, and a dog's demeanor at harnessing time could best be described as 'major excitement.' Dogs left behind will bark continuously and show other signs of distress.

Where did sled dog behavior come from? The selection and training procedure just described is similar to Jack London's description in his novel, *Call of the Wild.* London was right in making 'Buck' a cross between a Saint Bernard and Scottish sheepdog, stolen in one of the lower 48 states. Buck and his fellow captives were dog derivatives, not wolf derivatives. Sled dogs behave like dogs, not wolves. Lead dogs are in that position because the driver sees their ability to run fast, to stay out in front and to take commands. They are not chosen because they establish dominance over the rest of the team. The driver is not a substitute leader forcing the dogs to submit, but an observant manager trying to reduce wasted energy and to select dogs that perform well. The driver's chief job is to anticipate problems and prevent conditions that disrupt the performance. Dominance struggles are selected against and discouraged when they do appear. The one place striking a dog is allowed is in preventing or breaking up a fight. Dogs are too valuable to be allowed to hurt each other.

A further distinction between sled dogs and wolves is that sled dogs are not running as a pack. Wolves pack only as adults. As large juveniles, they remain at a rendezvous point, much as a pet dog sitting in the yard waits for the 'parent'–owner to come and feed it. There is very little tendency even for adult domestic dogs to pack (see Bradshaw & Brown, 1990, for discussion), and when they do (e.g. a pack of hounds) there is a continuous vocal display that stimulates cohesion, rather than the social hierarchy of a wolf pack.

When sled dogs run, they are playing. They are exhibiting non-functional or non-rewarding behavior, in the classical meaning of reward. They are not chasing food or even another team; they do not have the social organization of a pack. No serious sled dog driver would hybridize a sled dog with a wolf, because the offspring would have traits counterproductive to racing fast on a team; an increase in size, improper gait, unmanageable behavior, and little tendency to play at racing. MacRury (1991) presents good evidence that Inuit dogs are not hybrid wolves, and that these people show little desire to produce such hybrids.

People who work with functional dogs tend to believe that breed-specific behavior is inherited. Not inherited in the sense that the behavior is the result of a gene product, but inherited in the sense that behavior is a consequence of, and limited by, the animal's morphological and physiological structures. Thus, when it is said that dogs inherit behavior, it is not meant to imply that the behavior is passed from generation to generation. Rather, the arrangement of the nervous system, morphology and physiology is in some way a consequence of the genes that the animal inherited. Running and pulling behavior in sled dogs is a perfect example. The dog has the potential before training begins, and it is the handler's job to release that potential.

The implication is that any measurable differences in behavior are products of measurable differences in structure. Structure is any inherited pattern of molecules, including not only bones, muscles and nerves, but also support systems like blood, hormones, neurotransmitters and other biochemical components. The theory does not deny environmental influences nor does it deny an animal's ability to modify behavior through learning. But learning itself is heritable, in that the organism needs to have the structures with which to learn, and the type of learning is limited by those structures. The sled dog cannot learn to run at racing speed if its feet snowball or if it does not have the reach. Equally important, the fact that sled dogs can be trained is dependent on a lack of innate aggressive or predatory behavior that cannot be controlled, or other behavior patterns that might detract from the play routines. Malamutes, for example, make poor sled dogs not only because they are too big to run at top speed for very long, but also because they tend to be vociferous fighters. Hounds often do very well as racing sled dogs but their drivers are in perpetual fear of crossing a game trail and having the chase instinct elicited. Leaving the groomed track with 16 dogs in pursuit of a rabbit can be an exciting moment in the life of a driver.

The ability to perform a particular task is rooted deeply in trainability, with the proviso that the animal has the structure and disposition to be trained. The word 'disposition' is used in the sense of having, or not having, innate motor patterns that either facilitate or disrupt the trainable performance. Sheepdogs provide another illustration of this principle.

Sheepdogs

Behaviorally, there are two main types of sheepdog, each type represented by a number of breeds. The two types have been selected to work in similar habitats (grasslands) and respond to similar stimuli (livestock) but they perform very different tasks. The duty of livestock guarding dogs is not to disrupt the sheep but to live in their midst and disrupt their predators. The duty of herding dogs is to disrupt the behavior of livestock and conduct them, on command, to another place.

Livestock guarding dogs

A livestock guarding dog must be attentive, trustworthy and protective with its livestock. The quality and frequency of these traits are dependent on the emergence of innate motor patterns, and the reinforcement or prevention of innate behavior. As with sled dogs, size is also an important factor. In attaining typical weights of 30–40 kg and heights of 50–60 cm, guarding dogs are large enough to present a deterrent to most predators.

The critical socialization periods of dogs (Scott & Fuller, 1965; Fox & Bekoff, 1975; Serpell & Jagoe, Chapter 6) are extremely important in the training of a livestock guardian. During its first year a pup passes through several stages of receptivity when it is biologically and psychologically ready to learn a particular behavior. As it passes out of a stage, the chances of it learning the associated behavior diminish.

Attentive behavior is learned between four and about 14 weeks when the pup forms its social bonds (Fig. 3.4) and can be 'taught' to direct normal intraspecific social behavior (especially care-soliciting, including food-begging and other tactile responses) to another species (Fox, 1971). The trainer accomplishes this by allowing the pup to display such behavior to sheep or other livestock, and not to other species, especially humans.

To shape a young guardian's trustworthy behavior, it is necessary to prevent it from displaying any tendency to play with the stock. Such attempts may come at about five months, but are commonly associated with the emergence of predatory motor patterns, e.g. eye/stalk, chase or bite (Figs. 3.6 and 3.7 show these behavior patterns in border collies), at about seven months. This disruptive behavior can be diminished either by correction or by the removal of the stimulus that elicited the display. Once past the associative stage, extinction of the tendency to show these behavior patterns usually follows. By the time it is an adult, a guarding dog is an attentive, trustworthy part of the flock (Fig. 3.5).

Livestock guarding breeds may have a longer period of social bonding than most other breeds, and may even show high frequencies of juvenile motor patterns well into adulthood. This is important for their function as livestock guardians. In good working sheep-guarding dogs, predatory motor patterns never or only weakly emerge. It is rare to see eye, stalk or dissect, although chase and grab–bite do occur. Eating is not a predatory behavior; guarding dogs will eat exposed flesh whether the 'prey' is dead or not, but since they do not show dissect, a pen full of them will not eat an unopened carcass.

In livestock guarding dogs, any predatory, courtship or territorial motor patterns directed toward sheep are often described as play because they do not appear functional, and are disruptive of functional behavior. In the guardians, play behavior, such as chasing, biting or sexually mounting sheep, actually represent emergent adult motor patterns and are detrimental to a good working performance. The behavior of livestock guarding dogs appears to depend on correct early socialization and on a juvenilized adulthood. As adults, they play-wrestle with conspecifics but tend not to play with objects, nor will the good ones chase a ball. They sound and act like overgrown puppies.

Livestock guarding has little to do with the legendary brave companion fiercely protecting its master's property. Rather, guarding dogs protect by disrupting predators by means of behavior that is ambiguous or contextually inappropriate: barking, tail-wagging, social greeting, play behavior and, occasionally, aggression. Many species of predator

Fig. 3.4. Developing socialization in a livestock guarding pup. Photo by Lorna Coppinger.

will stop a hunting sequence if disrupted, and 'discovery' by a big dog is often enough to avert predation.

Livestock guarding dogs are an adjunct to the pastoral community. They 'work' continuously without commands and there is no need even to name them; their genealogy is not important to shepherds. What the shepherds selected for was behavior. What they got was big dogs with weakened adult stereotypical behavior sequences, which displayed early juvenile motor patterns longer during ontogeny. Since many sheep-raising communities have historically been transhumant (nomadic), gene flow occurs between dogs of different regions. Females in estrus are not usually penned, and so mate with other mountain dogs and also with non-working dogs. Mongrelization tends to disrupt stereotypic behavior, and thus does not necessarily decrease the working abilities (Coppinger, Smith & Miller, 1985), and may at the same time improve genetic health (hybrid vigor). In the southwestern United States, Navajos use a small mongrel very effectively with their sheep (Black & Green, 1985).

Regional phenotypes do exist and have been identified as breeds. But in such cases, selection is based on a preferred coat color within the guardian cohort. The Anatolian shepherd dog of Turkey often shows the tail carriage, sleek design and smooth coat of locally popular rabbit-hunting greyhounds. The Italian Maremmano–Abruzzese (Maremma) is usually white, as are the Hungarian komondor and kuvasz. The Yugoslavian šarplaninac comes in a wide range of coat colors from white to black, often with large spots, but is predominantly a dark grey or tan dog. In spite of these variations in color, stock guardians from countries of western Europe and all the way to eastern Asia are a similar size and shape, and no significant differences have been found in their guarding abilities (Coppinger *et al.*, 1988).

Recently, increased interest in pet and show dogs

Fig 3.5 The trustworthy, attentive behavior of an adult livestock guarding dog enables it to live with livestock and protect the animals from predators. Photo by Ray Coppinger.

in many of these countries, and the popularity of vacationing in the mountains, has led to a diminution of much of the guarding dog stock. When the mountain 'breeds' become a focus of national pride, they increase in value and are sold, taking their good working genes out of the region. In cities, people decide to keep written genealogies and registries, selection changes to companion or show animals and the functional guarding disposition suffers. The breed in one sense gains from this popularity, but the breed's gain is the shepherd's loss.

Herding dogs

Herding dogs conduct livestock from one place to another by causing fear-flocking and flight behavior. They are known in the trade as 'chase and bite' dogs. The various breeds have been selected for specific behavior patterns directed toward specific livestock. They are generally divided into headers, heelers and catch dogs depending on their tendency to circle livestock and bring them toward the handler, drive livestock away from the handler in a trailing pattern, or actually stop the livestock by bringing them to the ground. There are also breed differences in how far away from the handler they will work. Australian shepherd dogs and New Zealand huntaways work with their heads up and vocalize while they chase sheep. Border collies and kelpies are silent and use a circling out-run followed by a head-down eye, stalk and chase. Blue heelers bite at the hocks of cattle, a behavior that is generally not acceptable for dogs that herd the more vulnerable sheep. In the American

Fig. 3.6. Border collie showing 'eye' and 'stalk.' Photo by Ray Coppinger.

southwest, catch dogs are popular for wild range cattle. Catahoula leopard cowhog dogs are used in packs, forcing range cows into tight herds that are then driven by ranchers on horses, while the dogs prevent individuals from leaving the herd.

It is often assumed that herding dogs are using motor patterns homologous to the predatory behavior patterns of the ancestral wolf. These include the orientation posture, followed by eye, stalk, chase, grab–bite (that may be preceded by a forefoot stab), kill–bite (often called crush–bite and sometimes including a head-shake movement) and finally, dissect and consume (Figs. 3.6 and 3.7). In many wild carnivores predatory motor patterns are so stereotyped that a field biologist can look at a kill and know immediately the species of the predator.

The different breeds of herding dog display different motor patterns at different frequencies. Also, the intensity of the display varies to the extent that some breeds are easier to train to show (or not show) a particular behavior. It is difficult, for example, to train corgis and heelers not to nip at flying hocks and, therefore, they can be miserable companions for joggers.

During the early socialization period, herding dogs are allowed to socialize with people and other dogs but not to have continuous access to livestock. Although the emergence of predatory motor patterns is ontogenetically variable, training of herding dogs does not, and really cannot, begin until after the onset of eye, stalk and chase. The dog is encouraged to show these motor patterns toward livestock, and then is commanded off ('get down!') before it can continue with grab–bite, crush–bite and dissect. In Border collies the latter behavior patterns seem to be only weakly present, while in some of the catch dogs it behooves the handler to get to the scene and command the dog off before it does any damage to the restrained animals.

Experiments at Hampshire College showed that the release of the eye behavior in adult Border collies was stimulated in part by anticipation of movement of the 'prey.' When chickens were sedated just enough so that they stood motionless and inactive, Border collies were unable to hold the eye pattern and would resort to displacement activities such as play-bowing or barking at the birds. They resumed the eye posture at any suggestion of motion (Coppinger *et al.*, 1987).

Fig. 3.7. Border collie showing 'chase.' Photo by Ray Coppinger.

Showing of eye seems to provide its own reward. Once the onset of the motor pattern occurs ontogenetically, the only way to keep most collies from showing eye is to remove them from the stimulus. The same is true of other behavior patterns necessary for a good herding dog, because the dog seeks the rewards, and modifying the behavior is difficult. The solution to the problem of too much or too little correct behavior is often to get another dog.

Because the eye, stalk and out-run behavior patterns seem to be so strictly heritable, once the proper quality and frequency of display are established for a particular task, then the breed cannot be improved by crossing with other breeds, as can be done in guarding and sled dogs. Appropriately, discussions of genealogy are common among people who rely on their dogs to herd. Like sled dog drivers, sheep herders tend to get their dogs from other, well-known herders. Border collie breeders know exactly what a dog is like if it is related to MacKnight's Gael or Greenwood's Moss. Conversations between owners tend to be entirely about behavior and breeding, and scarcely ever about what the dog looks like. Because their job requires precise movement, herding dogs tend to be of a more uniform conformation than livestock guarding dogs, but not as uniform as racing sled dogs.

The development of breed-specific behavior

We have described three types of working dogs, selected to perform specific tasks. The effect of selection for performance (and ultimately for a 'breed') has been to change the onset of innate motor patterns which facilitate or disrupt that performance. Sled dogs are selected not to show hierarchical behavior, and that allows them to run as a team and the handler to switch dogs to different positions. Livestock guarding dogs cannot show the predatory behavior toward sheep because the dogs would not be trustworthy and could not be left to guard the sheep. Sheep-herding dogs, in contrast, must show the predatory eye, stalk and chase patterns, but it is a fault for them to show the grab–bite of the cattle heelers. Eye and stalk (or 'point') are virtues for pointers, but chase and crush–bite (called a 'hard mouth') are faults. One out of an original six Border collies bought in Scotland for the studies at Hampshire College did not show eye, and about 250 out of 1000 livestock guarding dogs displayed one or more disruptive behaviors toward sheep.

It appears that dogs are genetically programmed to behave like dogs, but different breeds and even different dogs do not display the program in the same way. One might imagine an inherited tape-recording with all the canine motor patterns programmed on that tape: search, locomote, attach and suck are on the early (neonatal) portion of the tape, while submission and food-begging are part of the juvenile section, followed by eye, stalk, chase, grab–bite, crush–bite and dissect. These latter are perfected in the adult portion, along with scent-marking, dominance, courtship, reproductive and parental behavior.

Dogs do not learn these behavior patterns. They are instinctual. They appear at the appropriate time, and are directed to the appropriate environmental stimulus. In one experiment, students played a tape-recording of the 'I'm lost' call of a newborn puppy. Border collie mothers retrieved the little tape recorder and placed it in the litter, but only during the 12 days after parturition. This motor pattern is available to females only during a specific period, and only in response to a specific signal. The isolation call is 'known' the moment the puppy is born, but is only given by puppies, and lost in the adult.

What a dog can learn is predetermined, governed by inherited motor patterns. Handlers train working dogs based upon inherited, breed-specific behavior. In contrast, obedience trainers tend to think of breeds as having 'constitutional differences' (Freedman, 1958) and teach with reinforcement using either instrumental or classical conditioning. O'Farrell (1986, p. 23) recognized that 'the definition [of reinforcement] has a certain circularity in that reinforcing events are those which promote learning.' Dog behaviorists seldom make the point that innate behavior is internally motivated, and that the display of such behavior is its own reward (Eibl-Eibesfeldt, 1975). The sled dog running the race, the bird dog retrieving the bird, the sheepdog showing eye or the guardian dog licking the face of a sheep are rewarded by the mere act of performing, and need not be further rewarded by the handler.

These breed-specific motor patterns are choreographed by selecting animals that display portions of the hypothetical tape and not others. One would expect the tape to vary between individuals, and that through careful observation of performance, the breeder could select the desired variants. The initial variation can be increased by hybridizing, as in the creation of racing sled dogs.

The genetic tape-recording raises a number of questions. Where did the tape come from? What are the processes of genetic change? Is it possible to obtain features that are not on the tape? Many observations, experiments and theories have examined these questions over the years in an attempt to understand the evolutionary relationship between domestic dogs and their wild ancestor.

The origin of dogs

The most widespread and accepted view places the wolf, *Canis lupus*, as the wild progenitor of the domestic dog (Zeuner, 1963; Scott & Fuller, 1965; Lorenz, 1975; Herre & Rohrs, 1977; Olsen & Olsen, 1977). Another hypothesis regards the wolf as the partial ancestor, but with some degree of hybridization with other members of the genus *Canis*, particularly jackals and coyotes (Darwin, 1875; Fiennes & Fiennes, 1970; Chiarelli, 1975; Clutton-Brock, 1977 and Chapter 2). A third possibility is that the domestic dog arose from a wild *Canis familiaris* type similar to those modern forms considered to be 'primitive,' such as the Australian dingo (*Canis familiaris dingo*), the New Guinea singing dog (*Canis familiaris hallstromi*), and the Asiatic and African pariah dogs (*Canis familiaris* spp.) (Macintosh, 1975; Oppenheimer & Oppenheimer, 1975; Brisbin, 1976, 1977). Additionally, there is the opinion that the origin of the domestic dog remains a mystery, and to speculate on its true ancestry is premature (Fox, 1978; Manwell & Baker, 1983).

Traditionally, researchers have used two methods to determine the origin of the domestic dog: either searching for archaeological evidence of domestication and relating it to wild species present at the time (e.g. Bökönyi, 1969), or comparing the domestic species with wild species existing now (e.g. Price, 1984; Wayne, 1986). Gaps in the fossil record and lack of assumed transitional forms have frustrated the process of creating accurate phylogenetic relationships.

Recently, using genetic and biochemical methods, researchers have shown domestic dogs to be virtually identical in many respects to other members of the genus. Researchers find little variation in karyotypes (the number and shape of the chromosomes) within the genus *Canis* (Chiarelli, 1975; Fischer, Putt & Hackel, 1976; Simonsen, 1976). Results using mtDNA (mitochondrial DNA) data also reveal startling similarities among canids (Vrana, 1988, unpublished BA thesis, Hampshire College). Mitochondrial

DNA is passed from mothers to daughters with no genetic recombination. This allows the evolutionary development of maternal lineages to be reconstructed. Figure 3.8 shows one of 20 possible arrangements of the data, the specific junctions depending on various options chosen for the analysis. All analyses, according to Braun (1990, unpublished paper, Hampshire College), who reworked Vrana's data, resulted in very similar arrangements. Greater mtDNA differences appeared within the single breeds of Doberman pinscher or poodle than between dogs and wolves. Eighteen breeds, which included dachshunds, dingoes and Great Danes, shared a common dog haplotype. Alaskan malamutes, Siberian huskies and Eskimo dogs also showed up in the common dog haplotype and were no closer to wolves than poodles and bulldogs. These data make wolves resemble another breed of dog.

To keep the results in perspective, it should be pointed out that there is less mtDNA difference between dogs, wolves and coyotes than there is between the various ethnic groups of human beings, which are recognized as belonging to a single species. The results are not surprising since, reproductively, wolves, coyotes, jackals and dogs are all interfertile, and cross-breeding still occurs in the wild between wolves, coyotes and dogs (Young & Goldman, 1944; Lehman *et al.*, 1991). In the strict biological sense, none of these animals are distinct species and 'new species never developed during domestication' (Kruska, 1988, p. 212). In other words, these canines are all working with virtually the same genetic tape and to say that wolves are more closely related to any given breed is misleading. The fact that a šarplaninac born in Yugoslavia, a Border collie born in Scotland and a wolf born in Canada can interbreed, and have virtually the same genetic sequence on their maternally-inherited mtDNA, is strong evidence that there is little phylogenetic distance between any of them. On the other hand, it is often said that no other group in the animal kingdom has achieved such a diversity of form in so short a time as *C. familiaris*. Variation within this species is greater than between all the rest of the canids. For example, Fig. 3.9 shows domestic dogs at the extremes of a cluster analysis of skull measurements, with the wild canids relatively closer together.

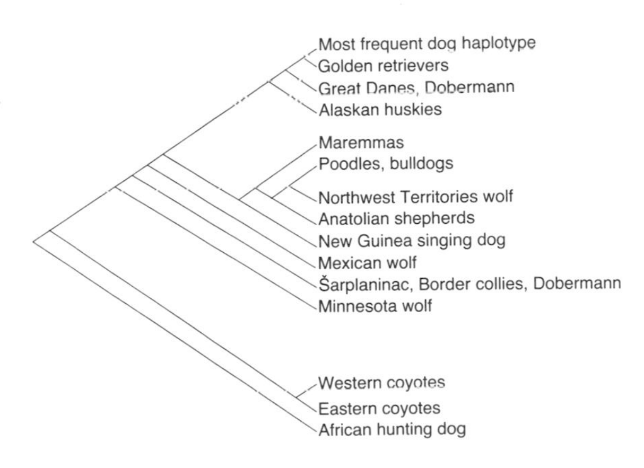

Fig. 3.8. This parsimonious tree of mtDNA analysis makes the wolf look like another breed of dog. From C. Braun (1990), unpublished paper, Hampshire College; based on data of P. Vrana (1988), unpublished BA thesis, Hampshire College.

Dogs range in size from Chihuahua to Saint Bernard, a 100-fold difference. However, that 100-fold difference in size does not imply a 100-fold difference in shape. Most dog skulls are actually quite similar in shape and proportion, except for the long-faced (dolichocephalic) breeds like the borzoi, or the short-faced (brachycephalic) pug-like breeds. Morey (1992) gives evidence that they have shown that same shape and proportion since prehistoric times. Eliminating size as a factor, most breeds are a consistent and predictable design. Under the jowly skin of the Saint Bernard or the sleek, tiny head of the Mexican dog reside skulls shaped very much like those of the wolf, jackal and coyote. It was not until we began to measure the skulls of dogs we had known ('Alas Poor Yorick. I knew him . . . '), that the real difference between dog and wolf skulls became visible.

The difference between wolf and dog is the result of a reduction in head, brain and tooth size. An adult dog which has a head the same size as an adult wolf has a 20% smaller brain (Fig. 3.10). But a dog that is the same weight as an adult wolf has a head about 20% smaller (Fig. 3.11). Thus, a 45 kg livestock guarding dog has a small head, and within that head, a brain the size of a three- to four-month-old wolf that still has most of its deciduous puppy teeth.

Some would argue that head size in dogs is a relative constant and it is simply body size that changes dramatically (Deacon, personal communication).

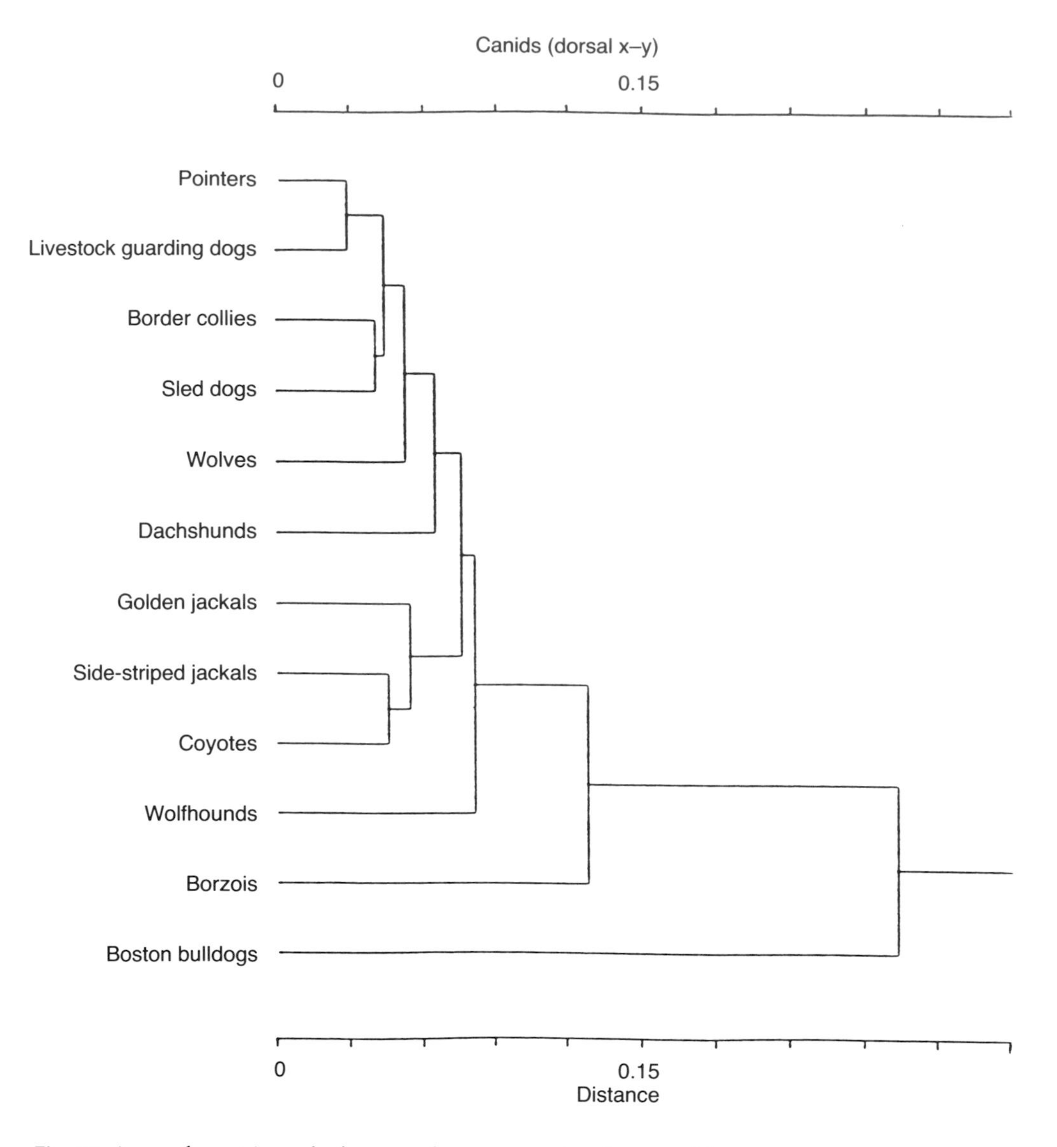

Fig. 3.9. A morphometric method was used to generate data from cranial landmark points to find out how the skulls of five species of canid differed in shape (size is scaled to a constant). Domestic dogs differed more from each other than they did from wild canids. This tree shows relationships based on similarity, not evolutionary history. There are interesting similarities between this tree and one produced by Clutton-Brock *et al.* (1976). From R. Schneider (1991), unpublished BA thesis, Hampshire College.

Using data on the ratio of brain-case length to width, Dahr (1941) concluded that dogs evolved from a single ancestral form smaller than the wolf but larger than the jackal. Until the matter is cleared up, we call the ancestor 'wolf' because it is the shortest word to type. Kruska (1988) argued that most domesticated animals have reduced brain to body proportions compared with their wild ancestors, and that there is a disproportionate reduction in the limbic system. He related this to the attenuation of aggressive behavior and cognitive changes in domestic species.

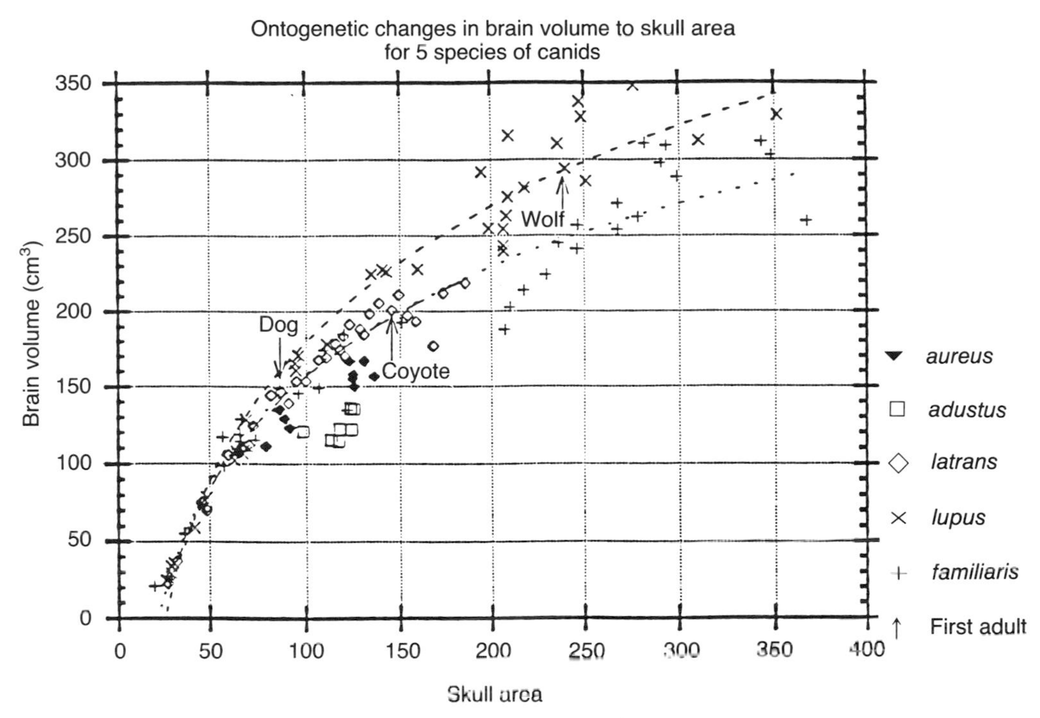

Fig. 3.10. Even though all members of the genus *Canis* start with the same size skull and brain, wolves develop bigger brain volumes than dogs for the same size head. From R. Schneider (1991), unpublished BA thesis, Hampshire College.

The behavioral origin of domestic dogs

The search for some anatomical similarity between dogs and the rest of the genus is perhaps asking the wrong question, since the real difference between them is behavioral: dogs are tameable (Scott, 1954) and trainable, wild canids are not. Anatomical details are fascinating, and give clues to understanding where the behavioral tapes came from and how the transformation was made between the wild canid and the tame dog. We would like to contend that the behavioral and anatomical commonalities of dogs suggest that the ancestor of breeds of dog was the dog itself.

It has been hypothesized that dogs descended more than once from several wild species (a polyphyletic evolution), and that the working behavior of the different breeds (e.g. hunting, guarding or sledding) were independently derived. Lorenz (1955) once proposed a theory (which he later rejected) that the spitz-type dogs were of wolf origin, whereas hunting dogs such as pointers were derived from jackals. One still sees references to these ideas in the popular press and there is still the notion that some palaeolithic hunter brought home a wolf puppy that grew to be a hunting companion and thence the founding sire of dogs.

Too often, writers forget that it is populations that evolve, not individuals, and that the evolving population has to be in some way isolated from the parent population. In many areas of the world, wild canids scavenge dumps and clean up human waste in villages. Boitani *et al.* (see Chapter 15) believe that wild wolves in Italy eating spaghetti at the dumps represent a step in the direction of domestication. The very act of creating refuse selects for those wild individuals that are less shy of people. Thus, even if the dog is pre-Neolithic, the evidence still points to it being a product of permanent settlement (Davis & Valla, 1978).

Meggit's (1965) observations suggest a model for the domestication process. Australian aboriginal people collected dingo pups from dens and although these animals were eaten only in times of distress,

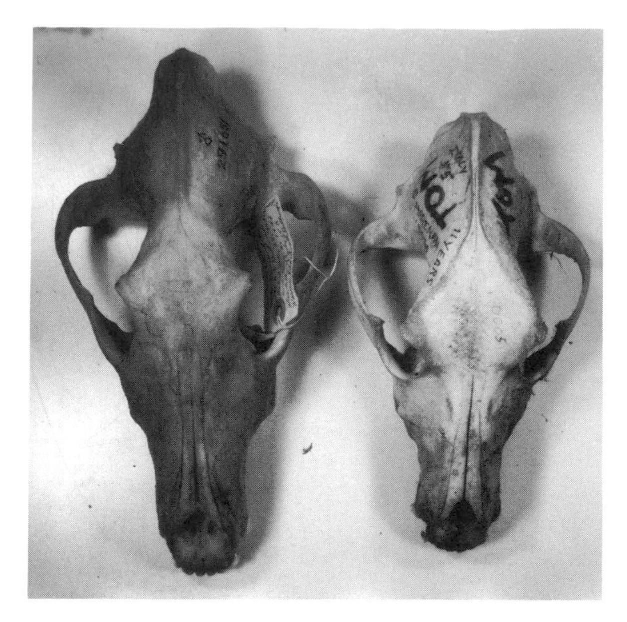

Fig. 3.11. The skulls of a 43 kg wolf (left) and a 43 kg dog. Wolf specimen NMNH 271651, courtesy of the US National Museum of Natural History; dog 5M77-B, Hampshire College. Photo by R. Coppinger.

they were fostered and allowed to scavenge around the settlement. After two years these dingoes would become sexually active and return to the wild to breed. If pups collected for village life have a better survival rate than their siblings left in the wild, then a number of selective forces could be postulated. Females that made dens and whelped near villages where their pups could be found would have a higher likelihood of reproductive success than more secretive nesters. In canids with a long maturation period, growth and development are limited by the provisioning capacity of the mother (Macdonald & Moehlman, 1983). Wolves and African hunting dogs solved the pup-feeding problem with packing behavior, in coyotes the male helps, and jackal pairs are assisted by the 'maiden aunt.' The tremendous success of the domestic dog is based on its ability to get people to raise its pups.

The advantage to people for this service is food and fur, and of course the clean-up function. Puppies on Pacific islands were a cherished source of protein (Titcomb, 1969), and made a good lunch on long canoe trips (Corbett, 1985). Eskimos of the Alaskan northwest commonly used dogs as food, and valued dog fur for garments. These animals were allowed to forage the village for rodents, wastes and human feces, and if fed, got the portions of game or rotten food that humans generally did not use (Lantis, 1980). Powers & Powers (1986) found that dogs were a highly prized food source, as well as a sacrificial object, for the Oglala peoples of North America. However, they noted that neither dogs that had been given names nor more chewy, adult animals were eaten. Dogs represented a means to turn the waste and surplus food of one season into the harvest of another. The evolution of the dog is probably identical to other domestic species that first began as village scavengers (Zeuner, 1963), became useful as food and fiber, and then, as they became tameable and trainable, were used for work.

What we are searching for is an evolutionary scenario, biologically beneficial to both humans and dogs, which would also include some allopatric or allochronic isolating mechanism. Isolation in space or time is necessary for speciation to occur. For a wild population to remain intact while the domestic population evolves is difficult to envision (see Mayr, 1963, for discussion). More plausible is the notion that the advent of village life and its changes on the immediate environment created enough differential mortality for adaptation to transform the wild wolf into the tame village dog.

The genetic tape we are looking for is not found in the wolf. By the time people began selecting for working breeds, a greatly modified wolf tape had already evolved into the small-headed, small-brained, small-toothed, tameable, trainable village pet. Three hypotheses are now offered to explain how the tape was converted into the ancestral dog and then altered for use by the various working breeds: direct trait-by-trait selection, neoteny and hybridization.

Direct trait-by-trait selection

A common assumption is that changes between dogs and their ancestors, or among breeds of dog, are the product of a process of trait-by-trait selection akin to natural selection. It is hard to imagine that many breed-specific traits (e.g. the short limbs of the dachshund, the brachycephalic face of the Pekingese) were ever initially selected for in terms of an adaptionist scheme. Rather, these traits were probably inadvertent outcomes of some other process, only to be subsequently reinforced through artificial selection.

Little evidence exists of breeders choosing among dogs for a gradual accumulation of some character, such as shorter and shorter legs, in order to produce a short-legged dog that could walk into a hole and extract a rabbit. Nor are the selective advantages of floppy ears, curly tails, jowls or variegated coat colors immediately obvious. Bemis (1984, p. 304) pointed out that 'the observation of evolutionary change by itself is insufficient evidence that adaptation actually occurred.'

When considering the diversity of dog breeds, it is unreasonable to propose 'rational adaptive explanations to account for each of these changes ... Rather, some, perhaps most of the changes are interpreted as the product of selection operating on a restricted set of characters' (Bemis, 1984, p. 303), effecting what Coppinger & Smith (1989, p. 23) called the 'whole package of behavior and morphology.' Darwin observed that 'if man goes on selecting, and thus augmenting, any peculiarity, he will almost certainly modify unintentionally other parts of the structure, owing to the mysterious laws of correlation' (cited in Løvtrup, 1987, p. 105). If there is selection for just one trait, many others may be altered in the process (Geist, 1971).

In an experiment to develop a more manageable animal, Belyaev (1979, p. 305) bred only those silver foxes that consistently displayed tame behavior and found that after 20 years of intense selection, 'quite new morphological characters appeared that are not found in wild animals but are quite characteristic of some breeds of dog: a peculiar position of the tail ... and finally, the drooping ears characteristic of young dogs.' Unlike wild types, some of these foxes had black-and-white piebald coats (Fig. 3.12) and even a diestrous breeding cycle. Careful to mitigate the effects of inbreeding, Belyaev selected from over 10 000 foxes and outbred to numerous farms. He also demonstrated that these traits were not due to simple Mendelian gene segregation.

Belyaev selected for a behavior: tameness. The resulting dog-like physical characteristics were not selected for. They could not have been selected for, since they did not exist in the ancestor, even as rare recessives. Some characteristics of dogs, such as piebald coats, droopy ears, curly tails or biannual estrus are not found in the wolves, i.e. there is no evolutionary homologue. It is difficult to see how neolithic breeders could have selected for them. Early selection for tameness, however, may have created the same leap in the dog that it did in Belyaev's foxes.

Arguments for direct selection do not account for the appearance of these novel characters. Coppinger & Feinstein (1991) attempted to determine the selective advantages of barking, but observed that this ubiquitous dog behavior was not common in the rest of the canines. Rarely in the adults, but more often in the juveniles of wild species, has barking been recorded (Lehner, 1978; Schassburger, 1987). Arguments that barking originated by natural selection, with specific benefits for particular breeds, do not recognize an initial developmental mechanism behind this ambiguous vocalization. Barking and the other features found in dogs seem to emerge as artifacts – independent of natural selection. Why would a neolithic breeder of wolf- or bear-hounds select for small teeth, small skulls and small brains? It seems more reasonable to assume that since all dogs have these concordant features, they are a consequence of some underlying process of domestication.

Even with a theory like 'punctuated equilibria' (Eldredge & Gould, 1972), there does not seem to be enough time in the dozen millennia since the first dogs appeared to collect all the mutations necessary to produce the hundreds of breeds existing today. Geist (1987, p. 1067) concurred with others (Waddington, 1957; Løvtrup, 1977, 1987) in the rejection of gradualism, stating that 'the gradualist model, in its selectionist forms, fails to address the fact that phenotypes, not genotypes, are the raw material of natural selection.' We would extend the concept to include the behavioral phenotype. In the case of our three working types (and Belyaev's foxes), it was the ability to perform that was selected for, with little cognitive notion of what the resulting dogs would or should look like.

That wolves and all breeds of dog have identical karyotypes and are interfertile, suggests that the only genetic differences are allelomorphic, that is, changes in the rates and times at which various developmental events occur. Differences in coat color and pattern, for example, are due to allelic differences in the production and distribution of melanin (Little, 1957; Burns, 1966). A white dog does not have genes for white. Rather it produces less of the pigment (melanin) that makes another dog black. But genes for the pigment melanin also affect other derivatives of the same tyrosine pathway, such as hormones (e.g.

Fig. 3.12. Selecting for tame behavior produced a number of dog-like characteristics in foxes. Photo courtesy of D. Belyaev.

adrenalin) and neurotransmitters (e.g. dopamine), all of which are involved in the tamable/trainable package. Thus, what seems to be selection for a single gene, or more properly for an allelic form of a gene, produces large morphological and behavioral effects.

Neoteny

After Darwin evoked 'the mysterious laws of correlation,' some authors postulated evolutionary mechanisms to explain why certain characters evolve together, or why saltatory events (leaps) sometimes occur in evolution. Belyaev selected for tameness and he also got characters that he had not selected for and did not even want.

A number of authors have been impressed that adult dogs look very much like wolf puppies. The short face of the dog, they thought, might be a juvenile characteristic inherited from the ancestor by retarding the developmental processes so that the descendent dog would have many of the morphological and behavioral features of the juvenile ancestor, i.e. the wolf puppy. It is easy to imagine that the universally tame quality of dogs stems from the care-soliciting behavior of the ancestor's juvenile, and that the small head/teeth/brain feature comprises a concordance of juvenile traits with tameness. Most juvenile mammals are reasonably tame, and their care-soliciting behavior tends not to be species-specific, the way courtship or care-giving behavior is. Juvenile behavior also tends to be dependent, whereas adult behavior is independent. Thus, the theory is that if early people had selected for wolves that retained puppy-like qualities long into adulthood, they would have ended up with something like dogs, just as Belyaev did when selecting for tameness in foxes.

The evolutionary process whereby an animal retains its youthful characters as an adult is called neoteny (see de Beer, 1958; Gould, 1977, for reviews). The idea that dogs evolved neotenically was proposed originally by Bolk (1926) on the basis of morphology. Dechambre (1949) postulated that breeds are differentially neotenic, while Fox (1965, 1978), Frank & Frank (1982) and Coppinger and associates (1982, 1983, 1985, 1987, 1989) extended the theory to include behavior. Coppinger & Coppinger (1982) proposed that breed-specific behavior is linked closely with distinct stages of ancestral ontogeny and can be predicted on the basis of head shape (see Fig. 3.13).

Coppinger *et al.* (1987) further hypothesized that the low frequency of predatory behavior found in livestock guarding dogs is a result of farmers selecting for animals that were retarded in their ontogenetic development. Similarly, herding dogs were arrested in a developmental stage where some of the predatory patterns (eye, stalk, chase) emerged, while others (bite) were less developed. Wayne (1986) and Gould (1986) reiterated the idea that stages of ontogeny provide the 'snapshot' upon which evolution operates.

Wild mammals pass through several distinct stages in their lives (ontogeny). Starting as eggs, they progress through the fetal stage, emerging as neonates, then go through adolescence and become adults. The popular conception is that as mammals grow, they develop. Incomplete as infants, they become complete as adults. The adult is seen as the niche-adapted animal. But it is important to remember that the fetus and neonate are fully developed specialists, perfectly adapted to their own special niches. The mammalian neonate is one of the marvels of evolution. It has its own organ system for feeding, including a unique portion of the brain and a highly specialized tongue, facial muscles and unique care-soliciting behavior. It has its own specific vocalizations, given only during the neonatal and early juvenile period. Growing up means substituting one set of organ and behavior systems for another, very much as the fetus substitutes

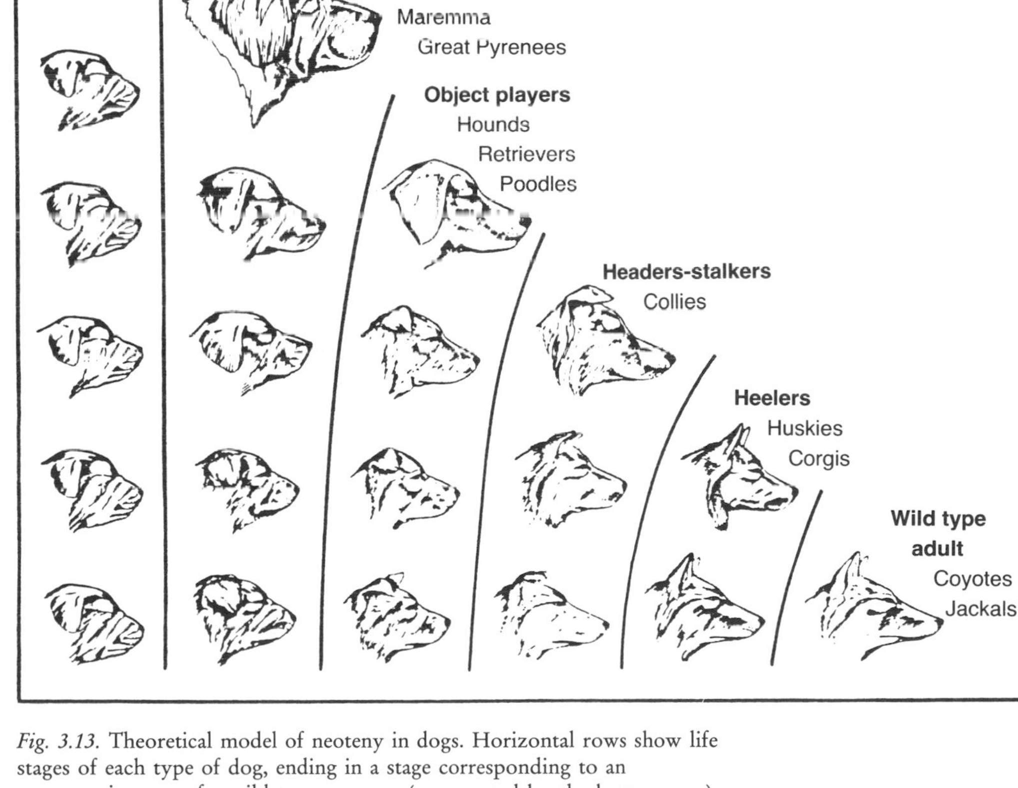

Fig. 3.13. Theoretical model of neoteny in dogs. Horizontal rows show life stages of each type of dog, ending in a stage corresponding to an ontogenetic stage of a wild-type ancestor (represented by the bottom row). Illustration by C. Lyon for Coppinger & Coppinger (1982); courtesy of *Smithsonian Magazine.*

the placenta for lungs as a way of getting oxygen from its environment.

'Growing up' is not a matter of growing bigger but rather remodeling (Enlow, 1975) or metamorphosing (Coppinger & Smith, 1989). Sucking, for example, does not grow or even develop into chewing. Sucking disappears and chewing appears ontogenetically (Hall & Williams, 1983). Behavior patterns appear, overlap and disappear as the organism leaves one ontogenetic stage and enters another.

The metamorphic period between the neonatal and the adult stages is usually referred to as the juvenile or adolescent period. The adolescent shows, in varying frequencies, both neonatal and adult motor patterns. What is important here is that the neonate cannot chew and the adult cannot suck, but the adolescent can and often does do both.

The neotenic argument holds that selection has favored those dogs that become reproductive in the adolescent or metamorphic period of the ancestral wolf. The reproductive dog does not mature to the relatively fixed behavioral repertoire of the niche-adapted adult wolf. The dog's stereotypic neonatal behavior beings to disappear, but the retardation of development prevents these animals from ever fully developing the ancestral adult's functional motor sequences.

In many ways this provides a perfect description of the adult domestic dog with neonatal and adult behavior mixed. Here is an animal that solicits care, begs for food and sits around at a rendezvous point waiting for 'parents' to show up with food, and at the same time will eye and chase a ball.

One of the theoretical consequences of this process is that the dog can use these inherited but non-stereotyped neonatal/adult motor patterns in novel ways, simply because during the metamorphic stage neither the neonatal nor the adult motor patterns are connected in functional sequences. The subject of using unsequenced micro-motor patterns as tools of learning is reviewed by Coppinger & Smith (1989). Dogs use bits of behavior, mix them or delete them, in ways that are not available to the adult wolf. When an animal mixes neonatal and adult behavior in non-functional sequences, the result is often called 'play.' 'Learning' takes place when an animal rearranges the motor patterns into functional sequences.

It is not surprising that juveniles of most mammalian species play more and learn better than adults. There may be some truth to the axiom, 'You can't teach an old dog new tricks.' It might be more accurate to say that you cannot teach new tricks to an adult animal that has rigidly sequenced motor patterns, which is why it is easier to teach the (neotenic) dog than the (adult) wolf. It is interesting to note that wolves have a more rapid cognitive development, even though the dog excels at being taught (Frank & Frank, 1982).

All the domestic animals are thought to have been derived from their ancestors by some form of neotenic retardation simply because they exhibit morphology and behavior throughout life that appear in the juvenile stages of their wild relatives (Zeuner, 1963; Ratner & Boice, 1975; Geist, 1978; Clutton-Brock, 1981; Price, 1984; Coppinger & Smith, 1989). If the theory has any validity, then searching for the ancestor of the domestic dog in adult populations of wild species is futile, for it does not exist there (see Geist, 1971; Northcutt, 1990).

The problem of postulating that specific motor patterns such as eye, stalk, chase and bite are homologous to the predatory behavior patterns of the wild ancestor (Holmes, 1966; Vines, 1981; McConnell & Bayliss, 1985) is illustrated in Fig. 3.14. Any hypothesis of homology needs to be examined with specific criteria (Roth, 1988) before a direct correspondence can be made. Even if these behavior patterns are homologues, the fact that the motor patterns can be isolated from the rest of the predatory motor patterns or that they can be combined with neonatal behavior suggests they are homologous to the unsequenced display of the ancestor's juvenile, rather than to the sequenced motor patterns of the adult.

A major problem with notions of evolutionary retardation is that they imply a gradual and global process acting upon several organ systems through phylogenetic time. Some authors feel that such a global process underestimates the dynamic nature of development and ignores the processes of ontogenetic change (Fink, 1982; Alberch, 1985; Atchley, 1987). It may have been that Belyaev was inadvertently selecting for retarded development, which is how he got the tame descendants, but the diestrous cycle certainly never appeared in the ancestor's

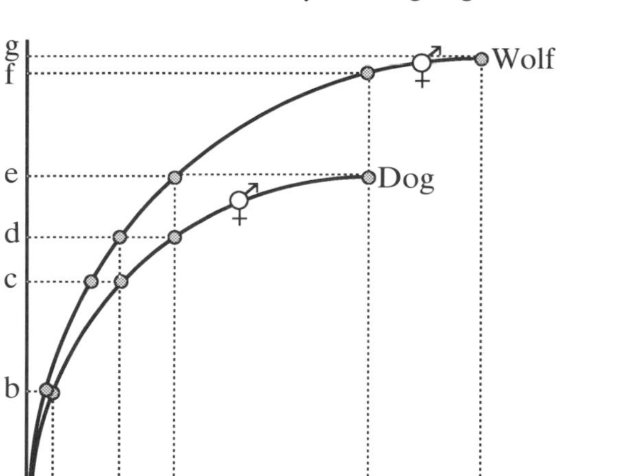

Fig. 3.14. Two theoretical ontogenetic sequences comparing 'homologous' characteristics of wolf and dog. Up to point 1b the trajectories are similar. From point b on, each character appears at a later ontogenetic stage in the dog descendant relative to the wolf ancestor. Even though the dog develops to stage 4 it is difficult to say whether character f is a true homologue since the developmental processes are so different. Character e is juvenile in the wolf and adult in the dog, while adult wolf character g never appears in the dog. Adapted from Alberch (1985, p. 48).

juvenile, and the piebald coat may have no phylogenetic history at all.

There is nothing wrong with the neoteny idea in theory, but it is difficult to find the morphological correlates or the concordance of characters necessary to substantiate the view portrayed in Fig. 3.13. Dogs are not short-faced wolves (see Olsen, 1985, for a discussion); in fact, members of the genus *Canis* are conservative in their skull-length proportions. Schneider (1991, unpublished BA thesis, Hampshire College), using a three dimensional digitizer system and a coordinate data analysis program, found no short-faced quality to dogs when compared with wolves, using robust statistical and morphometric methods; nor does Wayne (1986), who claimed that differences in the head shapes of dogs are in proportions of width and height rather than length. Similarly, Morey (1992) found the identical result in his examination of prehistoric dogs. The livestock guarding dogs, the herders and sled dogs all have palate to skull length proportions that are the same as wild members of the genus. The reduced skull and brain size might be neotenic characters, but of course early juveniles do not have permanent teeth, not even small ones.

The classical illustration of the neotenic process[1] is the salamander (*Ambystoma* spp.) that is arrested in a larval stage (de Beer, 1958). Becoming reproductive as a larva fits the global notion of carrying youthful characters into the adult period. But dogs are retarded in the metamorphic stage, which means that accommodation between the various features creates a lot of epigenetic noise. Just as the coat color of the foxes became piebald, so, metaphorically speaking, does the rest of the morphology and behavior.

The transformation from wild fox into dog-like creature that Belyaev made in a few generations simply by selecting for tameness, must mirror in some way the original transformation of wolf into dog. Belyaev thought that selection for tameness 'destabilized' the genome in such a way as to create evolutionary novelties. As we learn more about gene action and biochemistry we find that he was probably not very far from the truth. The diestrous heat cycle and the piebald coat may be results of a neotenic process, but they are not neotenic characters.

Arons & Shoemaker (1992) found livestock guarding dogs different from border collies and from sled dogs in both the distribution and amount of neurotransmitters in various sections of the brain. Guarding dogs have low levels of dopamine in the basal ganglia, whereas Border collies and huskies have higher levels. The distribution of dopamine in these three types is supportive of differential neoteny (fig. 3.13) since it correlates well with the findings that altricial neonates tend to have low levels of dopamine. One would expect that selection for the differential display of breed-specific predatory behavior would be reflected in various neurological patterns. Whether these neurological patterns are the result of direct selection for the single character or a pleiotropic effect cannot be determined without a number of clearly concordant features (Alberch *et al.*, 1979).

[1] A long discussion of which heterochronic process – paedogenesis, fetalization or neoteny – is most parsimonious would be inappropriate here, and we use neoteny to represent them all because it is the shorter and more pleasant word.

In our laboratory we continue to look for that concordance, because neoteny is such an appealing hypothesis, but it is a difficult hypothesis to test.

Hybridization

There is another way in which heterochronic shifting of developmental events can take place: hybridization or mongrelization. Of the 200–400 breeds (pick your favorite number), the majority have been created since the late nineteenth century by cross-breeding.

New forms or saltatory changes can be produced quickly by hybridization (Stebbins, 1959). 'Indeed, selection for tameness in the wild progenitor of dogs and other domestic species for the last 10 000–15 000 years may have been facilitated by hybridization and the resulting disruption of eco-specific behaviors' (Coppinger, Smith & Miller, 1985, p. 561). Not only does hybridization cause a disruption of eco-specific behavior in wild forms, but, more important, such systems as predation, courtship, territory, social hierarchy and pack formation fragment and sometimes disappear, in much the same way that they do under neotenic selection. Our feeling on the development of breeds is expressed by Haldane (1930, pp. 138), writing about the evolution of species, when he stated that there is 'every reason to believe that new species may arise quite suddenly, sometimes by hybridization, sometimes perhaps by other means. Such species do not arise, as Darwin thought, by natural selection. When they have arisen they must justify their existence before the tribunal of natural selection'.

As an evolutionary mechanism, hybridization can lead to morphological diversification and 'structural disharmonies' (Stockard, 1941, p. 17) – a concept reminiscent of Belyaev's 'destabilization'. Alberch (1982, p. 20) viewed diversification and adaptation as independent processes, with the former inherently preceding the latter. He stated,

> 'the role of development in evolutionary processes of morphological change is twofold. On one hand, the structure of the developmental program defines the realm of possible novelties . . . while, on the other, the regulatory interactions occurring during ontogeny can accommodate genetic and environmental perturbations and result in the production of an integrated phenotype.'

The limits of these structural disharmonies depend on the organism's ability to accommodate the changes. Accommodation is the fitting together of the various organ and behavioral systems. A simple example was given by Twitty (1966), an embryologist who transferred the eye germ cells from a big salamander species to a tiny species, resulting in little salamanders with great big eyes. Remarkably, the little salamanders grew eye sockets that fit their new eyeballs and added extra brain cells to 'accommodate' their large eyes.

As with the salamander, the dog's skull has features that have not been selected for but are simply the accommodation of one organ system to another. Evans & Christensen (1979, p. 1073–77) found that while 'there is considerable variation between breeds in regard to the position of the eyes, the size of the orbit, and the size and shape of the palpebral opening . . . the eyeball is nearly spherical, differing little in its sagittal, transverse, and vertical diameters.' The radius of the eyeball is usually 11 mm for most dogs, regardless of body size or breed. We suggest that measurements would support our initial findings that eyeball size is reasonably constant across the genus.

Figure 3.15(*a*) shows the relative differences (scaled to the same size) in skull shape between a newborn canine (see also Fig. 3.10) and a mesaticephalic adult (dachshund; see also Fig. 3.9). Most of the change is in the muzzle. In Fig. 3.15(*b*), the same newborn canine skull is compared with a brachycephalic adult (identified as a Boston 'bulldog'). Most of the change here is an accommodation of the zygomatic arch and the frontals to adult-sized eyeballs. The adult Boston bulldog has a skull the shape of a newborn puppy and eyeballs the same size as an adult wolf. Its brain volume is that of a 12-week-old wolf puppy. The wolf pup has its baby teeth, but the Boston bulldog has permanent, adult teeth stuck into a puppy jaw.

One can immediately see the problem with both the direct selection hypothesis and the neoteny hypothesis when examining skulls. During development, various parts of the skull have to encapsulate continuously a number of different organs and at the same time connect them together. Is the skull neotenic? Certainly the eyes and their sockets, and the teeth and their sockets are not. What about the brain? Parts of the brain might look neotenic and the size looks juvenile, but the visual cortex has neither the size nor shape of either the adult or the juvenile ancestor. It might be argued that the prominent

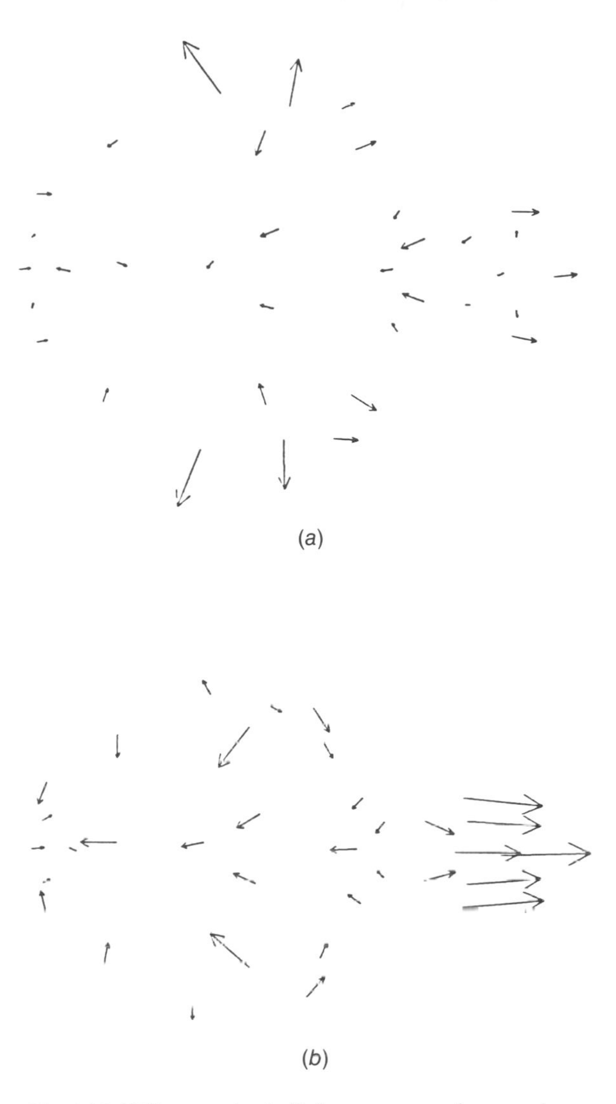

Fig. 3.15. Differences in skulls between a newborn canine pup and two types of adult dogs. Arrows indicate the vector magnitude and direction changes in the comparison of 35 dorsal cranial points. The factor of size has been removed and true shape changes are shown. As indicated, differences are greatest along the orbits and zygomatic arches. (*a*) Newborn pup and Boston 'bulldog' (brachycephalic); (*b*) newborn pup and dachshund (hypomesaticephalic). From R. Schneider (1991), unpublished BA thesis, Hampshire College.

'stop', often diagnostic in dogs with a large frontal sinus, is not a product of selection but is an epigenetic accommodation between the adult-sized eye and the puppy-sized brain.

Hybridization has the same effect on behavior that it does on morphology. Hybridization in many ways mimics neoteny in its effects on behavior (Coppinger *et al.*, 1985). Cross-breeds behave differently from their parents (Fox, 1978), theoretically producing phylogenetically novel combinations of onsets and offsets and sequencing of motor patterns. Pellis (1988) in his studies of play suggested that behavioral differences between closely related species are often the result of changes in ontogenetic timing. Several authors (Bekoff, 1972; Leyhausen, 1973; Eibl-Eiblsfeldt, 1975) have suggested that the trend away from hard-wired, inflexible behavior was one of the changes that gave modern mammals an adaptive advantage. Hybridization would appear to accomplish this metamorphosis much more rapidly, and perhaps much more radically, than the other systems of evolution we have referred to.

Conclusion

The concordance of morphological and behavioral features found in the domestic dog is best explained by the evolutionary mechanism of heterochrony (see Morty, 1992). Neoteny, in particular, may have played an important role in creating an organism suitable for domestication. Gould (1977, p. 348) stated that '. . . general juvenilization of behavior permits a more gregarious society.' Selection for juvenile behavioral characteristics would not only make for a more tractable animal, with juvenile care-soliciting behavior and lack of species recognition, but serendipitously would also retard the onset of dispersing motivations and adult inflexible motor patterns, resulting in a tameable and trainable companion.

Coppinger & Smith (1983) suggested that many of the ancestors of domestic animals were already neotenic and therefore predisposed behaviorally to interact socially with humans. Geist (1971, p. 346) showed that evolutionary changes during the colonization of new territory in the wake of glacial retreat could 'be adaptations not to the physical habitat but to changes in the social environment.' We have suggested that even if the dog is pre-neolithic, it was associated with permanent settlements, and selection favored those animals that were not wild or shy, which in turn allowed them to exploit energy resources associated with village life.

We believe that inadvertent selection for tameness is the cause of the observed morphological and behavioral divergence between dogs and their wild ancestor. Crossbreeding between individuals with novel and disparate features would create further novelties via epigenetic accommodation between organ and behavioral systems. These novelties become the bases of breed selection.

Dogs are trainable for the same reasons that they are tameable. Trainability is not particularly characteristic of either the neonate or the adult in the wild species, but it is a feature of adolescents and juveniles. The species that continue to learn easily as adults, such as dogs and humans, are thought to be neotenic. Adolescents also lack good species-recognition patterns, which might be a prime factor in fostering their social interaction.

Adolescents are not without inborn motor patterns, but those innate patterns tend to derive from two separate stages of ontogeny – the neonatal and the adult – and are not established in relatively fixed functional sequences. Thus, infant and adult behavior patterns are available to the adolescent for integration and rearrangement into novel patterns. Many motor patterns are internally motivated, and the display provides its own reward. Such behavior is easily arranged sequentially but can be eliminated only by removing the environmental stimulus that elicits it (Hart & Hart, 1985), by surgically modifying the animal (Anderson, 1970), or by getting another dog (Free, 1970).

Learning to perform a task is not based solely on behavioral plasticity, such that any breed can perform if the correct instrumental reinforcement is provided. Border collies must show eye before they can be trained to herd sheep. Livestock guarding dogs cannot perform if they do show eye. Furthermore, there is no known conditioning that can produce this motor pattern, and it is difficult to get rid of once it appears.

Efforts in our laboratory to raise Border collies and guarding dogs in controlled conditions have always produced aberrant adults. Either one type or the other is unable to work because the rearing conditions have been inappropriate for its normal development (Langeloh, 1984, unpublished BA thesis, Hampshire College). Nor is it possible to induce one type to perform the task of the other simply by raising it in the other's environment.

Sled dogs, guarding dogs and herding dogs provide classic examples of how working breeds of dog were created. We agree with Alberch (1982, p. 19) that '... diversification (characterized by changes in structural complexity) precedes adaptation.' Starting with inherited diversity created by hybridization, those dogs that perform best are the ones that survive reproductively. Selection is not for individual characteristics but for the accommodation of a multitude of characteristics. All dogs have the rudimentary 'talents' to perform particular tasks. Those talents are products of the genetic structures that make dogs tameable and trainable. Dog breeders capitalized on this by selecting animals for their performance, and created specialized breeds with distinct abilities to help people at work or at play.

Acknowledgments

Our thanks to James Serpell for getting this paper out of its neonatal form and into a creative metamorphic period, and to Lorna Coppinger who coaxed it to maturity. Mark Feinstein and Lynn Miller (Hampshire College), William Bemis (University of Massachusetts), Carl Phillips (Hofstra University) and Richard Thorington, Jr and Ralph Chapman (US National Museum of Natural History) all helped us reorganize our motor patterns in new and interesting ways: they teach well. David Macdonald (Oxford), James Serpell (Cambridge) and an anonymous reviewer showed us where the structural disharmonies were in the original draft. Unbeknown to them, Valerius Geist, Pere Alberch and Stephen J. Gould provided the mental saltations that keep us working on these problems. We also thank all those Hampshire College students who collected data, and all those dogs that 'behaved' for us.

References

Alberch, P., Gould, S. J., Oster, G. F. & Wake, D. B. (1979). Size and shape in ontogeny and phylogeny. *Paleobiology*, **5**, 296–317.

Alberch, P. (1982). The generative and regulatory roles of development in evolution. In *Environmental Adaptation and Evolution*, ed. D. Mossakowski & G. Roth, pp. 19–36. New York: Gustav Fischer.

Alberch, P. (1985). Problems with the interpretation of developmental sequences. *Systematic Zoology*, **34**, 46–58.

Anderson, A. C. (1970). *The Beagle as an Experimental Dog.* Ames: Iowa State University Press.

Arons, C. D. & Shoemaker, W. J. (1992). The distributions of catecholamines and beta-endorphin in the brains of three behaviorally distinct breeds of dogs and their F_1 hybrids. *Brain Research*, **594**, 31–9.
Atchley, W. R. (1987). Developmental quantitative genetics and the evolution of ontogenies. *Evolution*, **41**, 316–30.
Bekoff, M. (1972). The development of social interaction, play and metacommunication in mammals: an ethological perspective. *Quarterly Review of Biology*, **47**, 412–34.
Belyaev, D. K. (1979). Destabilizing selection as a factor in domestication. *Journal of Heredity*, **70**, 301–8.
Bemis, W. E. (1984). Paedomorphosis and the evolution of the Dipnoi. *Paleobiology*, **10**, 293–307.
Black, H. L. & Green, J. S. (1985). Navajo use of mixed-breed dogs for management of predators. *Journal of Range Management*, **38**, 11–15.
Bökönyi, S. (1969). Archaeological problems and methods of recognizing animal domestication. In *The Domestication and Exploitation of Plants and Animals*, ed. P. J. Ucko & G. W. Dimbleby, pp. 219–29. London: Gerald Duckworth.
Bolk, L. (1926). Das problem der Menschwerdung. Jena: Gustav Fischer.
Bradshaw, J. W. S. & Brown, S. L. (1990). Behavioural adaptations of dogs to domestication. In *Pets, Benefits and Practice*, Proceedings of Waltham Symposium 20, ed. I. H. Burger, pp. 18–24. London: BVA Publications.
Brisbin, I. L. Jr (1976). The domestication of the dog. *Purebred Dogs: American Kennel Club Gazette*, **93**, 22–9.
Brisbin, I. L. Jr (1977). The pariah: its ecology and importance to the origin, development and study of purebred dogs. *Purebred Dogs: American Kennel Club Gazette*, **95**, 22–7.
Burns, M. (1966). *Genetics of the Dog. Inheritance of Color and Hair Type*. Philadelphia, PA: J. P. Lippincott.
Chiarelli, A. B. (1975). The chromosomes of the Canidae. In *The Wild Canids*, ed. M. W. Fox, pp. 40–53. New York: Van Nostrand Reinhold..
Clutton-Brock, J. (1977). Man-made dogs. *Science*, **197**, 1340–2.
Clutton-Brock, J. (1981). *Domesticated Animals from Early Times.* Austin: University of Texas Press.
Clutton-Brock, J., Corbet, G. B. & Hills, M. (1976). A review of the family Canidae with a classification by numerical methods. *Bulletin of the British Museum (Natural History) (Zoology)*, **29**, 117–99.
Coppinger, L. (1977). *The World of Sled Dogs.* New York: Howell Book House.
Coppinger, L. & Coppinger, R. P. (1982). Livestock-guarding dogs that wear sheep's clothing. *Smithsonian Magazine*, April, 64–73.
Coppinger, R. P., Coppinger, L., Langeloh, G., Gettler, L. & Lorenz, J. (1988). A decade of use of livestock guarding dogs. In *Proceedings of the Vertebrate Pest Conference*, Vol. 13, ed. A. C. Crabb & R. E. Marsh, pp. 209–14. Davis: University of California.
Coppinger, R. P. & Feinstein, M. (1991). Why dogs bark. *Smithsonian Magazine*, January, 119–29.
Coppinger, R. P., Glendinning, J., Torop, E., Matthay, C., Sutherland, M. & Smith, C. (1987). Degree of behavioral neoteny differentiates canid polymorphs. *Ethology*, **75**, 89–108.
Coppinger, R. P. & Smith, C. K. (1983). The domestication of evolution. *Environmental Conservation*, **10**, 283–92.
Coppinger, R. P. & Smith, C. K. (1989). A model for understanding the evolution of mammalian behavior. In *Current Mammalogy*, Vol. 2, ed. H. Genoways, pp. 33–74. New York: Plenum.
Coppinger, R. P., Smith, C. K. & Miller, L. (1985). Observations on why mongrels may make effective livestock protecting dogs. *Journal of Range Management*, **38**, 560–1.
Corbett, L. K. (1985). Morphological comparisons of Australian and Thai dingoes: a reappraisal of dingo status, distribution and ancestry. *Proceedings of the Ecological Society of Australia*, **13**, 277–91.
Dahr, E. (1941). Über die Variation der Hirnschale bei wilden und zahmen Caniden. *Arkiv für Zoologii*, **33A**, 1–56.
Darwin, C. (1859). Letter to Asa Gray. *Journal of the Proceedings of the Linnean Society (Zoology)*, **3**, 50–3.
Darwin, C. (1875). *The Variation of Animals and Plants under Domestication*, 2nd edn, Vol. 1. London: John Murray.
Davis, S. J. M. & Valla, F. R. (1978). Evidence for domestication of the dog 12,000 years ago in the Natufian of Israel. *Nature*, **276**, 608–10.
de Beer, G. (1958). *Embryos and Ancestors.* Oxford: Oxford University Press.
Dechambre, E. (1949). La théorie de foetalization et la formation des races de chiens et de porc. *Mammalia*, **13**, 129–37.
Eibl-Eibesfeldt, I. (1975). *Ethology: The Biology of Behaviour*, 2nd edn. New York: Holt, Rinehart and Winston.
Eldredge, N. & Gould, S. J. (1972). Punctuated equilibria: an alternative to phyletic gradualism. In *Models in Paleobiology*, ed. T. J. M. Schopf, pp. 82–115. San Francisco, CA: Freeman Cooper.
Enlow, D. H. (1975). *Handbook of Facial Growth.* Philadelphia, PA: W. B. Saunders.
Evans, H. E. & Christensen, G. C. (1979). *Miller's Anatomy of the Dog*, 2nd edn. Philadelphia, PA: W. B. Saunders.
Fiennes, R. & Fiennes, A. (1970). *The Natural History of Dogs.* Garden City, NY: The Natural History Press.
Fink, W. L. (1982). The conceptual relationship between ontogeny and phylogeny. *Paleobiology*, **8**, 265–81.
Fischer, R. A., Putt, W. & Hackel, E. (1976). An

investigation of 53 gene loci in three species of wild Canidae: *Canis lupus*, *Canis latrans*, and *Canis familiaris*. *Biochemical Genetics*, **14**, 963–74.

Fox, M. W. (1965). *Canine Behavior.* Springfield, IL: Charles C. Thomas.

Fox, M. W. (1971). *Integrative Development of Brain and Behavior in the Dog.* Chicago: University of Chicago Press.

Fox, M. W. (1978). *The Dog: Its Domestication and Behavior.* New York: Garland STPM Press.

Fox, M. W. & Bekoff, M. (1975). The behaviour of dogs. In *The Behaviour of Domestic Animals*, 3rd edn, ed. E. S. E. Hafez, pp. 370–409. London: Baillière Tindall.

Frank, H. & Frank, M. G. (1982). On the effects of domestication on canine social development and behavior. *Applied Animal Ethology*, **8**, 507–25.

Free, J. L. (1970). *Training Your Retriever*, 4th edn. New York: Coward McCann.

Freedman, D. G. (1958). Constitutional and environmental interactions in rearing of four breeds of dogs. *Science*, **127**, 585–6.

Geist, V. (1971). *Mountain Sheep: A Study in Behavior and Evolution.* Chicago: University of Chicago Press.

Geist, V. (1978). *Life Strategies, Human Evolution, Environmental Design.* New York: Springer-Verlag.

Geist, V. (1987). On speciation in ice age mammals, with special reference to cervids and caprids. *Canadian Journal of Zoology*, **65**, 1067–84.

Gould, S. J. (1977). *Ontogeny and Phylogeny.* Cambridge, MA: Harvard University Press.

Gould, S. J. (1986). The egg-a-day barrier. *Natural History*, **95**, 16–24.

Haldane, J. B. S. (1930). *The Causes of Evolution.* London: Longmans Green.

Hall, W. G. & Williams, C. L. (1983). Suckling isn't feeding, or is it? A search for developmental continuities. *Advances in the Study of Behavior*, **13**, 19–254.

Hart, B. L. & Hart, L. A. (1985). *Canine and Feline Behavioral Therapy.* Philadelphia, PA: Lea & Febiger.

Herre, W. & Rohrs, M. (1977). Origins of agriculture. In *World Anthropology*, ed. C. A. Reed, pp. 254–79. The Hague: Mouton.

Holmes, J. (1966). *The Farmer's Dog.* London: Popular Dogs.

Kruska, D. (1988). Mammalian domestication and its effects on brain structure and behaviour. In *Intelligence and Evolutionary Biology*, ed. H. J. Jerison & I. Jerison, pp. 211–50. Berlin: Springer-Verlag.

Lantis, M. (1980). Changes in the Alaskan Eskimo relation of man to dog and their effect on two human diseases. *Arctic Anthropology*, **17**, 2–24.

Lehman, N., Eisenhawer, A., Hansen, K., Mech, L. D., Peterson, R. O., Gogan, P. J. P. & Wayne, R. K. (1991). Introgression of coyote mitochondrial DNA into sympatric North American gray wolf populations. *Evolution*, **45**, 104–19.

Lehner, P. N. (1978). Coyote vocalizations: a lexicon and comparisons with other canids. *Animal Behavior*, **26**, 712–22.

Leyhausen, P. (1973). On the function of the relative hierarchy of moods: as exemplified by the phylogenetic and ontogenetic development of prey-catching in carnivores. In *Motivation of Humans and Animal Behavior: An Ethological View*, ed. K. Lorenz & P. Leyhausen, pp. 144–247. New York: Van Nostrand.

Little, C. C. (1957). *The Inheritance of Coat Color in Dogs.* Ithaca, NY: Comstock.

Lorenz, K. Z. (1955). *Man Meets Dog.* London: Methuen.

Lorenz, K. Z. (1975). Foreword. In *The Wild Canids*, ed. M. W. Fox. New York: Van Nostrand Reinhold.

Løvtrup, S. (1977). *The Phylogeny of Vertebrata.* London: John Wiley.

Løvtrup, S. (1987). *Darwinism: The Refutation of a Myth.* London: Croom Helm.

Macdonald, D. W. & Moehlman, P. D. (1983). Cooperation, altruism and restraint in the reproduction of carnivores. In *Perspectives in Ethology*, Vol. 5, ed. P. P. G. Bateson & P. H. Klopfer, pp. 433–67. New York: Plenum.

Macintosh, N. W. G. (1975). The origin of the dingo: an enigma. In *The Wild Canids*, ed. M. W. Fox, pp. 87–106. New York: Van Nostrand Reinhold.

MacRury, I. K. (1991). The Inuit Dog: Its Provenance, Environment and History. Unpublished M.Phil. Thesis, Scott Polar Research Institute, University of Cambridge.

Manwell, C. & Baker, C. M. A. (1983). Origin of the dog: from wolf or wild *Canis familiaris*? *Speculations in Science and Technology*, **6**, 213–24.

Mayr, E. (1963). *Animal Species and Evolution.* Cambridge, MA: Harvard University Press.

McConnell, P. A. & Bayliss, J. R. (1985). Interspecific communication in cooperative herding: acoustic and visual signals from human shepherds and herding dogs. *Zeitschrift für Tierpsychologie*, **67**, 303–28.

Meggitt, M. J. (1965). The association between Australian aborigines and dingoes. In *Man, Culture, and Animals*, ed. A. Leeds & A. P. Vayda, pp. 17–29. Washington DC: American Association for the Advancement of Science.

Morey, D. F. (1992). Size, shape and development in the evolution of the domestic dog. *Journal of Archaeological Science*, **19**, 181–204.

Northcutt, R. G. (1990). Ontogeny and phylogeny: a re-evaluation of conceptual relationships and some applications. *Brain Behavior and Evolution*, **36**, 116–40.

O'Farrell, V. (1986). *Manual of Canine Behaviour.* Cheltenham, Glos.: British Small Animal Veterinary Association.

Olsen, S. J. (1985). *Origins of the Domestic Dog: The Fossil Record.* Tucson: University of Arizona Press.
Olsen, S. J. & Olsen, J. W. (1977). The Chinese wolf, ancestor of new world dogs. *Science*, **197**, 533–5.
Oppenheimer, E. C. & Oppenheimer, J. R. (1975). Certain behavioral features in the pariah dog (*Canis familiaris*) in West Bengal. *Applied Animal Ethology*, **2**, 81–92.
Pellis, S. M. (1988). Agonistic versus amicable targets of attack and defense: consequences for the origin, function, and descriptive classification of play-fighting. *Aggressive Behavior*, **14**, 85–104.
Phillips, C. J., Coppinger, R. P. & Schimel, D. S. (1981). Hyperthermia in running sled dogs. *Journal of Applied Physiology*, **51**, 135–42.
Powers, W. K. & Powers, M. N. (1986). Putting on the dog. *Natural History*, **95**, 6–16.
Price, E. O. (1984). Behavioral aspects of animal domestication. *Quarterly Review of Biology*, **59**, 1–32.
Ratner, S. C. & Boice, R. (1975). Effects of domestication on behaviour. In *The Behaviour of Domestic Animals*, 3rd edn, ed. E. S. E. Hafez, pp. 3–19. London: Baillière Tindall.
Roth, V. L. (1988). The biological basis of homology. In *Ontogeny and Systematics*, ed. C. J. Humphries, pp. 1–26. New York: Columbia University Press.
Sands, M. W., Coppinger, R. P. & Phillips, C. J. (1977). Comparisons of thermal sweating and histology of sweat glands of selected canids. *Journal of Mammalogy*, **58**, 74–8.
Schassburger, R. M. (1987). Wolf vocalization: an integrated model of structure, motivation, and ontogeny. In *Man and Wolf*, ed. H. Frank. Dordrecht, Netherlands: Dr W. Junk.
Scott, J. P. (1954). The effects of selection and domestication upon the behavior of the dog. *Journal of the National Cancer Institute*, **15**, 739–58.
Scott, J. P. & Fuller, J. L. (1965). *Genetics and the Social Behavior of the Dog.* Chicago: University of Chicago Press.
Simonsen, V. (1976). Electrophoretic studies on blood proteins of domestic dogs and other Canidae. *Hereditas*, **82**, 7–18.
Stebbins, G. L. (1959). The role of hybridization in evolution. *Proceedings of the American Philosophical Society*, **103**, 231–51.
Stockard, C. R. (1941). The genetic and endocrinic basis for differences in form and behavior. *American Anatomical Memoirs*, **19**. Philadelphia, PA: Wistar Institute of Anatomy and Biology.
Titcomb, M. (1969). *Dog and Man in the Ancient Pacific with Special Attention to Hawaii.* Honolulu, HI: Bernice P. Bishop Museum Special Publication 59.
Twitty, V. C. (1966). *Of Scientists and Salamanders.* San Francisco, CA: W. H. Freeman.
Vines, G. (1981). Wolves in dogs' clothing. *New Scientist*, **10**, 648–52.
Waddington, C. H. (1957). *The Strategy of the Genes.* London: George Allan and Unwin.
Wayne, R. K. (1986). Cranial morphology of domestic and wild canids: the influence of development on morphological change. *Evolution*, **40**, 243–61.
Young, S. P. & Goldman, E. A. (1944). *The Wolves of North America.* Washington, DC: The American Wildlife Institute.
Zeuner, F. E. (1963). *A History of Domesticated Animals.* New York: Harper and Row.

II Behaviour and behaviour problems

4 Genetic aspects of dog behaviour with particular reference to working ability

M. B. WILLIS

Introduction

Anyone using dogs as working animals has an obvious interest in knowing: (a) the degree to which specific features are inherited; (b) the magnitude of the differences which exist between breeds; and (c) the relationship, if any, between specific traits. Although the dog has had a longer association, and enjoys closer contacts, with humans than any other species, genetic studies on the domestic dog are not extensive. Most genetic work has been directed towards understanding the mode of inheritance of specific physical anomalies, whereas advanced studies on the inheritance of behaviour are relatively recent and stem largely from the pioneering work that led to the publication of Scott & Fuller (1965). Mackenzie, Oltenacu & Houpt (1986) provided a broad review of behaviour genetics of the dog, but the present review focuses specifically on the genetics of working behaviour, since it is in this field that behaviour studies are likely to have the greatest practical impact.

Background

As Mackenzie *et al.* (1986) have shown, studies on the inheritance of behaviour began around the turn of the century, but much of the early work was confined to well-established traits with a clear intention of finding Mendelian explanations. Thus we see Whitney (1929*b*) suggesting that, in foxhounds, vocal trailing is dominant to mute trailing. Similarly, Humphrey & Warner (1934) suggested that gunshyness in German Shepherd dogs (GSDs) was controlled by a simple gene series with two alleles: N causing under-sensitivity, and n causing over sensitivity. Thus NN animals were largely insensitive to sound, Nn were medium sensitive and nn oversensitive. In addition to auditory sensitivity, these workers also looked at sensitivity to touch and postulated a similar theory with SS being undersensitive, Ss medium sensitive and ss oversensitive. They considered that ss and nn animals were too sensitive to train effectively, and that the best animals to train were Nn/Ss. While most trainers would agree with the difficulties experienced in trying to train over or undersensitive animals, it is highly unlikely that such complex features as ear and body sensitivity are going to be inherited as simple Mendelian traits. Not only will such sensitivities be influenced by complex genetic factors, they will also be modified through interactions with the environment.

Humphrey & Warner (1934) calculated phenotypic correlations among 42 physical and 9 behavioural traits in GSDs, and found that 15 of the 378 possible correlations reached statistical significance. It is probable that the underlying genetic correlations were quite different and that the project was oversimplified in seeking Mendelian explanations. Nevertheless, the Fortunate Fields study (Humphrey & Warner, 1934) did achieve success in producing superior animals for guide dog/police work using this simplified system.

The present paper looks at behaviour in guide dogs, hunting dogs, livestock guarding dogs and police/service dogs. The importance to man of herding dogs is without question, but so few genetic data exist in this field that herding dogs are deliberately excluded from the present discussion.

Guide dogs for the blind

The role of the guide dog must rank as one of the most useful modern occupations for a working dog. For a long time the GSD was the breed of choice, but in more recent years Labrador and golden retrievers, and their crosses, have been commonly used (male GSDs being slightly too large for the task). Many guide dog organizations now breed their own stock (that in Britain being the largest dog owning body in the country), and studies from these organizations are now coming to fruition.

Most guide dog failures arise from character faults, particularly fearfulness (Scott & Bielfelt, 1976; Goddard & Beilharz, 1982). Clear sex differences have also been noted: females, for example, show more suspicion and fearfulness, while males show more initiative. When assessing the degree to which a character or trait might be inherited an important consideration is its heritability. Strictly, heritability is the proportion of the total variance observed in a trait that is due to additive effects. In lay terms, it might be best defined as that proportion of the parental superiority (over the population average) which is transmitted to the offspring. Thus, a heritability of 40% would suggest that only 40% of any parental superiority (or inferiority) would be passed on to the progeny. Heritability studies are usually based upon paternal half-sib correlations, i.e. correlations

Table 4.1. *Heritability estimates for guide dog traits (Australia)*

Trait	Sire		Dam		Combined	
	h^2	SE	h^2	SE	h^2	SE
Success	0.46	0.19	0.42	0.18	0.44	0.13
Fear	0.67	0.22	0.25	0.15	0.46	0.13
Dog distraction	−0.04	0.08	0.23	0.14	0.09	0.08
Excitability	0.00	0.09	0.17	0.13	0.09	0.08

Source: Goddard & Beilharz (1982).

between the performances of different progeny sired by a particular father. For such studies a series of sires needs to be used but the performance of the sire is not required, only that of his progeny. Alternatively, offspring–parent regressions can be made, in which case the performance of both offspring and parents needs to be known. These offspring–parent regressions can be based upon one parent (usually the sire) or the average of both parents (termed offspring–mid-parent regressions). A difficulty with such methods is that parents and offspring are, of necessity, assessed in different years, even if assessed at comparable ages.

Heritability studies are, strictly speaking, only relevant to the population from which they were derived and for the period of time when they were assessed. Nevertheless, they can provide a broad guide to inheritance. Because heritabilities are ratios they are susceptible to variation caused by environmental features. Failure to reduce environmental variation or to assess animals in a consistent fashion will tend to reduce heritability estimates and thus produce values that may be lower than the true figures. Despite these limitations, heritabilities are important because they give guidelines to the consequences of various selection procedures. Highly heritable traits should respond well to direct selection and performance testing, whereas low values may necessitate selection through progeny, or even the use of crossbreeding, to produce genetic advances.

Heritability estimates derived from 394 Australian guide dogs (Labrador retrievers) are shown in Table 4.1. They were generally high for 'success at becoming a guide dog' and for 'fear' (the principal cause of culling) (Goddard & Beilharz, 1982). Heritability estimates from American guide dogs (various breeds), based on over 700 males and over 1000 females, were produced by Bartlett (1976) and are given in Table 4.2. In many cases values did not differ significantly from zero, but some aspects of sensitivity were moderately heritable. Studies on Californian guide dogs (mainly GSD) have been published by Scott & Biefelt (1976) and are shown in Table 4.3. In 11 of the 13 traits, dam components attained higher heritabilities than sire components emphasizing the importance of maternal effects. However, most of the heritabilities were not significantly different from zero.

Both American studies suggest lower heritabilities for major traits than do the more recent Australian studies, but the reasons for this are unclear. The Australian figures derive from more sophisticated statistical analyses than the American ones, but one cannot

Table 4.2. *Heritability estimates for guide dog traits (USA)*

Trait	Male	Female	Both
Body sensitivity†	0.26	0.05	0.10
Ear sensitivity†	0.49	0.14	0.25
Fighting instinct†	−0.05	−0.08	−0.04
Protective instinct†	−0.21	−0.13	−0.12
Nose acuity††	0.30	0.05	0.12
Intelligence††	−0.17	−0.07	−0.06
Willingness††	−0.14	−0.04	−0.03
Energy††	−0.03	0.06	0.05
Confidence††	0.04	0.26	0.16
Self right†††	0.15	0.25	0.22

† low score least effect; †† low score greatest effect; ††† low score indicates most eager to submit to another.

Source: Barlett (1976)

Table 4.3. *Heritability estimates for guide dog traits (California)*

Trait	Feature	h^2
Sit	Forced sit with 3 repetitions	0.06
Come	Called with 5 repetitions	0.14
Fetch	Playful retrieving 3 repetitions	0.24
Trained response	Complex excitability, nervousness	0.08
Willingness	Responsiveness to tester	0.12
Body sensitivity	Complex reaction to pain	0.16
Ear sensitivity	Complex reaction to sound	0.00
New experience	Response to novel stimuli	0.06
Traffic	Reaction to moving cart	0.12
Footing crossing	Ability to identify surface underfoot	0.06
Closeness	How close passed to obstructions	0.04
Heel	Acceptance of leash training	0.10

Source: Scott & Bielfelt (1976)

exclude the possibility that American selection has gone on much longer and that this has resulted in the reduction in heritability estimates. It is also possible that the method of 'scoring' traits leads to marked differences in heritability estimates.

Genetic correlations between traits have been produced by Goddard & Beilharz (1983) for their Australian Labradors and by Bartlett (1976) for his American dogs. These are shown in Tables 4.4 and 4.5. The Australian work suggests high heritabilities for fear or 'nervousness', and strong positive correlations between this and 'sound shyness' and negative correlations with 'willingness'. Most practical breeders believe nervousness to be relatively strongly inherited, and there is empirical evidence in many breeds that breeding from nervous dogs leads to the production of increased proportions of nervous progeny (see below and Serpell & Jagoe, Chapter 6).

Hunting

Considering that man probably first used the dog as an aid to hunting (see Clutton-Brock, 1984 and Chapter 2), we have learned very little about hunting attributes in genetic terms. Whitney's (1929*a*, *b*) early work on hunting traits was largely confined to relatively simple behavioural features, whereas cross-breeding work by Marchlewski (cited in Burns & Frazer, 1966) showed that pointing behaviour was a complex trait with no indication that the progeny of superior field-trialists were any better than average. Sacher (1970) also showed that pointers' performance scores in field trials (on a 4-point system) were not normally distributed and that genetic influences could not be identified. Geiger (1972), working with German wirehaired pointers and a 12-point system, assessed 1463 progeny from 21 sires on four traits and obtained relatively high maternally derived heritabilities. However, he obtained insignificant sire values (Table 4.6) and no sex effects.

More recently, attempts have been made to examine the genetics of hunting potential in specific breeds in Scandinavia. Some of the principles involved in

Table 4.4. *Genetic correlations (below diagonal) and heritabilities (diagonal) in Labradors*

Trait	N	S	C	W	D	SS	B
Nervousness (N)	0.58						
Suspicion (S)	0.53	0.10					
Concentration (C)	−0.01	−0.31	0.28				
Willingness (W)	−0.57	−0.20	0.67	0.22			
Dog distraction (D)	0.11	0.63	−0.47	−0.41	0.08		
Sound-shy (SS)	0.89	0.47	0.33	−0.78	0.28	0.14	
Body sensitivity (B)	0.72	0.51	−0.29	−0.74	−0.21	0.59	0.33

Source: Goddard & Beilharz (1983).

Table 4.5. *Genetic correlations among temperament traits in American guide dogs*

Trait	BS	ES	OA	E	SR
Body sensitivity (BS)					
Ear sensitivity (ES)	1.00				
Olfactory acuity (OA)	0.75	0.58			
Energy (E)	−0.38	−0.77	−0.14		
Self-right (SR)	−0.15	−0.13	0.12	−0.14	
Confidence (C)	1.32	0.60	0.34	−0.04	−0.74

Source: Bartlett (1976).

Table 4.6. *Heritability estimates in German wirehaired pointers*

Trait	Sire	Dam
Hare tracking	0.03	0.46
Nose	0.01	0.39
Obedience	0.01	0.19
Seek	0.00	0.41

Source: Geiger (1972).

using tests as a basis for breeding work have been reviewed by Swenson (1987) and Vangen & Klemetsdal (1988). The latter looked at the English setter and Finnish spitz, both of which are used for hunting in Scandinavia more than they are in other countries. Heritabilities were calculated for various traits, as well as phenotypic and genetic correlations between traits. The researchers used 5285 English setter tests from 968 dogs by 224 sires, and 4864 Finnish spitz tests from 736 dogs by 212 sires. The results are shown in Tables 4.7 and 4.8.

Heritability estimates for English setter traits tended to be higher than those for those of Finnish spitz, but the authors emphasize that some traits did not show a normal distribution. For future work, they suggest that some traits be assessed as all-or-nothing, that identification by an experienced judge would be desirable and that adjustments for dog age are necessary (Vangen & Klemetsdal, 1988). For the Finnish spitz, breeding values[1] were calculated from the traits TS, HB and TI, and genetic progress[2] per year was shown to be +0.04%, −0.3% and +0.03% of the average score for the three traits, respectively. Vangen & Klemetsdal (1988) also recommended progeny testing, and they illustrated progeny test data for total scores that ranged from +7.83 to −4.26 but were based on too few progeny per sire to be meaningful.

Nevertheless the indications were that hunting traits were heritable, and that better systems of assessing such traits might lead to higher heritability figures and hence greater potential progress in selection. At present, failure to 'score' hunting traits accurately probably leads to lower estimates of heritability than may actually be the case. This will also lead to less accurate selection of breeding stock and hence reduced progress in hunting prowess.

Livestock guarding

In Central Europe, where the wolf survived longer than in places such as Britain, several large dog breeds evolved, which were intended to protect sheep flocks from predators and which were usually, though not exclusively, white in colour. In relatively recent times the effectiveness of such dogs as livestock protectors has been studied, primarily in USA. The first trials were conducted by Linhart *et al.* (1979), since which time it has been established that these dogs *can* deter coyote predation on sheep

[1] Breeding values for a sire can be defined as that sire's superiority over his contemporaries multiplied by the heritability of the trait being considered.

[2] Genetic progress is a measure of the advance in genetic quality that has occurred due to selection. Typically, it will be lower than the total improvement in quality, some of which will be due to environmental effects and/or training advances.

Table 4.7. *Genetic (above diagonal) and phenotypic (below diagonal) correlations together with heritabilities (diagonal) for English setters*

Trait	HE	SS	FW	CO	SI
Hunting eagerness (HE)	0.22	0.79	0.72	0.33	0.72
Style and speed (SS)	0.94	0.18	0.68	0.31	0.67
Field work (FW)	0.97	0.92	0.18	0.44	0.74
Cooperation (CO)	0.41	0.43	0.52	0.09	0.72
Selection index (SI)	0.80	0.74	0.64	0.61	0.17

Source: Vangen & Klemetsdal (1988).

Table 4.8. *Genetic (above diagonal) and phenotypic (below diagonal) correlations together with heritabilities (diagonal) for Finnish spitz*

Trait	TS	SA	FB	MK	BK	HB	FO	TI
Total score (TS)	0.11	0.48	0.51	0.57	0.48	0.66	0.60	0.72
Searching ability (SA)	0.61	0.07	0.15	0.30	0.35	0.22	0.48	0.43
Finding birds (FB)	0.94	0.79	0.11	0.13	0.10	0.17	0.16	0.28
Marking (MK)	0.77	0.97	1.00	0.04	0.48	0.35	0.33	0.47
Barking (BK)	0.46	–0.77	1.00	1.00	0.02	0.30	0.31	0.42
Holding birds (HB)	0.77	–0.01	0.31	0.55	–0.38	0.18	0.22	0.47
Following birds (FO)	0.59	1.00	0.55	0.37	–0.26	0.03	0.10	0.50
Total impression (TI)	0.83	–0.05	0.50	0.50	–0.14	1.00	0.13	0.09

Source: Vangen & Klemetsdal (1988).

(Green & Woodruff, 1983*a*, *b*). Ample evidence from questionnaires distributed to 399 livestock producers (763 dogs) demonstrated that these dogs were an economic asset: 71% of respondents suggested that their dogs were very effective, 21% somewhat effective and only 8% not effective (Green & Woodruff, 1988). Similar degrees of effectiveness have been reported by others (see review in Green & Woodruff, 1987). Bearing in mind the variable sources of the dogs and the differing environments to which they were exposed it is encouraging that they were successful at all.

In this area of behaviour there have been several studies on how livestock guarding dogs (LGDs) should be raised and trained (Coppinger *et al.*, 1983; Green & Woodruff, 1983*c*; McGrew & Andelt, 1985; Lorenz & Coppinger, 1986; see also Coppinger & Schneider, Chapter 3). Most studies involve raising LGDs with lambs from about eight weeks of age to encourage bonding, although the breeds studied have usually been those selected for this guarding trait. It has been argued that, over a long time, such breeds have been selected to show little or no predator behaviour at all (Coppinger, Smith & Miller, 1985), whereas herding breeds like the Border collie or kelpie have been selected to truncate the natural predatory sequence (see Coppinger & Schneider, Chapter 3). Breber (1977), however, has suggested that Maremmas only exhibit this loss of predatory behaviour if properly socialized to the sheep they guard, and only then if adequately fed. Coppinger *et al.* (1985) suggest that some mongrels may possess disrupted, missing or rearranged elements of the predatory sequence.

Bond-forming is well established in the dog (Scott & Fuller, 1965; Fox, 1978; Serpell & Jagoe, Chapter 6) and perhaps better exemplified in this species than in any other. It is argued that, because of the disruption of co-selected traits, mongrel dogs may be less likely to bond with alien species and be less protective of them than the Eurasian breeds selected for this affinity (Black, 1987). At the same

Table 4.9. *Comparison of five breeds of livestock guarding dog*

Breed[a]	No.	Effectiveness (%):			Aggression (%) to:	
		very	some	not	predators	dogs
Great Pyrenees	437	71	22	7	95	67
Komondor	138	69	1	12	94	77
Akbash	62	69	22	9	100	92
Anatolian	56	77	13	10	96	86
Maremma	20	70	20	10	94	94

[a] Other breeds omitted.
Source: Green & Woodruff (1988).

time, the smaller size of mongrels may prove less of a deterrent to predators. In contrast, Eurasian dogs are more trustworthy and large enough to deter many predators, although difficult and costly to obtain, and big enough to be a potential threat to humans.

Some studies on breed suitability have been hampered by lack of numbers. In their questionnaire study, Green & Woodruff (1988) looked at a variety of breeds but had sufficient numbers to compare only five. Their findings are shown in Table 4.9. Rates of success among the five breeds were not significantly different, although more komondors bit people than did Great Pyrenees, akbash or Anatolians. Similarly, fewer Great Pyrenees injured sheep than did komondors, akbash or Anatolians. It was observed that dogs reared with livestock from eight weeks or younger ($N = 280$) were more successful than dogs placed with livestock after eight weeks ($N = 227$), suggesting that bonding was easier prior to eight weeks of age.

A later study by Green (1989) used 100 purchased dogs placed with livestock breeders and showed a higher rating for Great Pyrenees than Anatolians. However, 40% of dogs injured livestock and 15% killed them. Significantly, more Anatolians than Great Pyrenees were involved in such attacks.

Recent data from Coppinger *et al.* (1988) were based on a ten-year study covering over 1000 dogs and using the co-operator questionnaire system. Using three purebreeds and two F1 crossbreeds they showed dogs to be effective at reducing predation. Average predation reduction was 64% with 53% of respondents claiming that predation was eliminated

Table 4.10. *Ratings of livestock guarding dogs (%)*

Breed[a]	Number	Good	Fair	Poor
Great Pyrenees	59	83	8	9
Anatolian	26	38	27	35

[a] other breeds deleted
Source: Green (1989)

entirely. Although reluctant to claim breed differences, on the ground that their dogs represented strains within breeds rather than breeds *per se*, the authors did demonstrate differences between breeds. In two years of the study, breed differences reached significance and in the traits 'attentiveness' and 'trustworthiness' Maremmas and F1 Maremma/Šarplaninac crosses scored better than either Anatolians, Šarplaninac or F1 Anatolian/Šarplaninac. Even if these represented strain rather than breed differences, they do show that real differences exist which could be exploited.

LGDs are clearly successful at protection work, and Coppinger *et al.* (1988) have argued that, in some areas, guarding dogs are essential for livestock farmers to remain in business. However, the underlying genetic basis for the behaviour needs further study to clarify the degree to which livestock protecting behaviour is inherited, and to further document breed differences and examine the extent to which selection could give greater success. It is not yet known the extent to which good LGDs transmit their superiority to their progeny. If this is high, then selection for superior protection dogs could prove as effective as the selection of superior guide dogs.

Police/armed service work

Studies at the Swedish army centre of Solleftea have undertaken temperament tests on German shepherd dogs at around 18 months of age, following a period of 'puppy walking' in private homes. Reuterwall & Ryman (1973) have shown that the proportion of additive genetic variance is quite small for all the temperament traits tested. Heritability estimates derived from their components of variance values were produced by Willis (1976) and are shown in Table 4.11. Only three values reach significance, although the total sample size (488 males and 438 females) is certainly adequate.

These disappointingly low values suggest a very low additive effect on these traits. However, as Mackenzie *et al.* (1986) suggest, the scoring system used was perhaps too complex, and assessments made at 18 months may not provide a true reflection of inherited differences. The effects of early experience may be important, particularly since dogs would have been 'walked' in very different situations and would have experienced varying environments (see Serpell & Jagoe, Chapter 6).

More encouraging values for inherited traits in army dogs have been produced by Falt, Swenson & Wilsson (1982, cited by Mackenzie *et al.*, 1986). These were based upon tests undertaken on eight-week-old GSD puppies and are given in Table 4.12.

Most dog trainers would agree that pursuing and picking up an object is easier for a dog to achieve than actually returning with that object. The heritability estimates obtained by Falt *et al.* (1982) tend to confirm this genetically. It is necessary, however, to know the degree to which early testing at eight weeks of age can accurately identify adult behaviour patterns. In genetic work early identification is desirable, but canine behaviour is constantly changing in early life and early selection must be highly correlated with adult performance if such selection is to be useful.

The Schutzhund working degree is widely used in Germany and elsewhere as the basis for testing and selecting working breeds of dog. Schutzhund is particularly associated with German shepherd dogs since, without a Schutzhund qualification, GSDs cannot be exhibited as adults.

Table 4.11. *Heritabilities (half-sib) of mental traits in GSDs*

Trait	Paternal half-sib values	
	males	females
Affability	0.17	0.09
Disposition for self-defence	–0.11	0.26**
Disposition for self-defence and defence of handler	0.04	0.16
Fighting disposition	0.16*	0.21*
Courage	0.05	0.13
Ability to meet sudden strong auditory disturbance	–0.04	0.15
Disposition for forgetting unpleasant incidents	0.10	0.17
Adaptiveness to different situations	0.00	0.04

*$P < 0.05$; **$P < 0.01$.
Source: Willis (1976) after Reuterwall & Ryman (1973).

Schutzhund comes in three degrees termed SchH I, SchH II and SchH III with increasing number associated with increasingly advanced tests. Essentially, the tests are divided into four broad components which might be termed: tracking, obedience, man-work (protection) and character (courage). Several thousand tests are undertaken in Germany each year, and Schutzhund groups have been set up in the USA, Britain and Eire. Despite the effort expended on such testing, little genetic work has appeared. Pfleiderer-Hogner (1979) analysed the SchH I results from 2046 tests on 1291 GSDs from 37 different sires, all the testees being born in 1973. She found no effect of dog age at test or month of trial, but she did find significant effects of sex and of the number of competitors in a test. Heritability values derived from sire and dam components and combined estimates are given in Table 4.13. None of the values reach significance.

Phenotypic correlations between the different features are shown in Table 4.14. Three of these (starred) were significant, that between man-work and character being the highest.

Table 4.12. *Heritability estimates for 8-week-old traits in GSDs*

Trait	Feature	Sire	Dam
Yelp	Time from first separation to yelp	0.66	0.73
Shriek	Time for serious cry of distress	0.22	0.71
Contact 1	Approach to stranger in strange place	0.77	1.01
Fetch	Pursue and pick up ball	0.73	0.10
Retrieve	Bring back ball	0.19	0.51
Reaction	To strange object in strange place	0.09	1.06
Social competition	Tug-of-war	0.11	0.76
Activity	Exploration in strange arena	0.43	0.76
Contact 2	Time spent near stranger	0.05	1.11
Exploration	Visits to strange objects in arena	0.31	0.83

Source: Falt *et al.* (1982). Values in excess of 1.00 are not strictly possible and will have resulted from insufficient numbers or will have high standard errors attached to them.

Table 4.13. *Heritability estimates for SchH I scores*

Trait	Sire	Dam	Combined
Tracking	0.01	0.20	0.10
Obedience	0.01	0.13	0.09
Man-work	0.04	0.07	0.06
Character	0.05	0.17	0.12

Source: Pfleiderer-Hogner (1979).

Table 4.14. *Phenotypic correlations between Schutzhund traits*

Trait	TR	OB	MW
Tracking (TR)			
Obedience (OB)	0.26**		
Man-work (MW)	0.11	0.20*	
Character (CH)	0.10	0.17	0.76***

*$P < 0.05$; **$P < 0.01$; ***$P < 0.001$.

Despite these findings, it is difficult to accept that Schutzhund testing is genetically a waste of time and one has to conclude that, if these traits have a genetic basis, either they are controlled by non-additive factors (i.e. interactions between genes) or the flaws in testing are sufficiently serious as to prevent statistical evaluation. It is highly probably that flaws in testing are large (the fact that numbers on test had a significant effect upon results is indicative of this), but either way the data would imply that selecting the best SchH I animals on the basis of current test results would not necessarily lead to progress in these traits. Pfleiderer-Hogner (1979) suggests the use of a performance test combined with sib selection, but the heritability values do not justify this. In addition, progeny testing, though time consuming, would be a necessary follow-up. A selection index based on tracking and man-work was also suggested, although more success might come from re-evaluating the testing procedure to more accurately reflect genetic aspects.

Temperament

Although most pedigree breeding stems from dogs exhibited in the show-ring the majority of dogs end up living as pets and this is true regardless of breed. The average owner requires an animal that is not nervous or aggressive but has a fairly easy-going and stable temperament. Clearly the character of a dog will be influenced by environmental features, in particular socialization (see Scott & Fuller, 1965; Serpell & Jagoe, Chapter 6). The nature of temperament, in its overall sense, is difficult to define and studies on its heritability are not readily available. Working with 575 US army GSDs, from 18 sires out of 71 dams, Mackenzie, Oltenacu & Leighton (1985) obtained a heritability of temperament of 0.51. Vague as this trait is, the figure is relatively high and suggests that selection for this feature would be successful. Most practical breeders would agree with this viewpoint, which is also in agreement with the Australian guide dog work.

Nervousness

Nervousness is a serious failing in most working dogs outside of sheep herding, and Thorne (1944) put forward the suggestion that extreme shyness is a dominant trait. His study was based on 178 dogs of which 83 were extremely shy. Forty-three of these animals descended from a single basset hound female of extremely shy nature, and he concluded from this that a dominant gene was at work (in contrast to the Fortunate Fields work suggesting a recessive trait). Certainly, Thorne's data were indicative of dominant inheritance, but this is unlikely to be true of all forms of shyness and may only apply to the specific situation he encountered. In their extensive work with pointers, Brown, Murphree & Newton (1978) developed a strain that was normal and friendly towards humans, and a second strain which exhibited extreme human-aversion. In the latter, it was concluded that much of the variability was additive. Character problems in German pointers in respect of gunshyness were reported by Kock (1984), but the heritability derived was only 0.06 ± 0.04 and not significant. Simple explanations for shyness/nervousness are unlikely to be valid and the feature is likely to be both complex in its mode of inheritance as well as in the environmental features which influence it.

In seeking to assess fearfulness in potential guide dogs, Goddard & Beilharz (1985) found Labradors to be less fearful than GSDs that were also more fearful than kelpies and boxers. It is generally held that sheep-herding breeds are more inclined to fearfulness and, despite its present role, the GSD was and still is to some degree a herding breed. Goddard & Beilharz (1985) found considerable within breed variation for fearfulness but no hybrid vigour due to cross-breeding, suggesting a largely additive trait, and this agrees with their high heritability figures (Table 4.1). This implies that within breed selection could be effective and they suggested this for their Labradors. However, they also suggested that selection against fearfulness was likely to be more effective in adults than in pups (Goddard & Beilharz, 1984, 1986).

That closed breeding programmes *can* succeed in enhancing behavioural traits is seen with the British Guide Dogs for the Blind Association where success rates over the past 30 years or so have risen from under 50 to over 90% (D. Freeman, 1988, personal communication). In the USA, Pfaffenberger (1963) described success rates in GSD guide dogs that rose from 9–90% in the 12 years from 1946. Similar closed programs seem desirable for police dogs rather than the current reliance upon 'gift' animals or those bred in 'show' kennels, since such animals may not have been selected for working features. Unfortunately, few British police forces, outside of the Metropolitan, actually have breeding programmes.

The present author has been judging GSDs in the show-ring since the 1950s. During the 1950–60s period it was normal to find up to 20% of animals nervous or fearful. Currently, one would expect to see nervousness only very rarely and 2% would be an approximate value. At first sight this appears to contradict the findings of Mugford & Gupta (1983), who, at Crufts dog show 1982, exposed 203 dogs from 15 breeds to a 60 second stare after which the dog was approached. They found marked levels of nervousness in GSDs. However, the GSD breed in Britain is split into two groups on physical type; one group is based largely on German bloodlines, and the other on 1950/60s British lines. It was the latter that were represented at the Crufts in question. In GSD breeding circles, it is believed that much of the improvement in the character of the breed stems from the use of German imports, all or most of which have been Schutzhund tested, and which tend to exhibit better temperament than the descendants of the 1950s British lines. Testing the British or even American lines would be more likely to reveal temperamental faults than would be the case in German-based bloodlines. Most studies critical of the GSD have been British or American based, rather than European. Researchers are thus assessing specific strains, between which large behavioural differences may exist.

Aggression

In respect of aggression, it is well established that males pose more problems than females (e.g. Beaver,

1983, Borchelt, 1983; Mugford, Chapter 10). Breed differences are claimed. The 'rage' syndrome of the red cocker spaniel is well known, though not actually confined to reds. There seems little doubt that this trait has a genetic basis, although the colour effect may reflect different bloodlines rather than any significance of colour itself. Cocker spaniel breeders do not mate parti-colours to self-colours so that the bloodlines of the two colour phases tend to be distinct. Behavioural differences would tend to reflect this. In foxes, links between certain coat colours and extreme or abnormal behaviour have been identified by Keeler *et al.* (1970) and Belyaev, Ruvinsky & Trut (1981). It seems unlikely, however, that any direct genetic relationships will be found between coat colour and temperament failings in the domestic dog.

Although never followed up, the sudden outbreak of unprovoked aggression among Bernese mountain dogs in Holland (Van der Velden *et al.*, 1976) provided evidence of polygenic inheritance of this trait (Willis, 1989). Lines of Swiss origin, noted for their aggressiveness exist in this breed. Similarly, several studies from the USA report high levels of protective or territorial aggression and fear-biting in GSDs (Beaver, 1983; Borchelt, 1983), though the finding is complicated by the numerical strength of this breed.

The degree to which aggression is acquired or inherited is both controversial and unclear. In the GSD, Wilsson (1985) has shown that social interactions between mothers and offspring during weaning have significant effects upon subsequent pup behaviour. Much trouble with aggression appears to stem from a failure to place *alpha* dogs[3] in the right hands. Most inexperienced owners do not know how to handle such animals, although, in this author's opinion, they often make the best companions as long as one is careful to establish one's authority early on in the relationship. Much of the recent media-alleged problems with Rottweilers may stem from the fact that this breed contains more than its share of *alpha* dogs. On the other hand, the results (unpublished) of this author's recent tests on Rottweilers revealed relatively few *alpha*-type dogs, but a great many owners who did not understand their animals and, in some cases, could not even play with them. Similarly, recent (unpublished) character evaluation studies of Bernese mountain dogs revealed only one *alpha*-type animal. About half the population examined were classified as easy-going and stable, and the rest were equally divided between the categories 'hesitant' and 'nervous/insecure'.

There is a clear difference between the dominant or assertive *alpha* dog and a fighting dog, such as the pit bull terrier. These animals have been deliberately selected for an ability and eagerness to fight, and to this degree are unlike most other breeds of dog in which, as with the wolf, intraspecific aggression tends to be ritualized more often than serious. It is possible that the behaviour patterns of pit bull terriers have been altered by selection to a degree that is not seen in other breeds, even those in which assertive animals are relatively common. Such alteration of behaviour would be a direct result of selection for fighting prowess. That this selection has been so successful in its objective implies that selection in the reverse direction could be equally successful, and that within a few generations pit bull terriers could be changed back into acceptable members of canine and human society.

It may also be important to distinguish between aggression directed against humans and aggression against other dogs. The best bloodlines for working ability in the GSD breed (Perry, 1980) stem from and are still dominated by Vello zd Sieben Faulen (born in 1956) and two important sons, Bernd and Bodo von Lierberg (born in 1962) from whose lines some inter-dog aggression has been apparent. Unfortunately, the most frequently seen bloodlines in the world outside of America now stem from Canto vd Wienerau (1968–72), who was recorded as not being ideal in character and from whom this flaw has doubtless been transmitted, although scientific documentation is not available (see Willis, 1991).

The future of dog breeding

Rightly or wrongly it is a fact that dog breeding in most countries is dominated by the show-ring, which makes it essential that judges in whatever breed discard nervous or aggressive animals, regardless of physical beauty. Hopefully, breeders will then follow this example since winning animals are the ones most

[3] I use the term *alpha* dog in this context to refer to a particular type of animal characterized by: (a) assertiveness in dominance contexts; (b) confidence and lack of nervousness when faced with novel and/or alarming situations or people; and (c) a tendency to protect its owner and owner's possessions. More often than not, these animals are males.

frequently used for breeding purposes. From this perspective, the sooner compulsory character assessments or working tests are introduced the better.

Although sheep-herding has not been discussed here, it is interesting to note that the Border collie has, in the past two decades, moved from the control of the International Sheepdog Society (ISDS) to the Kennel Club (KC), though the ISDS still controls working animals. Although it was decided that a working test would remain for KC registered dogs before they could become full champions, the sad truth is that few Border collies have taken this test and still fewer have passed it. With such failure to attend to essential features, it will be only a matter of time before the ill-named *Show* Border collie will have lost its ability to work. There, if not watched over by the breed societies, will go most working breeds, and this is more likely to happen in Britain and the United States than in continental Europe where breed clubs still control breeding procedures.

There is a need to further document the genetics of canine behaviour. Dogs are usually bred by people who have no training in either behaviour or genetics, despite having a wealth of hands-on practical experience. If canine behaviour is to be modified appropriately, then breeders will need to be able to understand the consequences of their actions, both in terms of breeding as well as husbandry and training. They will need to be able to make sense of new behavioural evidence and, increasingly, they will want to know the extent to which behaviour can be predicted early in life and, if so, how early and by what means.

From a scientific standpoint, dog breeding is likely to change markedly over the next decade, as Best Linear Unbiased Prediction (BLUP)[4] techniques are applied to the process, as they are already to the breeding of farm animals. BLUP techniques, which involve deriving a single breeding value prediction from a number of different information sources, are currently being used to predict such traits as hip dysplasia in dogs (Lingaas & Klemetsdal, 1990; R. Beuing, personal communication). In the future it would be feasible to see BLUP values derived for behaviour traits, although the problem of finding accurate and reliable scoring methods still remains.

Breeders usually select for physical as well as mental traits and, as Mackenzie *et al.* (1986) point out, the genetic relationships, if any, between physical and mental traits need to be documented. There is also a need to determine the degree to which the effects of learning can modify behaviour, and to document and define more accurately the influence of maternal effects.

[4] BLUP techniques take account of information on the dog itself, its parents, siblings, half-siblings and progeny, as well as similar data on the parents and the mates to which the dog was bred. The technique generates a single value – where 100 represents an 'average' animal and values of less than 100 represent superior breeding values and over 100 inferior values – which makes the use of BLUP values easy to apply even if the mathematics of their calculation is complex.

References

Bartlett, C. R. (1976). Heritabilities and Genetic Correlations between Hip Dysplasia and Temperament Traits of Seeing-eye Dogs. Masters thesis. Rutgers University, New Brunswick.

Beaver, B. V. (1983). Clinical classification of canine aggression. *Applied Animal Ethology*, **10**, 35–43.

Belyaev, D. K., Ruvinsky, A. O. & Trut, L. N. (1981). Inherited activation–inactivation of the star gene in foxes: its bearing on the problem of domestication. *Journal of Heredity*, **72**, 267–74.

Black, H. L. (1987). Dogs for coyote control: a behavioral perspective. In *Protecting Livestock from Coyotes.* ed. J. S. Green, pp. 67–75. Dubois, ID: US Sheep Experiment Station.

Borchelt, P. L. (1983). Aggressive behaviour of dogs kept as companion animals: classification and influence of sex, reproductive status and breed. *Applied Animal Ethology*, **10**, 45–61.

Breber, P. (1977). *Il Cane da Pastore Maremmano-Abruzzese.* Florence: Ed. Olimpia.

Brown, C. J., Murphree, O. D. & Newton, J. E. O. (1978). The effects of inbreeding on human aversion in pointer dogs. *Journal of Heredity*, **69**, 362–5.

Burns, M. & Fraser, M. N. (1966). *Genetics of the Dog. The Basis of Successful Breeding.* Edinburgh: Oliver & Boyd.

Clutton-Brock, J. (1984). Dog. In: *Evolution of Domesticated Animals*, ed. I. L. Mason, pp. 198–211. London: Longmans.

Coppinger, R., Coppinger, L., Langeloh, G., Gettler, L. & Lorenz, J. (1988). A decade of use of livestock guarding dogs. In: *Proceedings of The Vertebrate Pest Conference*, Vol. 13, ed. A. C. Crabb & R. E. Marsh, pp. 209–14. Davis: University of California.

Coppinger, R., Lorenz, J. R., Glendenning, J. & Pinardi, P. (1983). Attentiveness of guarding dogs for reducing predation on domestic sheep. *Journal of Range Management*, **36**, 275–9.

Coppinger, R., Smith, C. & Miller, L. (1985). Observations on why mongrels may make effective livestock protecting dogs. *Journal of Range Management*, **38**, 560–1.

Falt, L., Swenson, L. & Wilsson, E. (1982). Mentalbeskrivning av valpar. Battre Tjanstehundar, Projektrapport 11, Statens Hundskola, Sveriges Lantbruksuniversitet and Stockolms Universitet (unpublished report).

Fox, M. W. (1978). *The Dog: Its Domestication and Behaviour.* New York: Garland STMP Press.

Geiger, G. (1972). Prufungswesen und Leistungsvererbung beim Deutschen Drahthaaringen Vorstehhund. *Giessener Beitrage zur Erbpathologie und Zuchthyqiene*, **4**, 40–3.

Goddard, M. E. & Beilharz, R. G. (1982). Genetic and environmental factors affecting the suitability of dogs as guide dogs for the blind. *Theoretical and Applied Genetics*, **62**, 97–102.

Goddard, M. E. & Beilharz, R. G. (1983). Genetics of traits which determine the suitability of dogs as guide-dogs for the blind. *Applied Animal Ethology*, **9**, 299–315.

Goddard, M. E. & Beilharz, R. G. (1984). A factor analysis of fearfulness in potential guide dogs. *Applied Animal Behaviour Science*, **12**, 253–65.

Goddard, M. E. & Beilharz, R. G. (1985). A multivariate analysis of the genetics of fearfulness in potential guide dogs. *Behaviour Genetics*, **15**, 69–89.

Goddard, M. E. & Beilharz, R. G. (1986). Early prediction of adult behaviour in potential guide dogs. *Applied Animal Behaviour Science*, **15**, 247–60.

Green, J. S. (1989). APHIS Animal damage control livestock guarding dog program. *Proceedings of the 9th Great Plains Wildlife Animal Damage Control Workshop.* Fort Collins, CO.

Green, J. S. & Woodruff, R. A. (1983*a*). The use of three breeds of dog to protect rangeland sheep from predators. *Applied Animal Ethology*, **11**, 141–61.

Green, J. S. & Woodruff, R. A. (1983*b*). The use of Eurasian dogs to protect sheep from predators in North America: a summary of research at the US Sheep Experiment Station. In *Proceedings of the 1st Eastern Wildlife Damage Control Conference*, ed. D. J. Decker, pp. 119–24. Ithaca, NY: Cornell University Press.

Green, J. S. & Woodruff, R. A. (1983*c*). Guarding dogs protect sheep from predators. *USDA Bulletin*, 455.

Green, J. S. & Woodruff, R. A. (1987). Livestock guarding dogs for predator control. In *Protecting Livestock From Coyotes*, ed. J. S. Green, pp. 62–8. Dubois, ID: US Sheep Experiment Station. Idaho.

Green, J. S. & Woodruff, R. A. (1988). Breed comparisons and characteristics of use of livestock guarding dogs. *Journal of Range Management*, **41**, 249–51.

Humphrey, E. S. & Warner, L. (1934). *Working Dogs – an Attempt to Produce a Strain of German Shepherds Which Combine Working Ability and Beauty of Conformation.* Baltimore, MD: John Hopkins University Press.

Keeler, C., Mellinger, T., Fromm, E. & Wade, L. (1970). Melanin, adrenalin and the legacy of fear. *Journal of Heredity*, **61**, 81–8.

Kock, M. (1984). Statistische und erbanalytische Untersuchungen zur Zuchtsituation, zu Fehen und Wesensmerkmalen beim Deutsch-Langhaarigen Vorstehund. Doctoral thesis, Tierarztliche Hochschule, Hannover.

Lingaas, F. & Klemetsdal, G. (1990). Breeding values and genetic trend for hip dysplasia in the Norwegian Golden Retriever population. *Journal of Animal Breeding & Genetics*, **107**, 437–43.

Linhart, S. B., Sterner, R. T., Carrigan, T. C. & Henne, D. R. (1979). Komondor guard dogs reduce sheep losses to coyotes: a preliminary evaluation. *Journal of Range Management*, **32**, 238–41.

Lorenz, J. R. & Coppinger, L. (1986). Raising and training a livestock guarding dog. *Oregon State Univ. Extension Service, Extension circular*, 1224.

Mackenzie, S. A., Oltenacu, E. A. B. & Houpt, K. A. (1986). Canine behavioral genetics – a review. *Applied Animal Behaviour Science*, **15**, 365–93.

Mackenzie, S. A., Oltenacu, E. A. B. & Leighton, E. (1985). Heritability estimate for temperament scores in German Shepherd Dogs and its genetic correlation with hip dysplasia. *Behaviour Genetics*, **15**, 475–82.

McGrew, J. C. & Andelt, W. F. (1985). Livestock guardian dogs. *Kansas State University External Services Bulletin*, MF713.

Mugford, R. & Gupta, A. S. (1983). Genetics and behaviour problems in dogs. *Applied Animal Ethology*, **11**, 87 (Abstract).

Perry, W. E. (1980). Some notes on the SV Federal Sieger trials – 1949 to 1978. *Handbook, GSD League of Great Britain.* pp. 67–71.

Pfaffenberger, C. J. (1963). *The New Knowledge of Dog Behaviour.* New York: Howell Books.

Pfleiderer-Hogner, M. (1979). Moglichkeiten der Zuchtwertschatzung beim Deutschen Schaferhund anhand der Schutzhundprufung 1. Doctoral thesis Ludwig-Maximilians-Universität, Munich.

Reuterwall, C. & Ryman, N. (1973). An estimate of the magnitude of additive genetic variation of some mental characters in Alsatian dogs. *Hereditas*, **73**, 277–84.

Sacher, B. (1970). Statistische-genetische Auswertungen von Zuchtbuchunterlagen bei Kleinen Munsterlander Vorstehhunden mit Hilfe der EDV. Doctoral thesis, Geissen Universität, Geissen.

Scott, J. P. & Bielfelt, S. W. (1976). Analysis of the puppy testing program. In *Guide Dogs for the Blind: Their Selection, Development & Training*, ed. C. J. Pfaffenberger, J. P. Scott, J. L. Fuller, B. E. Ginsburg & S. W. Bielfelt, pp. 39–75. New York: Elsevier.

Scott, J. P. & Fuller, J. L. (1965). *Genetics and the Social Behavior of the Dog.* Chicago: University of Chicago Press.

Swenson, L. (1987). Hunting and performance test data as basis for breeding evaluation. Unpublished presentation, Uppsala Conference on Dog Breeding, Uppsala, Sweden. 5pp.

Thorne, F. C. (1944). The inheritance of shyness in dogs. *Journal of Genetical Psychology*, **65**, 275–9.

Van der Velden, N. A., de Weerdt, C. J., Brooymans-Schallenberg, J. H. C. & Tielen, A. M. (1976). An abnormal behavioural trait in Bernese Mountain Dogs (Berner Sennenhund). *Tijdschrift voor Diergenees-Kunde*, **101**, 403–7.

Vangen, O. & Klemetsdal, G. (1988). Genetic studies of Finnish and Norwegian test results in two breeds of hunting dog. *VI World Conference on Animal Production, Helsinki*, Paper 4.25.

Whitney, L. F. (1929*a*). Inherited mental aptitudes in dogs. *Eugenics*, **2**, 8–16.

Whitney, L. F. (1929*b*). Heredity of the trail-barking propensity in dogs. *Journal of Heredity*, **20**, 561–2.

Willis, M. B. (1976). *The German Shepherd Dog: Its History, Development and Genetics.* Leicester, UK: K. & R. Books.

Willis, M. B. (1989). *Genetics of the Dog.* London: H. F. & G. Witherby.

Willis, M. B. (1991). *The German Shepherd Dog: a Genetic History.* London: H. F. & G. Witherby.

Wilsson, E. (1985). The social interaction between mother and offspring during weaning in German Shepherd Dogs: individual differences between mothers and their effects on offspring. *Applied Animal Ethology*, **13**, 101–12.

5 Analysing breed and gender differences in behaviour

BENJAMIN L. HART

Introduction

The dog was the first animal apparently domesticated by humankind (see Clutton Brock, Chapter 2), and it is the species that has been most profoundly altered through selective breeding. Morphological differences among the various breeds of dogs are so diverse that, if we did not know better, it would be tempting to conclude that they originated from different species. Over the centuries, as dogs have been bred for various physical attributes, they have also undergone a variety of changes in behavior. Some breeds have been selected for behavioral changes that are useful in hunting, such as pointing at game birds, chasing foxes while vocalizing, or retrieving waterfowl shot by hunters. In other breeds, behavior associated with the performance of complex tasks, such as herding cattle or sheep, has been accentuated, while still others have been bred selectively to protect property (see Coppinger & Schneider, Chapter 3). These particular behavior functions that have come to characterize the different breeds of dogs are the outcome of the suppression or enhancement of existing 'native' canine behavioral characteristics, rather than the emergence of new behavior patterns (Scott & Fuller, 1965).

In modern Western societies, the practical functions of dogs are gradually diminishing in importance, while the behavioral attributes associated with the dog's companionship role in the human family are becoming increasingly relevant. Behavior of particular importance to people keeping dogs as pets includes expressions of dominance, territoriality, affection, sociability towards children and excitability. These characteristics are often very different from those that were selected for during the early histories of the different breeds, and there are obvious cases in which conflicts may arise between these two selective pressures. For example, dogs bred primarily for aggressive territorial protection seem to have become aggressive in an overall sense, acquiring a tendency to challenge their owners for dominance, especially when the latter are not sufficiently assertive. Similarly, dogs which were bred for chasing foxes and vocalizing are more predisposed to vocalize around their owners' houses and gardens than other breeds, even in the absence of foxes.

This chapter focuses on methods of measuring or evaluating those differences in breed-specific behavior that are of primary interest to those keeping dogs as pets. Virtually all behavioral traits that are of interest to dog owners will be influenced by both genetic and environmental factors, and the degree of each type of influence (or their interactions) will vary from trait to trait. In general, those traits that have the strongest genetic input will be more reliable at discriminating between breeds than traits more readily influenced by environment. One of the goals of the present research was to find a methodology that would not only reveal consistent breed differences in behavior, but also determine which traits, if any, discriminate best between breeds.

One previous approach to identifying breed differences in behavior is represented by the work of Scott & Fuller (1965), who conducted extensive laboratory experiments on six breeds of dogs (basenjis, beagles, cocker spaniels, Shetland sheepdogs and fox terriers) using tests that measured such traits as emotional reactivity, trainability and problem-solving ability. These investigators followed the principle that general traits must be tied down to specific behavioral tests, and that more than one specific test needed to be used before a judgement could be made with regard to a general behavioral category. Emotional reactivity, for example, was indicated by physiological measures, such as heart rate and respiratory rate, as well as by behavioral signs, such as distress vocalizations or tail-wagging. Tests of emotional reactivity included dogs' responses to experimenters who either approached them speaking softly, or who grabbed them by the muzzle and forced their heads from side to side. Using these sorts of tests, terriers, beagles and basenjis were judged significantly more reactive then shelties or cocker spaniels. Tests of trainability included leash training, and forcing the dog to remain on a stand and then to jump down upon command. Judging from the results of these tests, cocker spaniels were the easiest to train, and basenjis and beagles consistently the most difficult. Shelties and terriers varied, in the sense that on one task they might be easy to train and on another, difficult. Scott & Fuller's (1965) work revealed, among other things, the value of studying behavioral traits in the laboratory where the members of each breed can be raised in a uniform environment and with the same early experience.

Another possible method of studying breed differences in behavior would be to visit dogs of different breeds in their owners' homes and subject them to a standardized battery of behavioral tests. This approach might give an idea of the degree of variability within breeds as a function of differences in environment and early experience. Behavioral tests of different breeds, either in the laboratory or in the owner's home, would lend themselves particularly to analyses of such traits as watchdog behavior, territorial guarding, excitability and general activity. However, some traits are so much a function of the interaction between owner and dog that neither laboratory tests nor those administered in the owner's home would be likely to generate useful information. A dog with a tendency to be socially dominant, for example, may not reveal this trait in the presence of a relatively authoritarian owner. Similarly, a tendency to fight with other dogs may not be evident in a laboratory where encounters with other animals are restricted. In addition, laboratory or home-based tests on representative samples of many different breeds are likely to be prohibitively expensive to conduct.

A number of methods of assessing the behavioral characteristics of companion animals have involved surveys of pet owners or professionals familiar with the various breeds. These methods may be viewed as alternatives to laboratory or structured testing (Hart & Hart, 1984). An example of one such method involves obtaining narrative responses of selected informants, such as breeders, judges or dog owners. Most books on dog behavior make use of this sort of relatively-subjective consensus information. Another more quantifiable approach to obtaining information on dog behavior was employed by Serpell (1983), who invited dog owners to complete a questionnaire in which a variety of canine behavioral traits were represented as a series of visual analogue scales. Alternatively, questionnaires directed at pet owners can be used to obtain tabulated information on the frequency of occurrence of different behavior patterns, such as urine marking in the house, threats to the owner, or destructive episodes (Hart & Hart, 1984). Finally, one can make use of selected informants or 'experts' by inviting them to perform forced-choice ranking of breeds on different behavioral characteristics. The data yielded by this approach allows for more complex statistical manipulations, such as factor and cluster analysis.

Informants who are familiar with dogs kept under a variety of circumstances may be able to supply information that could not be obtained reliably from behavioral tests or other methods. An experienced obedience instructor, for example, is likely to become familiar with a particular breed's tendency to be dominant over the owner or aggressive towards other dogs.

A disadvantage in using the opinions of informants is that, while the opinions reflect information based on first hand experience with dogs, they are necessarily subjective and therefore prone to biases. These biases may be individual or they may reflect erroneous cultural conceptions about breeds. Such biases might be particularly prevalent with dogs because people are commonly emotionally attached to their dogs. It is impossible to eliminate the influence of biases, especially those of a general cultural nature, when informants are the source of data. A method of forced ranking, if done with breeds that are not the informants' favorite breeds, should limit the effect of such biases.

Survey of books

In examining the dog breed books, such as the official publication of the American Kennel Club and similar catalogues of breeds, one finds that the source of information is not given. Vague terms, such as 'loyal companion,' 'regal,' 'patient' and 'alert,' are used as behavioral descriptions. These terms say little specifically about a breed, let alone provide the basis for comparisons between breeds.

Two published volumes, Tortora (1980) and Howe (1976) compared all of the 121 breeds then registered by the American Kennel Club. Tortora, using a 5-point range (with intermediate scores), ranked breeds on indoor activity, outdoor activity, vigor, behavioral constancy, dominance to strange dogs, dominance to familiar people, territoriality, emotional stability, learning rate, obedience learning, problem solving learning, watchdog ability and guard-dog ability, sociability with family members, sociability with children, sociability with strangers and sociability with strange dogs, among other traits.

Both authors stated they relied upon breeders for information with input modified by their own opinions when assigning scores.

To get an idea of how well Tortora's and Howe's systems discriminate among breeds, the frequency with which the authors used a score in rating breeds was tabulated. In neither book were the breeds evenly distributed across the range of the scoring systems. Of 16 traits used by Tortora, the rating system used on 11 of the traits listed 60% or more of the breeds on two adjacent ranks and in two instances (dominance towards people; watchdog ability) two adjacent ranks accounted for 90% or more of breeds (ratings spanning two or more ranks were not counted). An even distribution would have been 40% at two adjacent ranks.

Of four traits used by Howe with a 3-point scale (learning, watchdog, friendliness, behavior with strange dog) a range of 53 to 71% of all breeds were assigned to one of the three ranks. An even distribution would be represented by 33% of breeds at each rank. Two traits, housebreaking ease and guard dog, were rated on a 2-point scale (essentially yes/no), and 70 and 75% of breeds were assigned to one of the two ranks. The absence of an even distribution across the rating scale may reflect an actual clustering of dog breeds at some extreme on the scale. For example, dog breeds may vary little in watchdog behavior and would legitimately be clustered around the highest level. Alternatively, dogs may vary considerably in watchdog behavior, but the person assigning the rating may erroneously feel that most breeds are high in this trait. At any rate, when breeds are clustered, as in Tortora's ranking of watchdog behavior and dominance, there is much weaker discrimination among breeds than with traits such as obedience learning that clustered only 60% of breeds on two adjacent ranks.

The project described next was an attempt to obtain breed rankings that provided maximum discrimination between breeds. Since the intention was to have breeds ranked against each other, informants who were in a position to compare diverse breeds were sought. It was felt that most breeders and dog owners would not only have individual biases but would not be in a position to compare many diverse breeds. For reasons explained below, obedience judges and small animal veterinarians were therefore chosen as informants.

Rationale of the project

The methods employed in the study were based on three assumptions: (1) that significant differences exist in many behavioral characteristics among the various breeds of dogs, although the magnitude of differences, and within-breed and between-breed variability, will differ from trait to trait; (2) that some of these behavioral differences are reflected in the perspectives of authorities who have extensive experience with various dog breeds and dog-owner relationships; and (3) that the behavioral information about breed differences that exists in the minds of the authorities can be obtained by interviewing large numbers of authorities with a data collection format that minimizes the opportunity for the authorities to emphasize the breeds in which they may have personal interest. This latter premise was addressed by deciding to ask authorities to rank a small number of randomly chosen breeds on various behavioral characteristics. The ranking method avoids use of absolute scoring that would be problematical because different authorities will have different ideas about what a score may mean. The extent of agreement between authorities in their rankings of the various breeds on a trait is an indication of the reliability of breed differences and should be reflected in statistical significance. An absence of breed differences on a particular behavioral trait would be indicated by large random variability in the rankings and a lack of statistical significance. This procedure admittedly measures the perceptions of authorities, not actual breed differences. The methodology does, however, control for consistency among these perspectives. There was no way of checking the validity of the ranking against an independent method of ranking the breeds such as a laboratory test. Perhaps the greatest danger in using the ranking system is to force a spread among breeds on a trait that is, in actuality, fairly uniform. With this type of error there should be greater variability in rankings and an absence of statistical significance.

Methodology

The selection of a group of authorities to be interviewed for ranking breeds turned out to be more difficult than expected. Four types of individuals with extensive experience with dogs were considered: obedience judges, dog show judges, professional handlers of dogs and veterinarians in small animal practice. For all groups, a national directory was available for randomly choosing names and for selecting equal numbers of men and women. Obedience judges that are listed in the national directory have had extensive experience in leading obedience classes. They are well acquainted with interactions between dogs and their owners, and have talked to people about the problems or attributes of their dogs. They also travel in social circles where there is frequent discussion of breeds. Similarly, dog show judges and dog handlers belong to a social milieu in which dog breeds are a frequent topic of conversation. Small animal veterinarians observe the interactions between owners and their animals directly in the examination rooms, they handle dogs in their hospital wards, and they listen to the complaints and boasting by clients about their dogs.

During preliminary interviews, we found that dog show judges were reluctant to rate breeds against each other with regard to behavioral traits, perhaps out of a feeling that the traits would have positive or negative value and they did not wish to rank some breeds above others. Dog handlers tended to handle just a few, usually closely-related breeds and did not have a broad comparative perspective. Obedience judges and small animal veterinarians were therefore selected as being the most familiar with all breeds in a variety of situations and the most willing to rank breeds against each other. Nevertheless, it could be argued that obedience judges and veterinarians experience dogs mainly in structured situations and do not have a perspective based upon familiarity with dog breeds in the home environment. Without a doubt there is no ideal group of authorities or informants and a critic of this project could justifiably propose a different authority group.

Thirteen behavioral characteristics were chosen as being of primary interest to prospective pet owners. Dog breeds were ranked by 48 obedience judges and 48 animal veterinarians who were selected randomly from a directory so as to represent males and females equally and to represent eastern, central and western United States equally. Each was asked to rank a list of seven breeds chosen from a master list of 56 on each of the characteristics presented in Table 5.1. So

Table 5.1. *Behavioral traits ranked in order of decreasing reliability in differentiating between dog breeds, as indicated by magnitude of F ratio*

Behavioral characteristic	*F* ratio
Excitability	9.6
General activity	9.5
Snapping at children	7.2
Excessive barking	6.9
Playfulness	6.7
Obedience training	6.6
Watchdog barking	5.1
Aggression to dogs	5.0
Dominance over owner	4.3
Territorial defense	4.1
Affection demand	3.6
Destructiveness	2.6
Housebreaking ease	1.8

as to make clear what a characteristic represented, a question was formulated that illustrated the usefulness of the behavior. For example, *excitability* was asked about in the following way: ‘A dog may normally be quite calm but can become very excitable when set off by such things as a ringing door bell or an owner’s movement towards a door. This characteristic may be very annoying to some people. Rank these breeds from least to most excitable.’ The trait *watchdog behavior* was approached by asking: ‘Now we would like to find out your opinion regarding watchdog capability of these breeds. A woman living alone in the city wants to get a dog that will sleep by her bed and frighten intruders by barking if anyone breaks into the house in the middle of the night. Rank these breeds from least to most as to which will consistently sound an alarm when it hears something unusual and will bark at intruders.’

The 56 breeds on the master list represented the 55 most frequently registered breeds of the American Kennel Club in 1978 plus the Australian shepherd, which is popular but not recognized until recently by the American Kennel Club. Other aspects of the methodology including the procedure of randomization for selecting breeds and order of questions are dealt with elsewhere (Hart & Miller, 1985). Data were processed by a computer program that was designed to create a ranking of 1 through 56 breeds for each trait based on a mean score of each breed derived from the 1–7 rankings. An analysis of variance was conducted to determine if there were significant breed differences on each trait.

Results

Analysis of variance revealed that there were statistically significant differences on all behavioral characteristics among the various breeds ($P < 0.005$) with an F ratio range of 1.84 to 9.56. The range in F ratio verified our suspicion that some characteristics provided a more reliable basis for distinguishing among breeds than did others. The F ratio assigned to each characteristic is given in Table 5.1. The trait of *excitability* had the highest F ratio and was therefore judged to discriminate the best among breeds, while *ease of housebreaking* had the lowest F ratio and was presumably the least reliable in distinguishing among breeds. In other words one could more reliably predict differences between breeds for the characteristic of *excitability* than for *ease of housebreaking*. The progressive decrease in F value reflects a decreasing reliability in breed difference. Another indication that some behavioral traits distinguished among breeds better than others was the number of breeds that would have to be spanned in the rankings to be assured of a significant difference at the $P < 0.01$ level. This measure was less for the traits with a high F ratio than traits with a low F ratio. This latter analysis was done with the Bonferroni correction that takes into account all the multiple comparisons being made at one time and is an overly conservative measure of where significance lies. With *excitability*, the span was 32 and for *ease of housebreaking* the span was 50. A practical application of the data would not require using a span of breeds that was so wide. When differences in the ranking of breeds were tested as a function of geographic region, type and sex of the authority there were five occasions in ranking the 56 breeds on each of 13 traits where these differences reached the $P < 0.001$ level and where at least half the members of an authority group were familiar enough with the breed to rank it. These occasions were: Old English sheepdog ranked 15 overall on general activity but ranked 5 and 35, respectively, by obedience judges and veterinarians; German shorthaired pointer ranked 23 overall on excessive barking but ranked 4 and 43, respectively, by obedience judges and veterinarians; Yorkshire terrier ranked 39 overall on dominance over owner but ranked 56, 52 and 1 respectively, by East, Central and West; toy poodle ranked 30 overall on dominance over owner but ranked 10 and 53, respectively, by obedience judges and veterinarians; Dalmatian ranked 33 overall on territorial defence but ranked 6 and 50, respectively, by obedience judges and veterinarians.

For the sake of making the ranking of the breeds on each trait more understandable, a decile grouping was developed in which each decile comprised five or six breeds (according to the following symmetrical formula: 6, 5, 6, 5, 6, 6, 5, 6, 5, 6). Table 5.2 shows the use of these decile rankings on the traits of excitability and watchdog barking. Behavioral profiles for each breed were also developed utilizing each of the 13 traits. Examples of four behavioral profiles of breeds are shown in Fig. 5.1.

As mentioned above, each trait was considered to be a reflection of one or more general underlying behavioral predispositions that would be revealed by factor analysis. The computer program (Dixon, 1981) performed a factor analysis by first extracting the principal components and then rotating the most important components into uncorrelated factors having the simplest definition in terms of the original 13 traits. The data for the factor analysis were the scores derived from the relative rankings of each of the breeds on the 13 traits. It was found that four principal components accounted for 88% of the variance. Increments in cumulative variance accounted for beyond the four principal components were considered too small to warrant further consideration. The percentage of variance assigned to each of the principal components was 43%, 24%, 14% and 7%. After rotation each of the 13 behavioral traits was assigned to one of the four components according to the highest score coefficient on each component. In Table 5.3 the traits are listed under each component in order of decreasing loading on the component to which they were assigned. The three major components are referred to as reactivity, aggression, and trainability. In the breed profiles, traits are grouped according to the component to which they were assigned and a component or factor decile envelope is superimposed over the traits. Correspondence between the trait decile rankings and the component

Table 5.2. *Decile rankings – from least to most – on the characteristics excitability and watchdog barking*

Decile rank	Excitability	Watchdog barking
1	Bloodhound	Bloodhound
	Basset hound	Newfoundland
	Australian shepherd	St Bernard
	Chesapeake Bay retriever	Basset hound
	Rottweiler	Vizsla
		Norwegian elkhound
2	St Bernard	Brittany spaniel
	Golden retriever	Bulldog
	Akita	Siberian husky
	Labrador retriever	Afghan hound
	Great Dane	Alaskan malamute
3	Alaskan malamute	Golden retriever
	Bulldog	German shorthaired pointer
	Doberman pinscher	Old English sheepdog
	Old English sheepdog	Pug
	Collie	Bichon Frise
	Norwegian elkhound	Cocker spaniel
4	Vizsla	Labrador retriever
	Boxer	Weimaraner
	Chow chow	Great Dane
	Keeshond	Chesapeake Bay retriever
	Samoyed	Beagle
5	Brittany spaniel	Australian shepherd
	Welsh corgi	Collie
	German shepherd	Chow chow
	Afghan hound	Keeshond
	Siberian husky	Irish setter
	Poodle (standard)	Dalmation
6	Dalmatian	English springer spaniel
	Cocker spaniel	Samoyed
	Weimaraner	Boxer
	English springer spaniel	Pekingese
	German shorthaired pointer	Maltese
	Pug	Akita
7	Dachsund	Lhasa apso
	Bichon frise	Shetland sheepdog
	Lhasa apso	Welsh corgi
	Airedale terrier	Poodle (toy)
	Pekingese	Pomeranian
8	Shetland sheepdog	Boston terrier
	Beagle	Shih tzu
	Poodle (miniature)	Poodle (miniature)
	Boston terrier	Dachsund
	Pomeranian	Silky terrier
	Poodle (toy)	Fox terrier
9	Maltese	Yorkshire terrier
	Chihuahua	Chihuahua
	Shih tzu	Cairn terrier
	Irish setter	Airedale terrier
	Cairn terrier	Poodle (standard)
10	Scottish terrier	Rottweiler
	Yorkshire terrier	German shepherd
	Silky terrier	Doberman pinscher
	Schnauzer (miniature)	Scottish terrier
	West Highland white terrier	West Highland white terrier
	Fox terrier	Schnauzer (miniature)

decile rankings was usually close. Of course, in reality each component undoubtedly contributes to all the traits but is expressed primarily in one trait. The behavioral profiles of 56 breeds of dogs are presented elsewhere in book form (Hart & Hart, 1988) where they are available to people wishing to adopt a dog or advise others wishing to select a dog on the basis of behavior.

The most powerful analysis performed in this project explored the relationship between gender and behavior. During the interviews each of the 96 authorities, in addition to being asked to rank breeds of dogs, was also asked to state whether or not they felt that either male dogs or female dogs (intact) displayed each trait more prominently. A statistical test was performed to identify the traits for which males ranked higher than females or females ranked higher than males. The formula for expressing the quantitative degree of difference between male and female dogs was calculated as follows: the number of authorities ranking males higher than females minus the number of the authorities ranking females higher than males, divided by the total number of authorities times 100. Thus, if all authorities ranked male

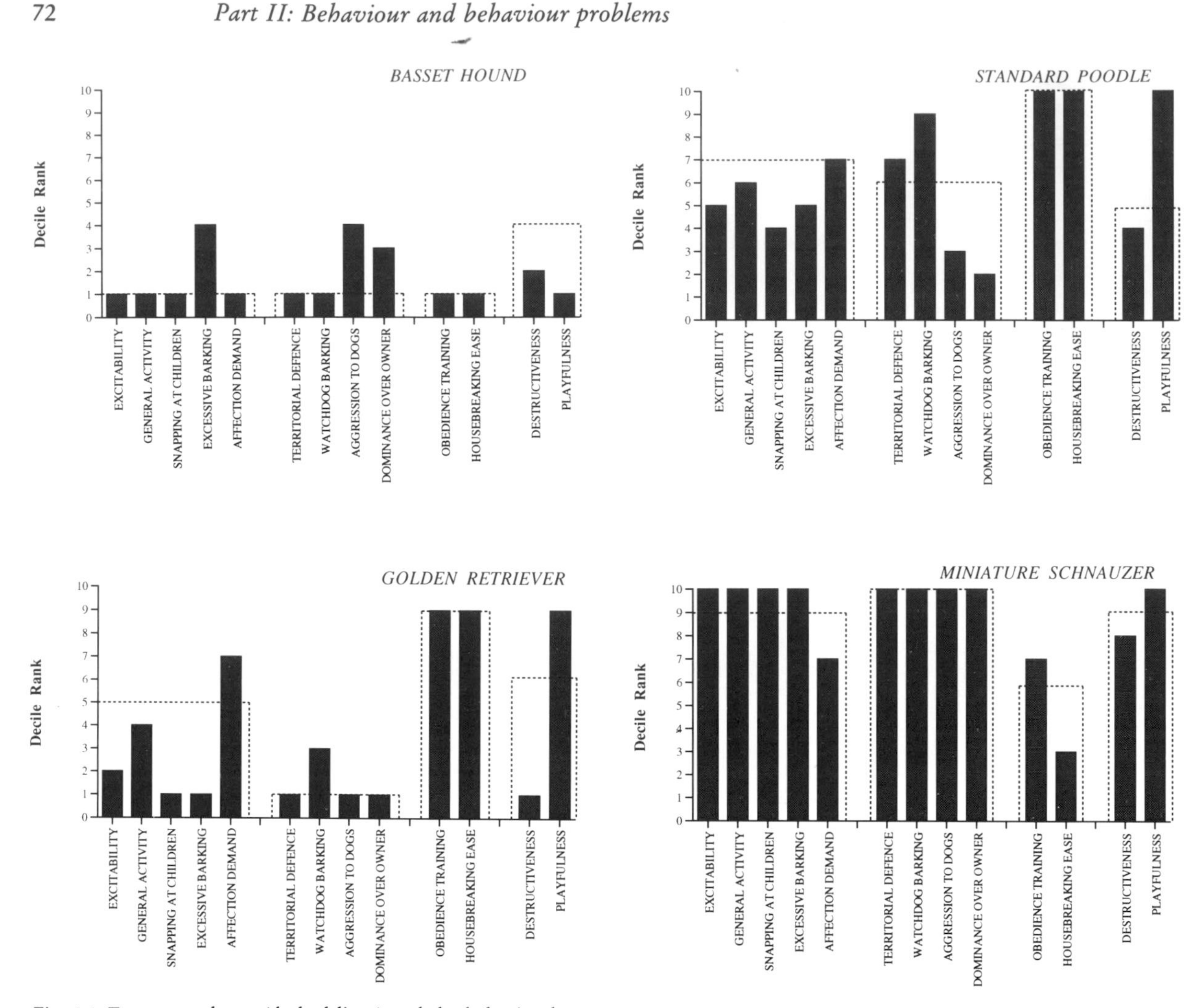

Fig. 5.1. Factor envelopes (dashed lines) and the behavioral traits assigned to the four factors for four breeds of dogs. Adapted from Hart & Hart (1985).

dogs higher than females, the score for males would be 100. Gender difference ($P < 0.0001$) was seen with all traits with the exception of *watchdog barking, excessive barking* and *excitability*. The data also provided information on the magnitude of gender differences in terms of the proportion of authorities ranking males higher than females or females higher than males. Figure 5.2 presents the results of the gender analysis. From a practical standpoint, those traits showing a percentage difference greater than 40 are most important. These are: *dominance over owner* and *aggression to other dogs*, in which males exceeded females, and *obedience training* and *housebreaking ease*, in which females exceeded males.

Returning to the premise that the data obtained from these authorities represent real differences between dogs, the analysis supports the idea that sexually dimorphic differences in behavior are graded rather than absolute. This is in correspondence with the results of laboratory studies where differences in frequency rather than absolute differences have been found to more accurately characterize behavior differences between males and females (Hart, 1985). Sexually dimorphic behavior patterns are also those

Table 5.3. *Behavioral traits listed in order of decreasing loading as assigned to the component for which they have the highest factor score coefficient*

Traits of component 1 (Reactivity)	Traits of component 2 (Aggressiveness)	Traits of component 3 (Trainability)	Traits of component 4
Affection demand	Territorial defense	Obedience training	Destructiveness
Excitability	Watchdog barking	Housebreaking ease	Playfulness
Excessive barking	Aggression to dogs		
Snapping at children	Dominance over owner		
General activity			

that tend to be altered by castration. According to one study, mounting, urine marking and aggression towards other dogs are altered by castration in 50 to 60% of animals (Hopkins, Schubert & Hart, 1976). The trait of *dominance over owner* is as strongly ranked as male-like as *aggression towards other dogs* in Fig. 5.2, although it is not known whether castration is as effective at suppressing this behavior as it is in the case of *aggression towards other dogs*.

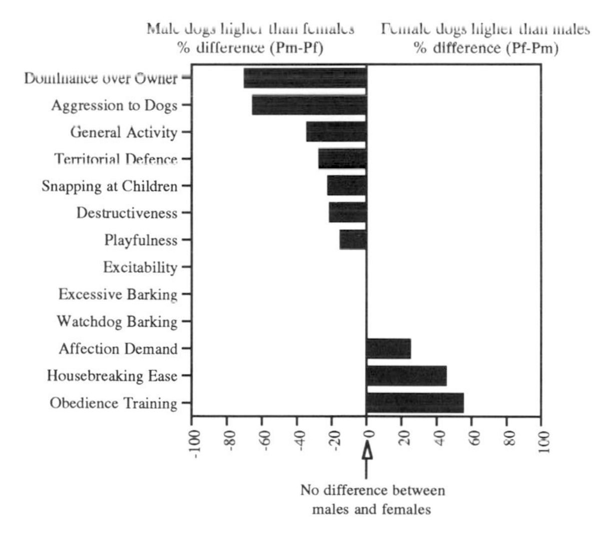

Fig. 5.2 Gender analysis of 13 behavioral traits. Solid bars represent significant differences between sexes ($P < 0.001$). Length of bar represents magnitude of difference. Adapted from Hart & Hart (1985).

Discussion

The project described above reveals that it is possible to obtain quantitative data reflecting the views of experts concerning behavioral differences between breeds of dogs and between genders. The experts or authorities who contributed the various rankings may, of course, have expressed biased individual opinions or harbored general misconceptions about some breeds. However, with the methodology used, no opinion about any particular breed was given more weight than another. The seven breeds presented to each authority for ranking were selected at random so there was also relatively little chance that authorities might be required to rank breeds that they personally liked or disliked.

Some important issues emerge from these data. It appears that some behavioral characteristics discriminate between breeds better than others; for example, the trait of *excitability* is better at discriminating than is the trait of *housebreaking ease*. It stands to reason there is less genetic influence, and more impact of training and the environment, on those traits that discriminate least well between breeds.

This project was not designed to obtain information about all registered breeds of dogs. When authorities were not sufficiently familiar with a breed

Table 5.4. *Obedience training: comparative ranking of breeds in three publications. Breeds chosen were from the lowest and highest deciles*

Decile and breed / Possible range (lowest to highest)	Ranking source		
	Hart & Hart (1988) 1–56	Tortora (1980) 1–5	Howe (1976) 1–3
Lowest decile			
Chow chow	1	1.2	2
Fox terrier	2	2.3	2
Afghan hound	3	2.0	1
English bulldog	4	2.0	1
Basset hound	5	2.8	1
Beagle	6	2.0	2
Highest decile[a]			
Minature poodle	51	5.0	3
German shepherd	52	4.5	3
Standard poodle	53	5.0	3
Shetland sheepdog	54	5.0	3
Doberman pinscher	55	5.0	3

[a]Australian shepherd ranked 56 and was not surveyed by other authors.
Source: Hart & Hart (1988).

Table 5.5. *General activity: comparative rankings of breeds in two publications. Breeds chosen were from the lowest and highest deciles*

Decile and Breed / Possible range (lowest to highest)	Ranking source	
	Hart & Hart (1988) 1–56	Tortora (1980) 1–5
Lowest decile		
Basset hound	1	1.6
Bloodhound	2	1.9
Bulldog	3	1.0
Newfoundland	4	2.0
Collie	5	2.0
St Bernard	6	2.0
Highest decile		
Silky terrier	51	4.0
Chihuahua	52	5.0
Miniature schnauzer	53	4.3
Fox terrier	54	5.0
Irish terrier	55	5.0
West Highland white terrier	56	5.0

Source: Hart & Hart (1988).

Table 5.6. *Snapping at children: comparative rankings of breeds in two publications. Breeds chosen were from the lowest and highest deciles*

Decile and breed / Possible range (lowest to highest)	Ranking source	
	Hart & Hart (1988) 1–56	Tortora (1980) 1–5
Lowest decile		
Golden retriever	1	1.0
Labrador retriever	2	1.3
Newfoundland	3	1.0
Bloodhound	4	1.0
Basset hound	5	1.0
Collie	6	1.5
Highest decile		
Scottish terrier	51	3.8
Miniature schnauzer	52	3.5
West Highland white terrier	53	3.2
Chow chow	54	4.9
Yorkshire terrier	55	3.0
Pomeranian	56	5.0

Source: Hart & Hart (1988).

Table 5.7. *Watchdog barking: comparative ranking of breeds in two publications. Breeds chosen were from the lowest and highest deciles*

Decile and Breed / Possible range (lowest to highest)	Ranking source		
	Hart & Hart (1988) 1–56	Tortora (1980) 1–5	Howe (1976) 1–3
Lowest decile			
Bloodhound	1	4.0	3
Newfoundland	2	4.0	1
St Bernard	3	5.0	3
Basset hound	4	4.0	2
Vizsla	5	4.0	2
Norwegian elkhound	6	5.0	3
Highest decile			
Rottweiler	51	5.0	3
German shepherd	52	5.0	3
Doberman pinscher	53	5.0	3
Scottish terrier	54	5.0	3
West Highland white terrier	55	4.0	2
Miniature schnauzer	56	5.0	3

Source: Hart & Hart (1988).

Table 5.8. *Dominance over owner: comparative rankings of breeds in two publications. Breeds chosen were from the lowest and highest deciles*

Decile and Breed / Possible range (lowest to highest)	Ranking source	
	Hart & Hart (1988) 1–56	Tortora (1980) 1–5
Lowest decile[a]		
Golden retriever	1	1.0
Shetland sheepdog	3	2.0
Collie	4	2.8
Brittany spaniel	5	3.0
Bloodhound	6	2.9
Highest decile		
Fox terrier	51	3.1
Siberian husky	52	2.8
Afghan hound	53	2.7
Miniature schnauzer	54	3.0
Chow chow	55	4.1
Scottish terrier	56	3.0

[a]Australian shepherd ranked 2 and was not surveyed by the other author.
Source: Hart & Hart (1988).

to rank it, the breed was given a letter coding (m) and the breed was given no rank. Had more breeds been used, the authorities would have been confronted with a higher percentage of less common breeds, and the (m) code would have been used more frequently, making all rankings less reliable.

Even assuming that these breed profiles represent an approximation of actual breed differences, it should not be assumed that all or most members of that breed will display the traits in the proportions characterized by the profile. Clearly, there are major, genetically-related differences between dogs within a breed. Also, the environment in which the dog is raised is an important influence (Serpell & Jagoe, Chapter 6), as is the training the adult animal receives.

Information on breed profiles and trait rankings has been presented in the form of a monograph elsewhere (Hart & Hart, 1988). Although such techniques may help potential dog owners to narrow down their selection to four or five breeds, the final selection should be based on what can be learned about the particular pedigree within a breed. When it comes to selecting an individual, the behavior of the dam, sire and siblings from previous litters, and the environment in which the puppy was raised are important considerations.

Comparison with other published breed rankings

Tortora (1980) and Howe (1976) have also presented rank order ratings of dogs on various behavioral traits. It is therefore interesting to compare the results of the present study with these earlier sources. For traits such as *obedience training*, *general activity*, and *snapping at children*, where breeds were not highly clustered together by either Tortora or Howe, there was a reasonable level of general correspondence between the ratings obtained by the different methodologies (Tables 5.4–5.6). However, with *watchdog barking* and *dominance over owner*, both Tortora's and Howe's methods failed to discriminate effectively between breeds, while the present study obtained statistically significant differences between breeds on these traits (Tables 5.7 and 5.8). If there are major differences between dog breeds in watchdog barking and dominance, then the forced ranking system, utilizing the full range of the ranking scale, provides a more useful guide to describing breed differences in behavior (when the opinions of authorities must be used), than do less structured methods employing the edited opinions of breeders.

Given the increasing interest in the behavior of companion dogs, and the importance of the dog's behavior for the interaction between owner and dog, there is a need for an objective system for rating dog breeds on behavioral characteristics. The analysis presented in this chapter represents an initial approach, and may help to focus the attention of some behavioral scientists on the feasibility and usefulness of such methods as a means of investigating breed-specific behavioral predispositions. The methodology presented here has also provided some predictions that could be tested by other direct or indirect methods of behavioral assessment.

References

Dixon, W. J. (1981). *BMDP Statistical Software*. Berkeley; University of California Press.

Hart, B. L. (1985). *Behavior of Domestic Animals*. New York: W. H. Freeman.

Hart, B. L. & Hart, L. A. (1984). Selecting the best companion animal: breed and gender specific behavioral profiles. In *The Pet Connection. Its Influence on Our Health and Quality of Life*, ed. R. K. Anderson, B. L. Hart & L. A. Hart, pp. 180–93. Minneapolis: CENSHARE, University of Minnesota.

Hart, B. L. & Hart, L. A. (1985). Selecting pet dogs on the basis of cluster analysis of breed behavior profiles and gender. *Journal of the American Veterinary Medical Association*, **186**, 1181–5.

Hart, B. L. & Hart, L. A. (1988). *The Perfect Puppy: How to Choose Your Dog by its Behavior*. New York: W. H. Freeman.

Hart, B. L. & Miller, M. F. (1985). Behavioral profiles of dog breeds. *Journal of the American Veterinary Medical Association*, **186**, 1175–80.

Hopkins, S. G., Schubert, T. A. & Hart, B. L. (1976). Castration of adult male dogs: Effects on roaming, aggression, urine marking, and mounting. *Journal of the American Veterinary Medical Association*, **168**, 1108–10.

Howe, J. L. (1976). *How to Choose your Dog*. New York: Harper and Row.

Serpell, J. A. (1983). The personality of the dog and its influence on the pet-owner bond. In *New Perspectives on Our Lives with Companion Animals*, ed. A. H. Katcher & A. M. Beck, pp. 57–63. Philadelphia: University of Pennsylvania Press.

Scott, J. P. & Fuller, J. L. (1965). *Genetics and the Social Behavior of the Dog*. Chicago: University of Chicago Press.

Tortora, D. L. (1986). *The Right Dog for You*. New York: Simon and Schuster.

6 Early experience and the development of behaviour

JAMES SERPELL AND J. A. JAGOE

Introduction

It is generally assumed in the literature on both human and nonhuman social development that early (infantile) experiences are more important in terms of their effect on later (adult) behaviour than those occurring at other stages of the life cycle (see Levine, 1962; Simmel & Baker, 1980; Bateson, 1981). During the last 50 years, studies of the domestic dog have played a major part in reinforcing this assumption. In 1945 an extensive programme of research was initiated at the Roscoe B. Jackson Memorial Laboratory at Bar Harbor, Maine, into the relationship between heredity and social behaviour in dogs. This work, which included detailed descriptive and experimental studies of behavioural ontogeny, led to the conclusion that there are particular periods in early development when puppies are unusually sensitive to environmental influences and, therefore, especially vulnerable to permanent psychological damage (see Scott & Marston, 1950; Freedman, King & Elliot, 1961; Elliot & Scott, 1961; Scott, 1962; Scott, 1963; Scott & Fuller, 1965; Fuller, 1967; Scott, Stewart & DeGhett, 1974).

In this chapter we review the evidence linking adult behaviour and temperament with the effects of early experience in dogs, particularly with regard to the development of so-called behaviour problems.

Stages of development

According to the findings of the Bar Harbor studies (see Scott & Fuller, 1965; Scott *et al.*, 1974), the early development of the dog can be divided into a series of four 'natural' stages or periods: (1) the *neonatal period*; (2) the *transition period*; (3) the *socialization period*; and (4) the *juvenile period*. To these should be added the *prenatal period*, since it appears that long-term effects on behavioural development may also be produced in some mammals by events occurring *in utero* (see Joffe, 1969).

The prenatal period

The prenatal period has been largely overlooked as a developmental stage in the ontogenesis of canid behaviour. Nevertheless, studies of rodents indicate that transplacental maternal influences may affect the subsequent behaviour of the offspring. For example, subjecting females to stressful experiences during pregnancy tends to render their offspring more emotional or reactive in test situations later in life (Thompson, 1957; Thompson, Watson & Charlesworth, 1962; De Fries, Weir & Hegmann, 1967), and more emotional females tend to give birth to more emotional offspring independent of genetic influences (Denenberg & Morton, 1962). Various preferences and species recognition patterns may also be influenced prenatally (see Joffe, 1969; Gottlieb, 1976).

In an altricial species such as the dog, the immature state of the fetal nervous system makes it unlikely that significant prenatal effects on behaviour result from learning. Any changes in emotionality, for instance, are probably caused by direct effects of maternal corticosteroid hormones on the development of the fetus's subsequent physiological responsiveness to stress (see Hinde, 1970). Androgens derived either from maternal circulation or from the proximity and preponderance of male littermates have also been shown to exert a prepartum 'masculinizing' influence on the subsequent behaviour of female rat and mouse pups (see Hart & Hart, 1985). Whether similar effects exist in canids is at present unknown, and it should be emphasized that placentation and blastocyst elongation patterns vary markedly between species.

The neonatal period

During the *neonatal period* the puppy is still comparatively helpless and dependent on the mother, and adapted to a life of suckling and care-soliciting. At this age, from birth to approximately two weeks, puppies are sensitive to tactile stimuli and certain tastes and, possibly, smells (see Thorne, Chapter 7), but their motor abilities are very limited, and neither their eyes nor their ear canals are open or functional. Because of the immature state of their neurosensory systems, it was originally assumed that canine neonates were largely incapable of associative learning (Scott & Marston, 1950). Subsequently, it has been shown that neonatal puppies can learn simple associations, although slowly compared with older pups, and only within the limits of their own rather specialized sensory and behavioural capacities (see Cornwell & Fuller, 1960; Stanley, 1970, 1972). On the basis of this specialization, Scott & Nagy (1980) concluded that learned effects of early experience in the neonatal period are unlikely to carry-over to later periods to any appreciable extent.

Nevertheless, it is well established that short periods of daily handling, as well as a variety of other strong or noxious physical stimuli, can have marked, long-term effects on the behavioural and physical development of some mammalian neonates, including puppies (see Meier, 1961; Levine, 1962; Whimbey & Denenberg, 1967; Denenberg, 1968; Fox, 1978). These effects include accelerated maturation of the nervous system, more rapid hair growth and weight gain, enhanced development of motor and problem solving skills, and earlier opening of the eyes. In behavioural terms, canine neonates exposed to varied stimulation from birth to five weeks of age were found to be more confident, exploratory and socially dominant when tested later in strange situations than unstimulated controls (Fox, 1978). To account for these sorts of effects, Levine (1967) suggested that early handling and stress produces an adaptive change in the animal's pituitary-adrenocortical system that enables it to cope more effectively with stressful situations later in life.

The extent to which such effects prevent or contribute to the development of behaviour problems in domestic dogs is at present unknown, although Fox (1978) reported that the adoption of early handling programmes by some dog breeders (including the US Army Veterinary Corps) achieved remarkable improvements in stress-resistance, emotional stability and learning capacity. Both Fox (1971) and Zimen (1987) have also suggested that wolf pups hand-reared from birth or six days of age are more reliable and friendlier towards humans than those hand-reared from 15 days or subsequently. Even allowing for species differences, this would suggest that puppies are almost certainly sensitive to some effects of maternal style and human handling long before the end of the neonatal period.

The transition period

As its name suggests, the *transition period* is marked by a period of rapid transition or transformation during which the patterns of behaviour associated with neonatal existence disappear and are replaced by those more typical of later puppyhood and adult life. The whole process takes no more than a week, beginning with the opening of the eyes at around 13 (±3) days, and ending at approximately 18–20 days with the opening of the ear canals and the first appearance of the auditory 'startle' response to loud noises. Electroencephalogram (EEG) readings of visual cortex activity also indicate a sudden increase in alpha waves at about three weeks of age, although brain wave patterns and visual acuity do not attain adult levels until about eight weeks. Puppies also show a number of changes in behaviour during this transitional phase. They show an ability to crawl backwards as well as forwards, and begin to stand and walk, albeit clumsily. They start to defecate and urinate outside the nest, and anogenital licking by the mother is no longer required to stimulate elimination. Puppies begin showing an interest in solid food at this time, and they also start play fighting with littermates and displaying social signals such as growling and tail-wagging. Patterns of distress vocalization also change. Whereas, neonatal puppies yelp primarily in response to cold or hunger, a three-week-old puppy will also yelp if it finds itself outside the nest in an unfamiliar environment, even if it is otherwise warm and well-fed (Scott & Fuller, 1965; Fox, 1971).

Most of these behavioural changes make sense when considered in the context of development under non-domestic conditions. Among wolf pups living in the wild, 2–3 weeks represents the age at which they first emerge from the dark interior of the breeding den. Evidence from comparative studies suggests, however, that this period of transition begins slightly earlier and is completed more rapidly in wolves than in domestic dogs (Frank & Frank, 1982, 1985; Zimen, 1987).

In terms of learning and the effects of early experience, the transition period more or less resembles a continuation of the neonatal stage. Puppies' performances on both classical and operant conditioning tasks show a steady improvement at this age, although rates of learning and the stability of conditioned responses do not reach adult levels until 4–5 weeks of age (Scott & Fuller, 1965).

The socialization and juvenile periods

The socialization period in puppies was first described as a 'critical period' for the formation of primary social relationships or social attachments (Scott, 1962; Scott *et al.*, 1974). The concept of 'critical periods' was borrowed originally from embryology by Lorenz (1935), who used it to account for the phenomena of filial and sexual imprinting in precocial birds (i.e. those which are able to move about and feed themselves immediately after hatching). In this context, the *critical* period was seen as a narrow and clearly defined developmental window during which specific stimuli produced long-term and irreversible effects on behaviour. According to some interpretations, appropriate stimulation within the critical period was also necessary for normal development to proceed (e.g. Fox, 1978). More recent evidence suggests, however, that the boundaries of such periods tend to be gradual rather than sudden, and that behaviour or preferences acquired within them can usually be modified or reversed at later stages, albeit with varying degrees of difficulty. As a result, most authorities now favour the term 'sensitive periods' – that is, less distinct periods or phases in development when particular responses or preferences are acquired more readily than at other times (see Hinde, 1970; Immelmann & Suomi, 1981; Bateson 1979, 1981).

Primary socialization in puppies has been found to be largely independent of associated rewards or punishments, although emotionally arousing stimuli – both positive and negative – seem to accelerate the process (Scott, 1963; Scott & Fuller, 1965; Scott *et al.*, 1974). Among wolf pups reared under natural conditions, the process of socialization ensures that the young animals form their primary social attachments for their littermates, parents and other pack members. In the case of the domestic dog, it enables puppies to form non-conspecific attachments for humans or other animals encountered socially during the same period. For example, cross-fostered puppies raised throughout the socialization period with only kitten littermates were subsequently found to reserve all positive social behaviour for cats and kittens, and to avoid interacting with strange puppies. The kittens, on the contrary, engaged in positive social interactions with strange puppies, something normally-reared kittens show no desire to do (see Fox, 1969). Such experiments demonstrate that the character of the socialization experience not only determines the young animal's future social partners but also defines the species to which it effectively belongs. It has also been shown that early, non-conspecific encounters do not need to be particularly frequent or protracted for socialization to occur. Fuller (1967) found that puppies could be socialized to humans during this period with as little as two 20-minute sessions of exposure per week, and Wolfle (1990) has described a programme that achieved 'adequate socialization' of laboratory beagles with less than five minutes of human social contact per pup per week.

During the sensitive period for socialization, puppies also appear to form attachments for particular places, a phenomenon referred to as 'localization' by Scott & Fuller (1965), although 'site attachment' might be a more appropriate term to use. According to Scott & Fuller, the consequences of socialization and localization are so similar that they may represent the same process applied to different objects: 'this would mean that the puppy becomes attached to both the living and non-living parts of its environment at this age' (1965, p. 112).

The upper and lower boundaries of the socialization period have been determined by laboratory experiments in which puppies' social contacts have been observed and manipulated at different points and for different periods in early development. In what is often regarded as the definitive study, Freedman *et al.* (1961) reared eight litters of cocker spaniel and beagle pups in isolation (from humans, but not from their mothers or littermates) until 14 weeks of age, during which time each pup received one week of moderately intensive human testing and handling before being returned to the litter. Some pups received this week of human socialization at two weeks, others at three, five, seven or nine weeks of age, respectively. Five 'control' pups remained unsocialized until 14 weeks of age. At 14 weeks, all the pups were tested (or retested) for their responses to a 'passive' human handler, to being walked on a leash, and to being strapped into a physiological harness and subjected to various arousing or unpleasant stimuli. Those pups socialized between five and nine weeks approached a passive handler most readily, and were more easily trained to walk on a leash. Those socialized at seven weeks obtained the most favourable scores in terms of their reactions to being tested in harness. Control pups remained uniformly fearful and intractable even after many weeks of careful handling and petting. On the basis of these findings, the authors concluded that '$2\frac{1}{2}$ to 9–13 weeks of age approximates a critical period for socialization' (Freedman *et al*, 1961, p. 1017).

Behavioural observations of naive puppies' responses to human handlers at different ages tended to confirm these findings. Although initially fearful in the presence of an 'active' human handler, young puppies show a rapid increase in their tendency to approach and make social contact with an unfamiliar person between the ages of three and five weeks. Thereafter, this tendency declines. Conversely, from 3–5 weeks most pups show little or no fear of a 'passive' handler, but they then become increasingly wary or fearful of strange individuals or situations beyond this age. On the basis of this kind of evidence, Scott & Fuller (1965) concluded that the primary socialization period ran from about the third to the twelfth week after birth, with a peak of sensitivity between six and eight weeks. Below three weeks of age, they argued, a puppy's neurosensory systems are too underdeveloped to permit socialization, and beyond 12 weeks its growing tendency to react fearfully to novel persons or situations puts an effective upper limit on further socialization. Between six and eight weeks, however, a pup's social motivation to approach and make contact with a stranger outweighs its natural wariness, hence the view that this period represents the optimum time for socialization.

Based on the results of conditioned aversion experiments with beagle pups, Fox & Stelzner (1966) also identified a period at around eight weeks when pups are hypersensitive to distressing psychological or physical stimuli. Puppies' heart rates and rates of distress vocalization in strange situations also tend to show developmental peaks during this 6th–8th week period (Elliot & Scott, 1961; Scott & Fuller, 1965).

Subsequent studies and observations indicate that the upper boundary of the socialization period is far less clearcut than originally suggested, hence the decision to merge the socialization and juvenile periods in this review. Good evidence exists, for example, that young wolves – and many young dogs – that are well socialized at three months will, nevertheless, regress and become fearful again in the absence of periodic social reinforcement until the age of 6–8 months (Woolpy & Ginsberg, 1967; Woolpy, 1968; Fox, 1971, 1978). Once properly socialized, however, adult wolves appear to remain so despite long periods of isolation from human contact (Woolpy & Ginsberg, 1967). Anecdotal evidence that young wolves experience a second, sudden-onset phase of heightened sensitivity to fear-arousing stimuli at around 4–6 months of age may account for these phenomena (see Fentress, 1967; Mech, 1970; Fox, 1971). Conversely, it has also been established that adult, untamed wolves and inadequately socialized dogs can still be socialized to humans, although the process requires considerable patience and, in the case of wolves, involves isolating the animal from everything but human contact for periods of 6–7 months (Woolpy & Ginsberg, 1967; Niebuhr *et al.*, 1980). Finally, most authorities would probably agree that substantial individual and breed differences exist in the precise timing and quality of the socialization process. Zimen (1987, p. 290) attributed some of this variation to the variable and conflicting expression of 'two genetically independent motivational systems': the motivation to make social approaches to strangers and the motivation to flee from novel stimuli. In a study of social development in wolves, poodles and their hybrids, he found that these two tendencies tended to segregate out among the F_2 backcrosses (Zimen, 1987). Other work on wolves and domestic foxes (*Vulpes vulpes*) suggests that differences in developing aggressive tendencies may also affect the upper age limit for socialization (Woolpy & Ginsberg, 1967; Plyusina, Oskina & Trut, 1991).

The results of the Bar Harbor studies of socialization, and subsequent, related investigations by Guide Dogs for the Blind (see Pfaffenberger *et al.*, 1976) gave rise to various practical recommendations regarding the husbandry and training of domestic dogs. In particular, two basic rules for producing a well-balanced and well-adjusted dog were proposed. First, that the ideal time to produce a close social relationship between a young dog and its human owner is between six and eight weeks of age, and that this is therefore 'the optimal time to remove a puppy from the litter and make it into a house pet' (Scott & Fuller, 1965, p. 385). Second, that puppies should be introduced, at least in a preliminary way, to the circumstances and conditions they are likely to encounter as adults, preferably by eight weeks, and certainly no later than 12 weeks of age (Scott & Fuller, 1965; Pfaffenberger & Scott, 1976). The latter recommendation, as well as the basic idea of a 6th–8th week 'optimum period' for socialization, has never been seriously challenged in the literature. However, Slabbert & Rasa (1993) have recently criticized the recommended practice of removing pups from their maternal environment at six weeks of age. In a study of socialization and development in German shepherd puppies, these authors found that pups separated from their mothers and nest sites (but not their littermates) at six weeks exhibited loss of appetite and weight, and increased distress, mortality and susceptibility to disease compared with pups that remained at home with their mothers until 12 weeks of age. Both groups, however, showed the same degree of socialization towards their human handlers. Although the rates of morbidity and mortality observed in this study appear to be abnormally high, Slabbert & Rasa (1993, p. 7) nevertheless concluded that conventional rehoming procedures for puppies are deleterious to the animals' welfare, and that the evidence on primary socialization in dogs has been wrongly interpreted to mean that 'exclusive access to the desired bonding partner' is actually *necessary* at this time in order to achieve correct socialization.

Development of behaviour problems

Detailed and exhaustive coverage of the development of all, or even most, canine behaviour problems is not possible at this stage, since many problems have never been subjected to empirical research and so little useful information is available as to their probable ontogeny. In this section we will therefore focus on a few, common categories of behaviour problem for which at least some developmental data are available.

Aggression

Aggression is the most commonly reported category of behaviour problems in domestic dogs (Hart & Hart, 1985), and one that has received an inordinate amount of public and media attention in recent years (see Lockwood, Chapter 9; Serpell, Chapter 16). Despite the popular Lorenzian idea of aggressiveness being a single temperament trait (Lorenz, 1966), most modern ethologists would accept that canid aggression, like aggression in every other species, is likely to be context dependent, and that a dog which responds aggressively in one situation is not necessarily likely to do so in others. Unfortunately, wide disparities in methods of classifying aggressive behaviour problems now exist in the relevant literature (see e.g. Moyer, 1968; Borchelt & Voith, 1982*a*; Borchelt, 1983*a*; O'Farrell, 1986), and this has tended to hamper efforts to understand the ontogeny of this important category of problems. In this section we will consider only two forms of aggressive behaviour problem: aggression in relation to home-range or territory, and aggression associated with social dominance.

Territorial aggression

Many dogs, like wolves, display a tendency to react aggressively to unfamiliar intruders within their home ranges. In practice, this home range or territory usually comprises the immediate vicinity of the owner's home, but may also include other areas where the dog is regularly walked or confined. As with most traits, there are marked individual and breed differences in the tendency to display territorial behaviour, and it would appear that elements of this behaviour have been amplified by human selection, particularly in certain guarding breeds (see Hart & Hart, 1985; Adams & Johnson, 1993; Hart, Chapter 5). Extreme territorial aggression is also one of the more common forms of behaviour problem in dogs. One large North American survey reported that over 18% of owners classified their dogs as being 'overprotective' in a territorial sense (Campbell, 1986).

There appear to have been few, if any, studies of the development of territorial behaviour in either domestic dogs or wild canids. According to Borchelt & Voith (1982*a*), dogs with a behaviour problem are usually from 1–3 years of age, although this presumably reflects the age when the behaviour became a problem rather than the age when it first developed. Anecdotally, a number of authors describe the first appearance of overt hostility towards unfamiliar intruders in wolf pups at around 16–20 weeks of age, coinciding with a sudden phase of heightened sensitivity to novel or fear-evoking stimuli (Fentress, 1967; Mech, 1970; Fox, 1971). Whether this represents some sort of sensitive period for the acquisition of territorial behaviour is not known, although Mech (1970) pointed out that this is about the age when young wolves start moving away from the familiar den and rendezvous sites, and when they are therefore more likely to encounter strange or hostile territorial intruders. It has not been established whether domestic dogs show a similar increase in reaction to either dog or human strangers at this age, or whether particular experiences at this time contribute in any way to the onset and intensity of adult territorial behaviour.

Dominance-related aggression

In the literature on canine behaviour problems, the labels 'dominance aggression' and 'dominance-related aggression' are applied interchangeably to the tendency of some dogs to react aggressively to apparent challenges to their positions within the social hierarchy. These circumstances, it is said, include those in which the owner is apparently treated as a competitor for resources – e.g. food, space, sleeping position, etc. – or in response to supposedly 'dominant' gestures by the owner, such as holding, petting, grooming, restraining, punishing or pushing past the animal, staring or yelling at it, or even leaning over it. In general, dominance-related aggression is characterized by threats or attacks directed at the owner or a member of the owner's family rather than strangers. This form of aggression is also more commonly reported in intact males and neutered females, and it is one of the most frequent problems seen by behaviour therapists and trainers (see Campbell, 1975; Borchelt, 1983*a*; Voith & Borchelt, 1982; Hart & Hart, 1985; Line & Voith, 1986; O'Farrell, 1986; Wright & Nesselrote, 1987; Mugford, Chapter 10). O'Farrell (1986) suggested that socially dominant dogs are more likely to behave aggressively in contexts that are not directly concerned with the maintenance of social rank. If true, however, this observation would tend to undermine the whole dominance concept.

The concept of dominance originated with Schjelderup-Ebbe's (1922) early work on linear 'peck-order' hierarchies in domestic fowl, and is based on the notion that animals sometimes compete aggressively for status or social rank much as they do for other resources, such as food or mates. The idea has since been used (and misused) extensively to explain the organization and maintenance of social structure in many group-living animals (see Hinde, 1974). Although its usefulness as an explanatory construct has been criticised in relation to both primate and canid social organization (see e.g. Rowell, 1966; Lockwood, 1979; Bernstein, 1981), various observations suggest that the concept does provide a useful explanation for some aspects of group social dynamics, at least in wild canids such as wolves. Most observers of wolf behaviour have identified separate male and female linear dominance hierarchies in social groups of wolves, each headed by a so-called *alpha* animal. With the exception of the *alpha* female, who appears to share a similar status to that of the *alpha* male, males tend to be dominant over females (see Schenkel, 1967; Woolpy, 1968; Fox, 1971; Zimen, 1981). Using a more detailed analysis of behavioural interactions, van Hooff & Wensing (1987) detected a single, more-or-less linear dominance hierarchy in their wolf pack, although dominance relationships tended to be expressed in different ways among males and females. High-ranking males (and the *alpha* female) could be recognized by the preponderance of 'high' or dominant postures they adopted during interactions, particularly with males of similar rank. In contrast, the dominant partners in interactions between females, and between males and females (apart from the *alpha*) could be inferred from the relative preponderance of 'low' or submissive postures they received. Dominance relationships among male wolves might therefore be characterized as *dominance asserting*[1], while those among females, and between males and subordinate females, might be better described as *dominance acknowledging* (see van Hooff & Wensing, 1987).

Both Lockwood (1979) and van Hoof & Wensing (1987) concluded that neither the direction nor the frequency of aggressive threats or attacks were reliable indicators of dominance relationships in wolf packs. Although dominant individuals enjoy unquestioned priority in most cases, they do not invariably pull rank in competitive situations. Where the dominant individual has relatively little interest in pursuing a claim on a particular resource, subordinate animals may be able to gain access by means of threats or attacks. In other words, subordinates can be 'situationally dominant without upsetting the formal dominance' (van Hooff & Wensing, 1987, p. 248). These authors also suggest that the terms *dominant* and *dominance* should be confined to describing relationships, and that a term such as '*assertiveness*' should be used to describe an individual's propensity to strive for a dominant position within such relationships. Although still somewhat controversial, recent studies of deer mice (*Peromyscus maniculatus*) suggest that the tendency

[1] van Hooff & Wensing (1987) actually use the phrases *dominance confirming* and *subordinance acknowledging*. However, we consider our terminology less likely to create misunderstandings.

to become socially dominant or subordinate can be inherited from fathers to sons (Dewsbury, 1990). However, since dominance is not in itself a trait but rather an emergent property of a relationship (see Barrette, 1993), it is not yet clear what phenotypic characters are involved.

Experimental studies of the development of dominance and dominance-related behaviour in canids have been fraught with methodological problems. Despite strong theoretical reasons for rejecting this approach (see e.g. Hinde, 1974), dominance has usually been assessed experimentally by placing two (or more) littermates together in an artificial competitive situation – the so-called *bone-in-pen test* – and ranking them according to their ability to gain and keep possession of some desired object, such as a bone or toy, for a fixed period of time (see Scott & Fuller, 1965; Fox, 1972). This technique ignores the possible effects of transient motivational differences between individuals at the time of testing, or the influence of individual temperamental factors, such as overall persistence or confidence, which may or may not be related to social dominance.

In wolf pups, both Fox (1972) and MacDonald (1987) found that paired contests were unreliable as a means of assessing dominance. In this context, pups tended to share the bone rather than compete for it. When littermates were tested *en masse*, however, both authors regarded the outcome of such contests as a reliable indicator of the most socially dominant individual(s) within each litter. According to Fox (1972) the highest ranking individuals in this test also tended to be more active and exploratory when confronted by novel objects, and were more adept at killing prey (live rats) than low ranking pups. He also found that test results obtained at eight weeks of age correlated reasonably well with dominance ratings ten months later. MacDonald (1987) tested a litter of five male wolf pups *en masse* at frequent intervals between the ages of 17–180 days, and found that by the fifth week one pup was consistently outcompeting the others for possession of the bone. Between 15 and 20 weeks, this pup was sporadically beaten by various littermates, but thereafter he won consistently until the end of the study. By about six weeks, this pup also ranked consistently highest in tests of boldness or 'leadership' in unfamiliar situations. These findings led the author to conclude that the tendency to become socially dominant in wolves is one aspect of a relatively stable personality trait that can be detected as early as six or seven weeks of age (MacDonald, 1987).

The results of competition tests involving domestic dog puppies are generally more confusing, and are characterised by marked breed differences. Scott & Fuller (1965) based their assessments of dominance on standard, paired bone-in-pen style contests conducted at 5, 11, 15 and 52 weeks of age. They also allowed their puppies weekly 'training periods' with the bone between the ages of two and ten weeks. At five weeks they found little evidence of consistent dominance (defined as the ability to possess the bone for more than 80% of a ten minute test period) in any of the five breeds they studied. By 11 weeks, all breeds showed a large increase in the number of completely dominant relationships. Beyond this point, however, wire-haired fox terriers, Shetland sheepdogs, and basenjis showed a continuing increase until 52 weeks – though at different rates – while dominant relationships among beagles and cocker spaniels tended to decrease.

Breeds also differed in their tendency to threaten or attack each other in the test situation. Beagles and cocker spaniels almost never fought for possession of the bone and appeared to show very low levels of aggression at any age. Fox terriers also hardly ever fought in dyadic encounters, not because they were unaggressive (at around seven weeks aggression became such a problem in some litters that the pups needed to be separated) but because dominance relationships were apparently established at an early age in this breed. In contrast, shelties fought or attacked each other quite frequently at five weeks, but almost never after this, while basenjis continued to fight in competitive situations until they were a year old. The different breeds also showed sex differences in their tendency to establish dominant relationships. In the more aggressive breeds, such as fox terriers and basenjis, males tended to become completely dominant over females by 15 weeks or even earlier. In beagles and cocker spaniels there was no discernible tendency for males to be dominant over females at any age and, while male shelties showed no particular tendency to dominate females in competition for bones, other observations suggested that they were completely dominant in contexts unrelated to food competition. On the basis of these findings, Scott & Fuller (1965) concluded that the capacity to establish stable dominance relationships is ultimately a product of inherited aggressive tendencies, but that an individual's likelihood of becoming socially dominant is strongly dependent on previous experience of interactions with other members of the litter or social group.

More recently, Wright (1980) has challenged the view that bone competition tests provide an accurate measure of social dominance in puppies. He investigated dyadic bone-in-pen test outcomes, neophobic responses in a modified 'open field test', and social interactions in the rearing environment in a litter of five German shepherd puppies at $5\frac{1}{2}$, $8\frac{1}{2}$ and $11\frac{1}{2}$ weeks, and found that puppies' apparent dominance positions within the litter environment bore little or no relation to their likelihood of monopolizing a bone in a test situation. As in the studies of wolf litters described above, however, competitive ability in the bone-in-pen tests was correlated with confidence and exploratory behaviour in the unfamiliar open field. Wright (1980) concluded that neophobic responses to the unfamiliar test situation prevented otherwise socially dominant pups from competing successfully in the bone-in-pen test, and that these tests actually measure competitive tendencies rather than social dominance. He also suggested that the social dominance hierarchy within the litter did not appear to stabilize until $11\frac{1}{2}$ weeks.

The results of all these various studies suggest that human selection under domestication has had marked effects on the expression of dominance-related aggressiveness or assertiveness in *Canis familiaris*. Evidence from wolves indicates that dominance relationships within litters of pups may become established any time between four and eight weeks of age, at least as far as the highest ranking individuals are concerned. The tendency to win in competition for a limited resource may be part of an inherited or pre-established temperament trait associated with overall confidence in unfamiliar situations (Mech, 1970; Fox, 1972; Bekoff, 1974; Fox *et al.*, 1976; MacDonald, 1987). However, it is still unclear whether these two tendencies are linked or related, and serious doubts persist regarding what the various tests of dominance actually measure. In dogs, breeds may differ with respect to the contexts in which dominance behaviour is exhibited. In some it may be obvious in relation to food competition, while in others it may only be expressed during conflicts over space or the attentions of the owner. Some dog breeds, particularly certain terriers, appear to establish dominance relationships at least as early as wolf pups, while others, such as cocker spaniels and beagles, may never develop stable dominance relationships regardless of circumstances (Fox, 1972; Scott & Fuller, 1965; see also Bradshaw & Nott, Chapter 8). Even in beagles, however, postures associated with dominance displays, such as *standing over*, first start appearing in the behavioural repertoire at around four weeks of age during bouts of social play (Fox *et al.*, 1976). This would suggest that, if there is a sensitive period for the acquisition of dominance-related assertiveness in dogs, it is likely to peak during the first few weeks of the socialization period.

With regard to the possible effects of early experience on the development of dominance-related behaviour problems, the evidence is sparse and complicated by the inability to separate out the confounding and interdependent effects of overall *confidence*, *competitiveness* with respect to particular resources, and *assertiveness* in relation to the acquisition of social rank. Common sense would suggest, however, that prenatal environment, early handling effects, and early agonistic interactions and experience with the mother, littermates and with human care-givers may all exert a modifying influence on the development of these traits. The results of early social isolation experiments with Scottish terriers suggested that restricted social experience during the socialization and juvenile periods rendered these pups more or less incapable of competing aggressively for food in test encounters with normally-reared puppies (Clarke *et al.*, 1951; Melzack & Thompson, 1956). However, it is not clear from these findings whether the outcomes were the result of specific early experience deficits, or simply a lack of prior experience in general. More recently, Wilsson (1984) has provided largely anecdotal evidence that German shepherd bitches who adopt a relatively disciplinarian mothering style during the weaning period may encourage the development of submissive behaviour in their puppies. He also speculates that this might have long-term effects on these dogs' trainability later in life.

Fears and phobias

Nervous or fearful responses to strangers and unfamiliar situations are another common source of behaviour problems in dogs. Analysis of reasons for rejecting dogs from the US Guide Dogs for the Blind programme revealed that, out of 600 animals, 19% were frightened of loud noises, 15% were afraid of cars or farm machinery and 12% evinced fear of other animals. Eighteen per cent also displayed fear in a range of miscellaneous or unspecified situations (Tuber, Hothersall & Peters, 1982). Fearfulness is also cited as the most common reason for rejecting potential guide dogs by Goddard & Beilharz (1982, 1984). Campbell (1986) reported a 20% incidence of 'fear of noises' in his survey of 1422 American dog owners, and data from one behaviour problem referral clinic suggests that roughly one-third of all behaviour consultations involve fear-related behaviour problems, although it is not clear whether this figure includes separation-related anxieties (Shull-Selcer & Stagg, 1991).

Interpreting the available information on the development of canine fearfulness is rendered more difficult by the use of widely varying systems of classifying aversive or fearful behaviour. On the one hand, different fears or phobias are sometimes regarded as being more-or-less distinct (Hart & Hart, 1985), while on the other they may be lumped together with more generalized anxieties (Tuber *et al.*, 1982), or treated as symptomatic of some global temperament trait, such as 'emotionality' (Scott & Fuller, 1965; Scott & Bielfelt, 1976) or 'stimulus reactivity' (Wright & Nesselrote, 1987). Regardless of how it is classified, however, it is apparent that there is a strong genetic basis to fearful behaviour (see Serpell, 1987; Willis, Chapter 4). In 1944, investigations into the high prevalence of abnormally shy or fearful dogs in one laboratory colony revealed that 52% of the nervous animals were directly descended from a single bassett hound bitch who was a notorious fear-biter. It was concluded that 'shyness' is a dominant characteristic in dogs that is normally strongly selected against in the pet dog population (Thorne, 1944). Scott & Fuller's (1965) Bar Harbor studies seemed to confirm this assessment. Their findings with regard to puppies' fearful responses to being approached and handled by humans among basenjis, cocker spaniels and their various hybrids and backcrosses were, in their own words, consistent with the action of 'a single dominant gene causing wildness in the basenji . . . and a contrasting [recessive] gene for tameness in the cockers' (1965, p. 268). More recent evidence from guide dog breeding programmes demonstrates that the 'fearfulness' trait is moderately heritable (Goddard & Beilharz, 1982), and efforts to breed strains of abnormally fearful or 'nervous' pointer dogs, to serve as research models of human anxiety disorders, have also been highly successful (see Dykman, Mack & Ackerman, 1965; Dykman, Murphree & Ackerman, 1966; Dykman, Murphree & Reese, 1979; Murphree & Dykman, 1965; Murphree, 1973; Murphree *et al.*, 1977; Newton & Lucas, 1982).

In addition to genetic factors, it appears that neonatal (and possibly prenatal) environment and experiences may influence puppies' subsequent reactions to stressful or frightening situations through their direct effects on the development of pituitary-adrenocortical responsiveness (Levine, 1962; Whimbey & Denenberg, 1967; Denenberg, 1968; Fox, 1978). Although allowances must be made for species differences in placentation, evidence from studies of rodents would imply that the offspring of bitches stressed during pregnancy might be more nervous in strange situations than normal puppies (Hinde, 1970). In contrast, exposing puppies or fox cubs to handling or other mild stressors during the neonatal period tends to produce more phlegmatic and less easily stressed or frightened individuals (Fox & Stelzner, 1966; Fox, 1978; Pedersen & Jeppesen, 1990). Other, as yet unidentified, aspects of maternal style and behaviour may also affect puppies' confidence in stressful situations (see Scott & Bielfelt, 1976; Wilsson 1984).

Abundant evidence also exists for the learned acquisition of fearful and phobic behaviour. Various early isolation experiments have demonstrated that puppies reared in restricted, visually-isolated or environmentally-impoverished conditions from weaning until around 12–14 weeks of age exhibit varying degrees of neophobia when placed in unfamiliar situations (Clarke *et al.*, 1951; Melzack & Thompson, 1956; Fuller, 1967). Isolation effects are also thought to account for so-called 'kennel dog syndrome', i.e. the effect of leaving dogs in relatively restricted kennel environments beyond 12 weeks of age, resulting in animals that exhibit abnormal levels of timidity towards novel situations (Pfaffenburger & Scott, 1976). Similarly, pups reared with little or no human contact for the duration of the socialization period tend to develop a generalized fear of humans that is difficult, if not entirely impossible, to overcome subsequently (Freedman *et al.*, 1961; Elliot & Scott, 1961; Scott & Fuller, 1965). Such findings are consistent with the idea that young dogs preferentially 'imprint' on, or at least familiarize themselves with, certain biologically important aspects of their environment during the sensitive period for socialization, and that once these preferences and attachments are formed they are reinforced by the subsequent development of increasingly fearful and avoidant responses to anything novel or unfamiliar (see Bateson, 1979).

The results of conditioned aversion experiments further suggest that there may be relatively narrow periods of maximum sensitivity to frightening stimuli within the socialization period. By training beagle pups to associate human contact with electric shocks at 5, 8 and 12 weeks of age, Fox & Stelzner (1966) were able to demonstrate a short period at approximately eight weeks when pups were hypersensitive to distressing psychological or physical stimuli, and during which a single unpleasant experience could produce long-term aversive or abnormal effects. They concluded from this that, below five weeks of age, the effects of conditioning were unstable and quickly 'forgotten', while at 12 weeks the aversive effects were completely over-ridden by positive affiliative tendencies towards humans established during the socialization period. At around eight weeks, however, conditioning is stable and effective but strong social bonds are not yet fully established, hence the vulnerability of these puppies to psychological trauma at this time (Fox & Stelzner, 1966).

Anecdotal evidence from observations of tame or captive wolves indicate that at least some individuals go through a second period of intense sensitivity to fear-evoking situations at around 4–5 months of age (see Fentress, 1967; Mech, 1970; Fox, 1971; MacDonald, 1983). It is not known, however, if an equivalent period exists in dogs or whether frightening events occurring at this age have stronger or more durable effects on subsequent behaviour.

Fearful behaviour in dogs may be relatively stimulus-specific. For example, although most dogs are wary of unfamiliar objects in their home environment, particularly if the object is large or moves suddenly, it is apparent that some animals are far more aversive in these contexts than others (Melzack, 1954; Voith & Borchelt, 1985*a*). Similarly, noise phobic or 'gun-shy' dogs are a well-established category of problem animals (Hart & Hart, 1985; Stur, 1987; Shull-Selcer & Stagg, 1991), and specific 'anthropophobic' responses have been described in inbred strains of 'nervous' pointer dogs (Dykman *et al*., 1979). Individual differences in response to specific fear-evoking stimuli may also exist in wild canids, such as wolves. MacDonald (1983), for example, detected an inverse relationship between fear of strange people and fear of unfamiliar objects in a sample of five wolf pups. The most plausible explanation for all of these observations would be that most dogs (and wolves) are born with a 'biological "preparedness" to learn to fear certain evolutionarily relevant or prepotent stimuli' (Shull-Selcer & Stagg, 1991, p. 355), but that early experience, especially during the latter half of the socialization period (and possibly later), plays a major part in determining which fears are acquired and how strongly they are expressed in adult life.

Separation-related problems

'Separation-related behaviour' and 'separation anxiety' are terms commonly used in the behaviour problem literature to describe a particular class of problematical behaviour patterns in dogs that occur only in response to separation from the owner. Such problems include: *separation-related destructiveness* – biting, chewing and scratching household furniture and fittings, often near the site of the owner's most recent departure; *separation-related vocalizing* – barking, whining or howling; and *separation-related defecation* and *urination*. The latter may be symptomatic of generalized anxiety (although house-training problems, marking behaviour and pathophysiological disorders cannot be ruled out), while the two former behaviour problems are most easily interpreted as attempts by the dog to restore contact with the owner, either by escaping from confinement and following him, or by maintaining vocal contact (Borchelt & Voith, 1982*b*, Voith & Borchelt, 1985*b*; McCrave, 1991). Separation-related behaviour problems have a high prevalence in the pet dog population. According to McCrave (1991) such problems represent roughly 20% of the caseloads of behaviour consultants in the United States, although some report much higher figures (see e.g. Borchelt, 1983*b*). Marked breed differences in prevalence have also been reported. Mugford (1985) mentions unusually high prevalences in Labrador retrievers, German shepherd dogs and English cocker spaniels, but also states that crossbred dogs are far more prone to these problems than any pure breed. This statement is confirmed by the findings of McCrave (1991), who attributed the bias entirely to the fact that crossbred dogs are significantly more likely to be obtained from animal shelters.

Theories concerning the development of separation-related problems in dogs have been influenced, to some extent, by the extensive psychological literature on the effects of attachment and separation in human and nonhuman primates (e.g. Harlow & Harlow, 1966; Bowlby, 1973; Ainsworth *et al.*, 1978). According to the primate model of secure attachment, the mother or primary care-giver provides the 'secure base' from which the infant learns to explore its world and acquires confidence and stability in its relationships with others. Mothers who are anxious or ambivalent, however, disrupt the normal attachment process and tend to give rise to clingy, over-dependent infants who are abnormally distressed by separation. These ideas find parallels in the dog behaviour problem literature. For instance, the most commonly reported predisposing factor in the etiology of separation-related problems in dogs is said to be a sudden episode of enforced separation from the owner *preceded* by a period of prolonged and relatively constant and exclusive contact. The theory is that such dogs have either never learned to cope with separation from the primary attachment figure, and/or have developed an overly-intense attachment as a result of experiencing a long phase of constant and, presumably, highly-rewarding companionship. It is also suggested that dogs who experience the loss of a primary attachment figure, are more likely to develop insecure attachments to subsequent owners (Borchelt & Voith, 1982*b*; Voith & Borchelt, 1985*b*; McCrave, 1991), and this has been proposed as an explanation for the unusually high incidence of separation-related problems in animals adopted from animal homes and shelters (McCrave, 1991). In general, this tendency of some dogs to react badly to separation from the owner has been interpreted as a side-effect of unconscious human selection for increasingly affectionate, socially-dependent and infantilized pets (see Fox, 1978; Serpell, 1983; Mugford, 1985). In practice, however, it may be difficult to distinguish the genetic effects of human selection from learned patterns of behaviour acquired in response to the unusually dependent, childlike roles which many dogs are expected to play.

Regarding the possible impact of early experience on the development of separation-related behaviour, there has been surprisingly little research, although Borchelt (1984) stressed the importance of gradually introducing young puppies to separation in their new homes. Experiments at Bar Harbor revealed that, from about three weeks of age, puppies become extremely distressed if placed alone in a strange situation, away from the mother, littermates and nest sites (Elliot & Scott, 1961; Scott, 1962). The level of distress, as measured by frequency of distress vocalizations, rises to a peak at around 6–7 weeks of age after which it steadily declines, although animals which have experienced no separations until 9 or 12 weeks exhibit more distress when first tested than previously-tested puppies of the same age (Elliot & Scott, 1961). In their work with German shepherd puppies, Slabbert & Rasa (1993) argued that separation from the *mother* at six weeks of age has adverse effects on puppies' overall health and welfare, although it is not clear from their data whether pups react specifically to separation from the mother or from the nest site. From a biological standpoint, it would be inappropriate for puppies beyond the age of 3–4 weeks to be greatly distressed by periodic separation from their mothers, at least under natural conditions. Unlike primate infants who rely on their mothers (or other care-givers) for protection, young wild canids retreat to the safety of the den when danger threatens. Moreover, field observations of wolves have shown that, when pups are about 3–4 weeks old, the mother begins leaving them at the den site for periods ranging from 2–18 hours daily (Ballard *et al.*, 1991). Pups remain at the den continuously and do not begin to follow the adult members of the pack until they are at least 10–12 weeks old, and usually somewhat older (Gray, 1993). Evidence from wild canids would therefore suggest that removal from the familiar nest location is the most probable cause of the distress which accompanies rehoming in 6–8 week old domestic dog puppies, rather than the loss of any primary attachment figure(s). This distress may, however, have the effect of accelerating and intensifying the bond that is concurrently established with the new owner (see e.g. Scott *et al.*, 1974), and this may in turn exacerbate the effects of any subsequent separations.

Behaviour problem prevalence in relation to early experience: a recent survey

To investigate the possible long-term effects of early experience on the development of dog behaviour problems, Jagoe (1994) conducted a retrospective survey of the owners of 737 adult dogs. Subjects were recruited with the help of practising animal behaviour therapists ($N = 451$ 'problem' dogs), and via veterinary practices, a veterinary teaching hospital and random door-to-door inquiries ($N = 286$ 'control' dogs). Each dog owner completed a questionnaire[2] that was divided roughly into three sections. In the first part, owners were asked to provide background information about themselves and their household, and in the second they were invited to supply as much information as possible about the dog's early experiences and environment from birth–16 weeks of age (e.g. details of any early health problems, time left alone as a puppy during the day, the puppy's age when acquired, where it was acquired from, its age at first vaccination and the age when it was first taken out into public areas on a regular basis). Finally, owners were asked to indicate on a series of 5-point rating scales (ranging from 'almost never' to 'almost always') how frequently the dog displayed any out of a list of 40 possible behaviour problems arranged either according to their contexts of occurrence (various forms of aggression, fears and phobias, excitability and other miscellaneous problems), or their outcomes (separation-related behaviour). The rating scales were subsequently reduced to a more conservative, present or absent scoring system (1 to 2 – 'absent'; 3 to 5 = 'present') for the purposes of statistical analysis.

The results of this analysis are presented and discussed in greater detail elsewhere (see Jagoe, 1994). However, a subset of some of the more significant findings are provided here primarily to illustrate the apparent association between certain early events and experiences and the prevalence of adult behaviour problems in this sample of dogs.

Effects of source

According to their owners, the dogs in the study originated from six different possible sources: breeders, animal shelters, pet shops, friends or relatives, found or rescued off the streets and home bred (i.e. bred and reared in the current owner's home). When the 'problem' and 'control' dog groups were compared, a highly significant difference between them emerged in the proportions of animals derived from these different sources ($\chi^2 = 38.2$, $P < 0.0001$). In general, dogs obtained from pet shops, animal shelters and those rescued off the streets were over-represented in the 'problem' group, while those obtained from breeders, friends or relatives, or bred at home were under-represented (see Fig. 6.1).

When 'problem' and 'control' dogs were combined or pooled together, it was found that specific behaviour problems were significantly more prevalent among dogs originating from certain sources. In particular, dominance-type aggression and social fears[3] were significantly more prevalent than expected among dogs acquired from petshops (see Fig. 6.2(*a*), (*b*)), while coprophagia (eating faeces) was more common than expected among animals found unowned or rescued off the streets (Fig. 6.3). With regard to dominance-type aggression, subsequent post-hoc analyses revealed that the main contrasts were between dogs from pet shops, breeders and found unowned, in which the prevalence of this behaviour problem was high, and dogs from friends or relatives, shelters and home bred, in which it was low. With respect to social fears, the only significant contrast was between dogs from pet shops and shelters, on the one hand, and dogs acquired from friends or relatives, on the other.

The fact that dogs acquired from pet shops rated poorly in terms of behaviour problems is of interest, since pet shop puppies are often the result of mass production in so-called puppy-farms or puppy-mills

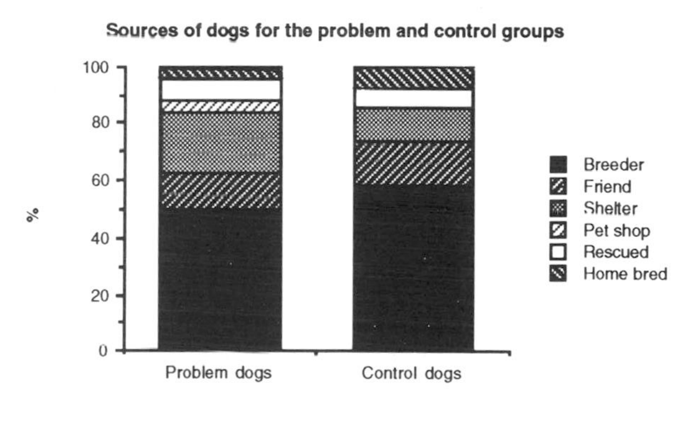

Fig. 6.1. Relative percentages of 'problem' and 'control' dogs acquired from different sources.

[2] Copies of the original questionnaire are available on request from the first author.

[3] Fear of strangers, children and unfamiliar dogs.

(*a*) Dominance-type aggression

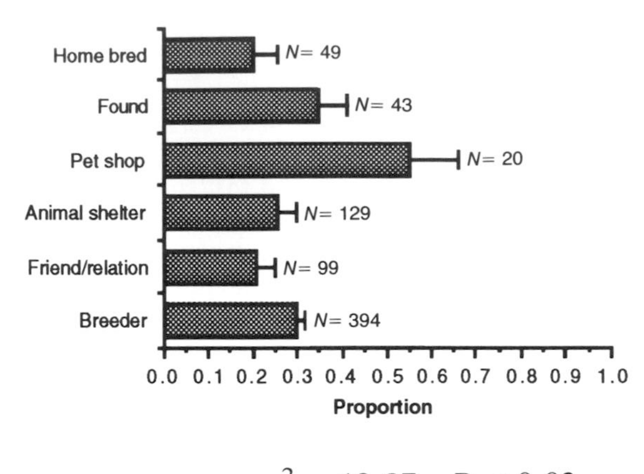

$\chi^2 = 12.87 \quad P < 0.03$

(*b*) Social fears

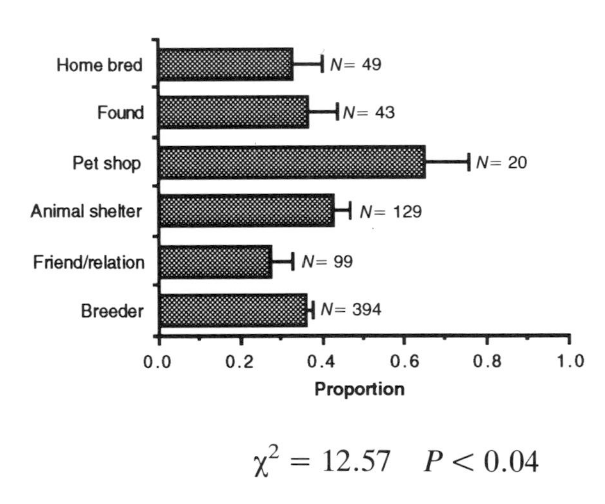

$\chi^2 = 12.57 \quad P < 0.04$

Fig. 6.2. Relationship between source of dog and prevalence of (*a*) dominance-type aggression and (*b*) social fears.

with little regard for their temperamental characteristics (see Lockwood, Chapter 9). Such animals may also undergo inadequate early socialization and a range of abnormal or traumatic early experiences that could predispose them to develop inappropriate adult behaviour. The preponderance of shelter animals and unowned strays in the 'problem' group is more difficult to interpret since presumably many dogs are abandoned or disowned as a result of having developed behaviour problems (see e.g. McCrave, 1991). The higher incidence of coprophagy in former strays may represent the adoption of scavenging habits by animals deprived of more reliable and palatable sources of food.

Coprophagia

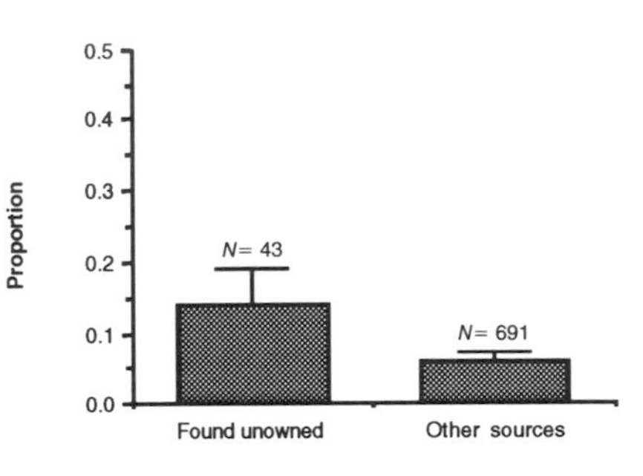

Fisher Exact Test, $P < 0.04$

Fig. 6.3. Relationship between source of dog and prevalence of coprophagia.

Effects of early illness

Of the 551 dogs for which owners supplied the relevant information, 73 (13%) were reported as having developed some form of illness during puppyhood (0–16 weeks of age). Owners were asked to state whether the illness had been serious or non-serious, but no information was requested on the precise timing of the illness or its duration. A number of statistically significant associations emerged between puppyhood illness and the prevalence of various adult behaviour problems. For example, dogs that had been ill as puppies were significantly more likely to display dominance-type aggression, aggression towards strangers, fear of strangers, fear of children, separation-related barking and abnormal or inappropriate sexual behaviour[4]. These results are illustrated in Fig. 6.4(*a*)–(*f*).

Although cause and effect cannot be determined from these data, most of these findings are consistent with the postulated effects of inadequate or restricted early socialization. Interpreted in this light, the fear-

[4] In practice, only one form of abnormal sexual behaviour was ever reported: sexual mounting of persons by male dogs.

(*a*) Dominance-type aggression

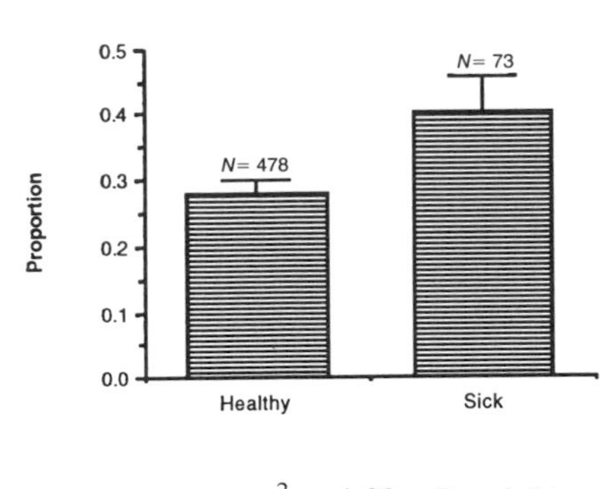

$\chi^2 = 4.32 \quad P < 0.04$

(*b*) Aggression towards strangers

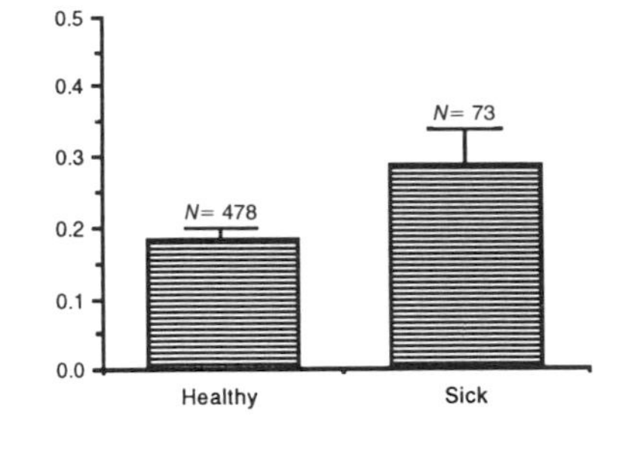

$\chi^2 = 4.7 \quad P < 0.03$

(*c*) Fear of strangers

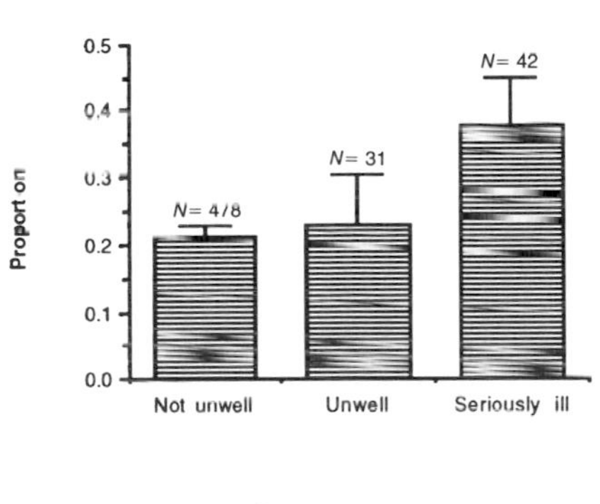

$\chi^2 = 6.18 \quad P < 0.04$

(*d*) Fear of children

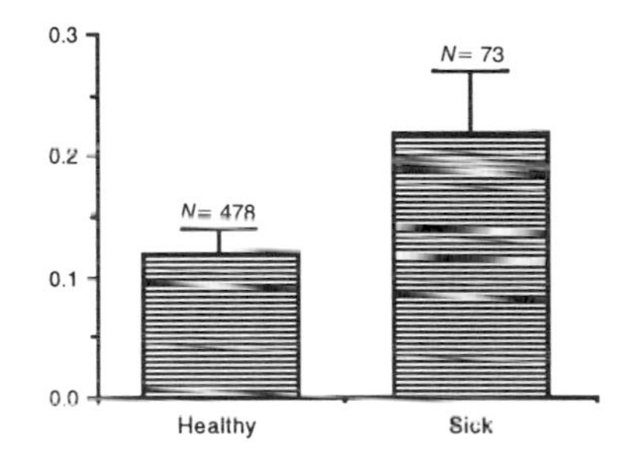

$\chi^2 = 4.94 \quad P < 0.03$

(*e*) Separation-related barking

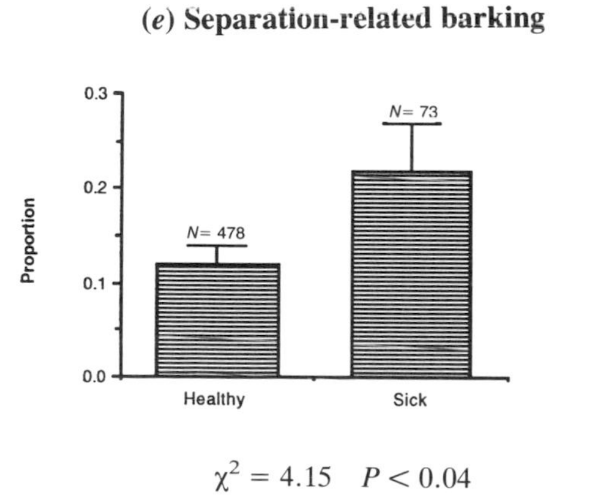

$\chi^2 = 4.15 \quad P < 0.04$

(*f*) Abnormal sexual behaviour

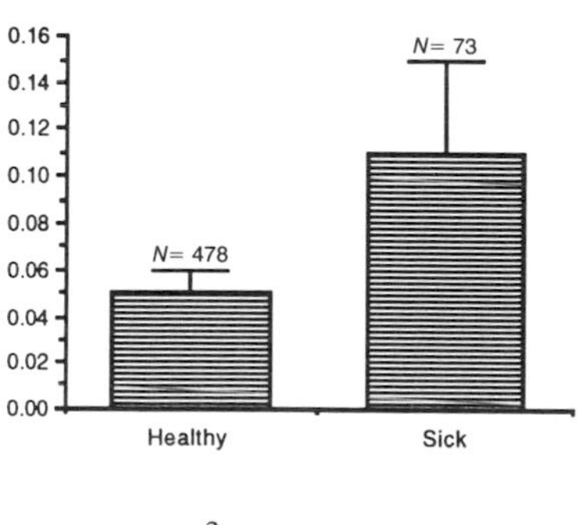

$\chi^2 = 3.69 \quad P < 0.05$

Fig. 6.4. Relationship between illness as a puppy and prevalence of (*a*) dominance-type aggression, (*b*) aggression towards strangers, (*c*) fear of strangers, (*d*) fear of children, (*e*) separation-related barking and (*f*) abnormal sexual behaviour.

ful and aggressive behaviour towards children and strangers would reflect limited early exposure to unfamiliar persons as a consequence of early isolation when ill, while dominance-type aggression, separation-related barking and abnormal sexual behaviour could all be regarded as symptoms of overly exclusive and constant early care and attention during the socialization period resulting in an overly dependent and 'human-socialized' dog. Sickly puppies presumably also experience more frequent and potentially aversive visits to veterinary surgeons, and this may conceivably help to account for the higher prevalence of aggression towards, and/or fear of, strangers in these animals.

Time left alone

Based on the 517 dogs for which owner's supplied appropriate information, time left alone as a puppy gave rise to relatively few associations with adult behaviour problems. There was a statistically significant linear tendency for separation-related destructiveness to increase with time left alone as a puppy, but other separation-related behaviour problems showed no association. Somewhat anomalously, increased prevalence of excessive barking was associated with both short (0–2 hours) and long (6–8 hours) periods of being left alone as a puppy. These effects are shown in Fig. 6.5(*a*), (*b*).

Given the very long periods of time that wild wolf pups are often left alone at den and rendezvous sites (Ballard *et al.*, 1991; Gray, 1993), it is in some ways surprising that dog puppies should exhibit any long-term adverse effects of being left alone during the day. The relatively neotenous, socially-dependent character of some breeds (Fox, 1978; Serpell, 1983; Mugford, 1985, see also Coppinger & Schneider, Chapter 3) may, however, render them more sensitive to this form of early social deprivation. Also, these survey data provide no information on the precise circumstances surrounding these early separations, some of which may be more important causative agents than the separations themselves.

(*a*) Separation-related destructiveness

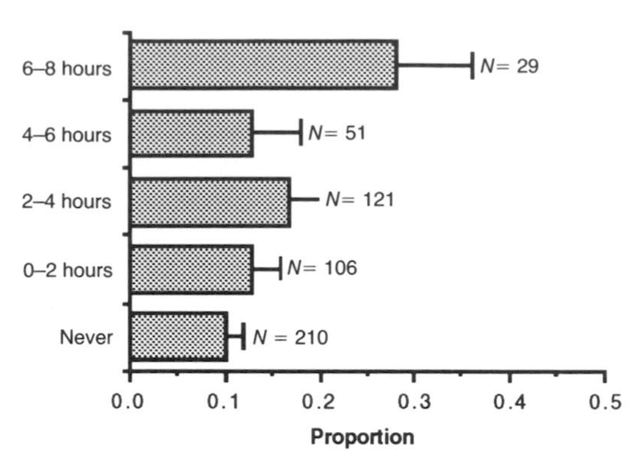

Mantel-Haenszel $\chi^2 = 5.54 \quad P < 0.01$

(*b*) Excessive barking

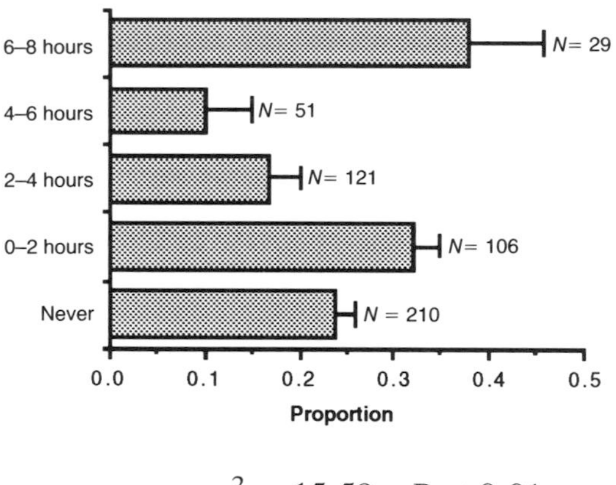

$\chi^2 = 15.58 \quad P < 0.01$

Fig. 6.5. Relationship between time left alone as a puppy and prevalence of (*a*) separation-related destructiveness and (*b*) excessive barking.

Age-related effects

Age at time of acquisition

Dogs acquired at different ages differed in their tendencies to display certain behaviour problems. In particular, fear of other dogs and fear of traffic both increased in a linear fashion with age of acquisition up to 24 weeks (dogs acquired after 24 weeks were excluded from this analysis). These trends are shown in Fig. 6.6(*a*), (*b*).

Since the bulk of animals acquired before 24 weeks of age came from breeders, this result is probably a consequence of the development of so-called 'kennel dog syndrome' or 'kennel-shyness' (Fox, 1968;

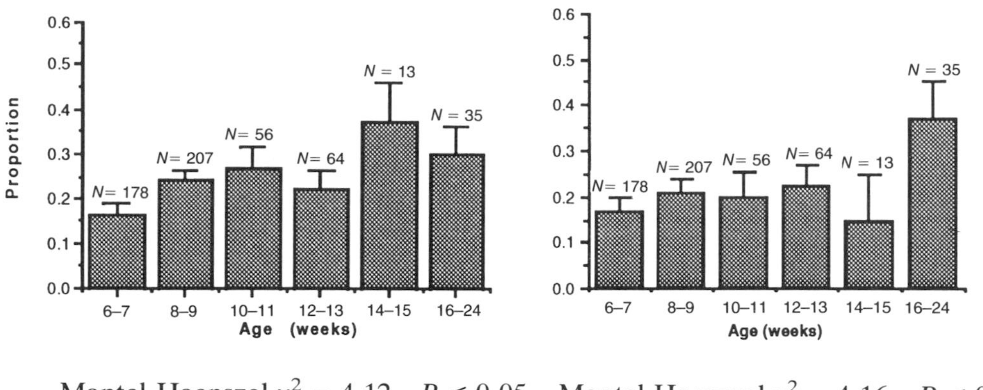

Mantel-Haenszel $\chi^2 = 4.12$ $P < 0.05$ Mantel-Haenszel $\chi^2 = 4.16$ $P < 0.05$

Fig. 6.6. Relationship between age at the time of acquisition and the prevalence of (*a*) fear of other dogs and (*b*) fear of traffic.

Scott & Bielfelt, 1976). It is also possible that prolonged social contact with littermates may potentiate fear of other dogs in socially subordinate individuals, particularly among the more 'assertive' breeds.

Age at first vaccination

Information on time of first vaccination was obtained for 471 dogs. Among these animals there were some striking and unexpected associations between vaccination time and the prevalence of behaviour problems. In the case of territorial-type aggression, a linear relationship suggested that this problem generally increases in prevalence with increasing vaccination age, at least up to the age of 20 weeks (Fig. 6.7). However, in the case of possessive or 'jealous' aggression[5], dominance-type aggression, social fears, separation-related destructiveness and over-excitability there was some evidence of an 8–9 week threshold point below which first vaccination was associated with reduced prevalence of behaviour problems (Fig. 6.8(*a*)–(*f*)).

The overall increases in behaviour problem prevalence with age at first vaccination can probably be accounted for in terms of restricted early socialization experience. Normal vaccination regimens recommend immunization at 8–10 weeks of age with the final vaccine being administered at 12 weeks. Initial

Territorial-type aggression

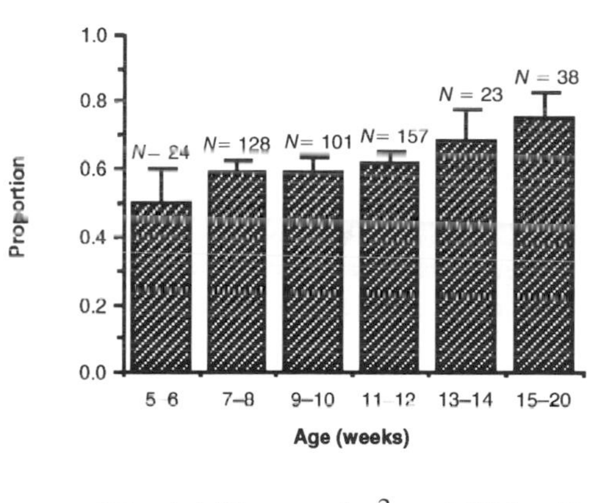

Mantel-Haenszel $\chi^2 = 4.884$ $P < 0.03$

Fig. 6.7. Relationship between age at first vaccination and the prevalence of territorial-type aggression.

vaccination at greater than ten weeks of age must be followed by an interval of two weeks before final vaccination. The unavoidable lag between maternally-derived and acquired immunity has resulted in veterinary surgeons recommending isolation from possible sources of infection until 7–10 days after this final vaccination. Inadvertently, this recommendation may be encouraging inadequate socialization

[5] Aggression when owner gives attention to a third party (either person or animal).

in puppies who receive their first vaccination above the age of ten weeks, since these animals receive their final vaccination at above 12 weeks of age, towards the end of the socialization period. Such animals run a greater risk of developing neophobic responses to strangers and novel situations, and may also become overly attached to their owners through relatively prolonged and exclusive early social contact.

As far as the apparent 8–9 week threshold effect is concerned, two possible explanations suggest themselves. First, because the majority of puppies are rehomed by eight weeks of age, most of the animals immunized below this age will have received their first vaccination at the breeder, in, or at least close to, the familiar litter and maternal environment. This type of vaccination experience is likely to be less traumatic overall than later vaccination, and may therefore result in fewer adverse effects on later behaviour.

Alternatively, or perhaps in addition, it is possible that puppies are more sensitive to the developmental effects of traumatic early experiences (such as vaccination) at certain points in puppyhood. Fox & Stelzner's (1966) conditioned aversion experiments with beagle pups suggested that puppies become unusually sensitive to distressing experiences at around eight weeks of age, and Fox (1968, p. 335) even referred to a case of 'eight-week vaccination trauma' in a dog that exhibited extreme avoidance of entering automobiles after being taken by car to the veterinarian for vaccination at eight weeks of age.

Age when first taken out into public areas

If the above findings were largely a consequence of isolation or restriction during the vaccination/socialization period, a similar range of associations would be expected between the prevalence of behaviour problems and the ages when puppies were first taken out into public areas on a regular basis. In practice, owners provided information on age when first taken out for a total of 446 dogs, but few associations with behaviour problem prevalence were identified.

This negative result is surprising since neophobic responses to the outside world would be expected in animals kept at home until the end of the socialization period or beyond (see e.g. Melzack & Thompson, 1956; Melzack & Scott, 1957; Scott & Fuller, 1965). It should be emphasized, however, that even infrequent and irregular exposure to novel persons and situations may be sufficient to effect adequate socialization in some dogs (see Fuller, 1967; Wolfle, 1990), and owners' recollections of the precise ages when pups were first taken out regularly may have been too inaccurate to constitute a reliable source of data.

Interactions between age-related effects

Interpretation of the possible effects of age-related experiences in puppyhood is complicated by the fact that some of the variables involved are related. For instance, age at first vaccination is inevitably correlated with age at acquisition ($r = 0.3283$, $P < 0.01$, $N = 445$), since owners tend to seek vaccination for new puppies soon after acquiring them. The effects of possible interactions between variables can be reduced or eliminated by effectively holding one variable constant. In the case of the effects of vaccination age, it was possible to control for possible confounding effects of age at acquisition by confining the analysis only to animals acquired by eight weeks of age ($N = 317$). When this analysis was performed, significant, positive linear relationships were found between age at first vaccination and the prevalence of over-excitability, general aggression[6] and separation-related problems (see Fig. 6.9). Although, in all three cases, there seemed to be some indication of an 8–9 week threshold effect, this could no longer be demonstrated statistically.

As above, these findings may reflect inadequate early socialization due to restrictions imposed on puppies by late vaccination regimens. The absence of any increase in social fears would suggest, however, that the earlier association obtained between vaccination age and social fear prevalence (see above) was an artifact of owners acquiring older, unvaccinated, kennel-shy puppies. Although difficult to interpret, the effective disappearance of the 8–9 week threshold effect in animals acquired by eight weeks might tend to imply that the trauma of rehoming, and of post eight-week vaccination, act synergistically to promote the development of later behaviour problems. However, clarification of the relative importance of these various factors will need to await the results of more detailed, prospective studies.

[6] A combination of all categories of aggressive behaviour.

(*a*) Possessive or 'jealous' aggression

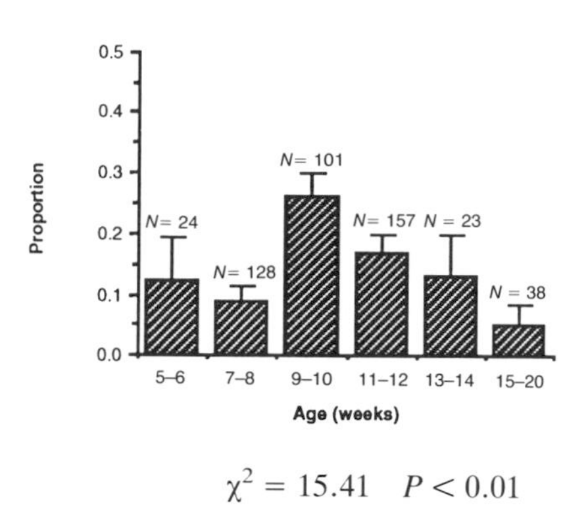

$\chi^2 = 15.41 \quad P < 0.01$

(*b*) Dominance-type aggression

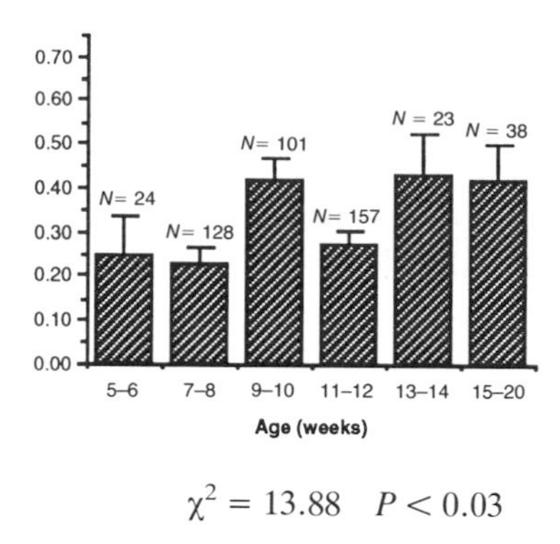

$\chi^2 = 13.88 \quad P < 0.03$

(*c*) Social fears

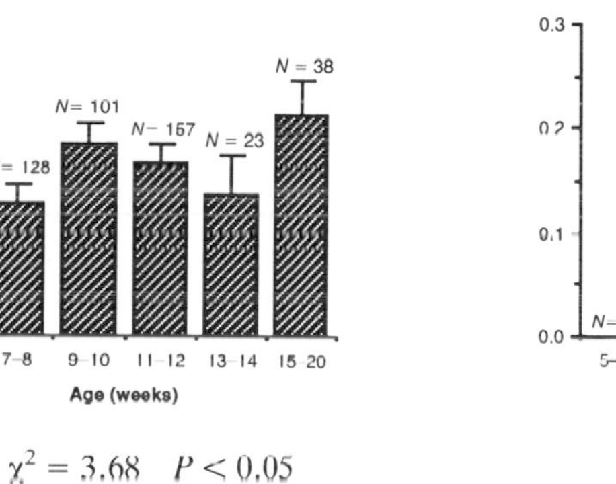

$\chi^2 = 3.68 \quad P < 0.05$

(*d*) Separation-related destructiveness

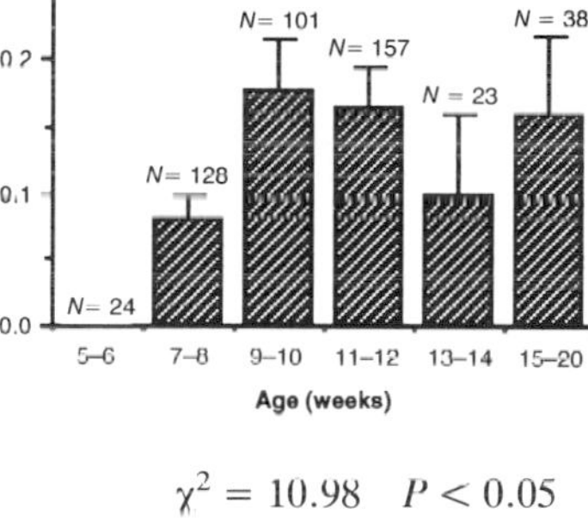

$\chi^2 = 10.98 \quad P < 0.05$

(*e*) General over-excitability

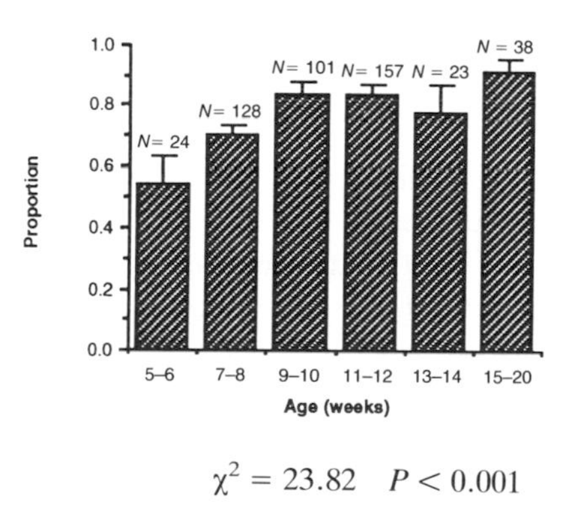

$\chi^2 = 23.82 \quad P < 0.001$

(*f*) Excessive barking

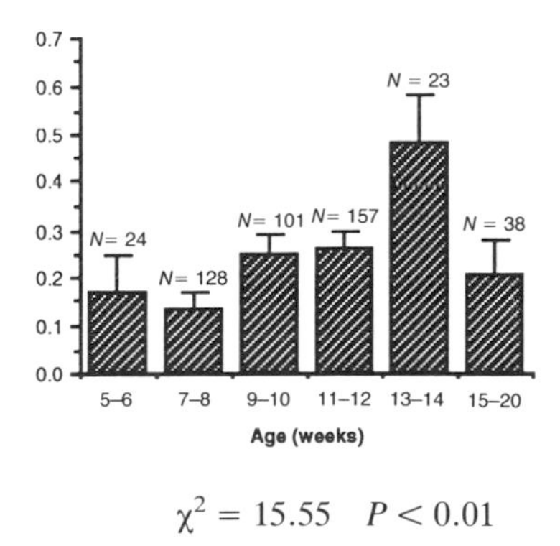

$\chi^2 = 15.55 \quad P < 0.01$

Fig. 6.8. Relationship between age at first vaccination and the prevalence of (*a*) possessive or 'jealous' aggression, (*b*) dominance-type aggression, (*c*) social fears, (*d*) separation-related destructiveness, (*e*) general over-excitability and (*f*) excessive barking.

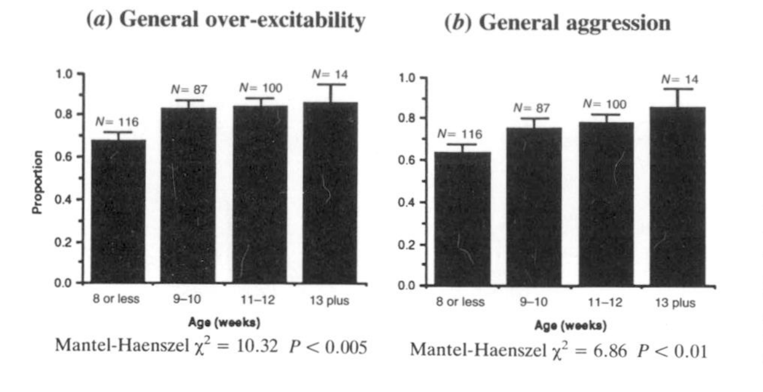

(c) Separation-related problems

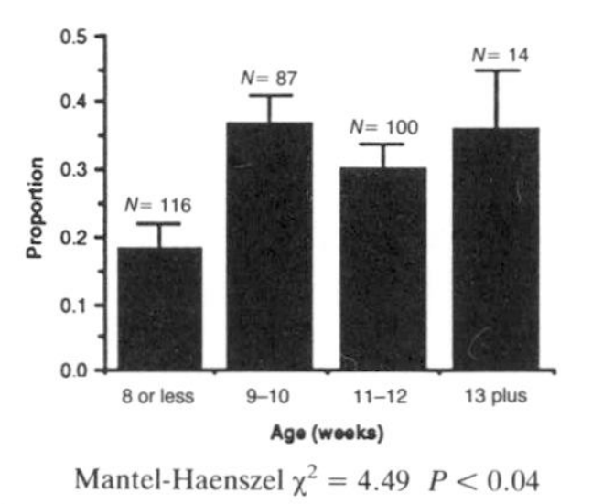

Fig. 6.9. Relationship between age at first vaccination and the prevalence of (*a*) general over-excitability, (*b*) general aggression and (*c*) separation-related problems (animals acquired by eight weeks of age only).

Conclusions

Although the conclusions to be drawn from retrospective reports are inevitably somewhat limited, the results of the above survey do appear to support the view that certain events and experiences, occurring during the socialization period, can have long-term deleterious effects on the behaviour of domestic dogs. At least in a general sense, this tends to bear out Scott & Fuller's (1965) original assertion that, in emotional terms, puppies are more sensitive and impressionable at this age, and are therefore also abnormally vulnerable to psychological injury.

For example, a period of relative isolation during the socialization period due to puppyhood illness appears to precipitate a long-term fear of strangers in some dogs. It also appears to intensify some dogs' attachments for their owners leading to separation-related problems and abnormal sexual imprinting. Long periods of daily social isolation or abandonment by the owner may also provoke adult separation problems and excessive barking. The existence of the phenomenon known as 'kennel dog syndrome' (Scott & Bielfelt, 1976) was also tentatively confirmed by the positive association between the prevalence of both social and non-social fears and the ages at which puppies were acquired. More surprisingly, the associations between behaviour problem prevalence and age at the time of first vaccination suggested that even relatively transient events in a puppy's early life may have long-term behavioural consequences, and that the precise timing of such events may be critical. Although it seems to have been ignored or forgotten in more recent literature, Fox & Stelzner's (1966) theory of a narrow, hypersensitive phase within the socialization period at around eight weeks of age may be relevant to these findings. Considering how close it is to the age when many puppies are traditionally rehomed, this putative 'critical' point would probably repay more detailed investigation, particularly in the light of Slabbert & Rasa's (1993) findings concerning the traumatic effects of removing puppies from their natal surroundings at six weeks of age.

A further lesson to be learned from this survey and from the wider review of the literature is that there are large individual and breed differences in the propensity to develop particular behaviour problems. It may be convenient to explain all canine behaviour in terms of supposedly ancestral lupine patterns, or to lump all dog breeds together as if the processes underlying their development were identical, but the truth is that each breed is behaviourally unique: a pattern or process that appears to hold true for Rottweilers may fall apart entirely when applied to Shih Tzus.

A number of specific questions or issues concerning the development of behaviour problems have also been highlighted by this review. In view of their potential importance with regard to puppies' initial responses to stressful or alarming situations, there has been a surprising dearth of research into the long-term effects of variation in prenatal and neonatal environment. At present, the possible long-term behavioural consequences for the developing fetus of stressing the mother during pregnancy have not been investigated. Similarly, the ways in which mother-neonate and human-neonate interactions can affect puppies' subsequent coping abilities remain largely unexplored. Both these areas deserve more detailed study. Anecdotal evidence for a second phase of heightened sensitivity to fear-evoking and territorial stimuli in wolves at around 4–5 months age (see Fentress, 1967; Mech, 1970; Fox, 1971) should also be re-examined in the domestic dog. It may be that the ongoing preoccupation with the traditional 3–12 week socialization period has led investigators to ignore or overlook the possible consequences of later exposure to biologically relevant events and stimuli. Finally, it may be time to revise the established view that correct socialization necessarily requires removing a puppy to its new home during the peak of the sensitive period. This idea is based on a misinterpretation of the available evidence, and it is one that may have adverse effects on puppy welfare and development. Separation from the natal environment at this vulnerable age is known to cause considerable distress in some puppies (see Elliot & Scott, 1961; Slabbert & Rasa, 1993), and it may also contribute to the development of adult behaviour problems. More effort should be given to exploring methods of adequately socializing puppies during this period without subjecting them to the trauma of rehoming until they are old enough to cope more easily with this difficult transition.

Perhaps the most outstanding conclusion to be drawn from this review is how little we actually know about the development of most behaviour problems in dogs, despite the obvious value of such knowledge in the struggle to reduce or eliminate these problems in the dog population. The systematic collection of information on the individual early histories of problem dogs by both clinics and consultants would go a long way towards plugging the gaps in our current knowledge. Clearly, however, there is also a need for focused, empirical studies to address some of the more specific developmental questions as they arise.

Acknowledgements

We wish to thank Intervet UK Ltd and Trinity College Cambridge for supporting JAJ's doctoral studies. JAJ would also like to express his thanks to David Appleby, Roger Mugford, Peter Neville, John Rogerson and all the other individuals and groups who helped him recruit the subjects for his research. We are also grateful to Marc Bekoff and Anthony Podberscek for reviewing an earlier draft of this manuscript.

References

Adams, G. J. & Johnson, K. G. (1993). Sleep-wake cycles and other night-time behaviours of the domestic dog, *Canis familiaris. Applied Animal Behaviour Science*, **36**, 233–48.

Ainsworth, M. D. S., Blehar, C. S., Waters, E. & Wall, S. (1978). *Patterns of Attachment: A Psychological Study of the Strange Situation*. Hillsdale, NJ: Erlbaum.

Ballard, W. B., Ayres, L. A., Gardner, C. L. & Foster, J. W. (1991). Den site activity patterns of grey wolves, *Canis lupus*, in southcentral Alaska. *Canadian Field-Naturalist*, **105**, 497–504.

Barrette, C. (1993). The 'inheritance of dominance', or an aptitude to dominate. *Animal Behaviour*, **46**, 591–3.

Bateson, P. (1979). How do sensitive periods arise and what are they for? *Animal Behaviour*, **27**, 470–86.

Bateson, P. (1981). Control of sensitivity to the environment during development. In *Behavioural Development*, ed. K. Immelmann, G. W. Barlow, L. Petrovich & M. Main, pp. 433–53. Cambridge: Cambridge University Press.

Bekoff, M. (1974). Social play and play-soliciting by infant canids. *American Zoologist*, **14**, 323–40.

Bernstein, I. S. (1981). Dominance: the baby and the bathwater. *Behavioral and Brain Sciences*, **4**, 419–57.

Borchelt, P. L. (1983*a*). Aggressive behavior of dogs kept as companion animals: classification and influence of sex, reproductive status and breed. *Applied Animal Ethology*, 10, 45–61.

Borchelt, P. L. (1983*b*). Separation-elicited behavior problems in dogs. In *New Perspectives on Our Lives with Companion Animals*, ed. A. H. Katcher & A. M. Beck, pp. 187–96. Philadelphia: University of Pennsylvania Press.

Borchelt, P. L. (1984). Behavioural development of the puppy. In *Nutrition and Behaviour in Dogs and Cats*, ed. R. S. Anderson, pp. 165–74. Oxford: Pergamon Press.

Borchelt, P. L. & Voith, V. L. (1982*a*). Classification of animal behavior problems. *Veterinary Clinics of North America: Small Animal Practice*, **12**, 571–85.

Borchelt, P. L. & Voith, V. L. (1982*b*). Diagnosis and treatment of separation-related behavior problems in dogs. *Veterinary Clinics of North America: Small Animal Practice*, **12**, 625–35.

Bowlby, J. (1973). *Attachment and Loss, Vol. 2: Separation.* New York: Basic Books.

Campbell, W. E. (1975). *Behavior Problems in Dogs.* Santa Barbara, CA: American Veterinary Publications.

Campbell, W. E. (1986). The prevalence of behavioral problems in American dogs. *Modern Veterinary Practice*, **67**, 28–31.

Clarke, R. S., Heron, W., Fetherstonhaugh, M. L., Forgays, D. G. & Hebb, D. O. (1951). Individual differences in dogs: preliminary report on the effects of early experience. *Canadian Journal of Psychology*, **5**, 150–6.

Cornwell, A. C. & Fuller, J. L. (1960). Conditioned

responses in young puppies. *Journal of comparative and physiological Psychology*, **54**, 13–15.

DeFries, J. C., Weir, M. W. & Hegmann, J. P. (1967). Differential effects of prenatal maternal stress on offspring behavior in mice as a function of genotype and stress. *Journal of comparative and physiological Psychology*, **63**, 332–4.

Denenberg, V. H. (1968). A consideration of the usefulness of the critical period hypothesis as applied to the stimulation of rodents in infancy. In *Early Experience and Behaviour*, ed. G. Newton & S. Levine, pp. 142–67. Springfield, IL: Charles Thomas.

Denenberg, V. H. & Morton, J. R. C. (1962). Effects of environmental complexity and social groupings upon modification of emotional behavior. *Journal of comparative and physiological Psychology*, **55**, 242–6.

Dewsbury, D. A. (1990). Fathers and sons: genetic factors and social dominance in deer mice, *Peromyscus maniculatus. Animal Behaviour*, **39**, 284–9.

Dykman, R. A., Mack, R. L. & Ackerman, P. T. (1965). The evaluation of autonomic and motor components of the nonavoidance conditioned response in the dog. *Psychophysiology*, **1**, 209–30.

Dykman, R. A., Murphree, O. D. & Ackerman, P. T. (1966). Litter patterns in the offspring of nervous and stable dogs. II. Autonomic and motor conditioning. *Journal of Nervous and Mental Disease*, **141**, 419–31.

Dykman, R. A., Murphree, O. D. & Reese, W. G. (1979). Familial anthropophobia in pointer dogs? *Archives of General Psychiatry*, **36**, 988–93.

Elliot, O. & Scott, J. P. (1961). The development of emotional distress reactions to separation in puppies. *Journal of Genetic Psychology*, **99**, 3–22.

Fentress, J. C. (1967). Observations on the behavioral development of a hand-reared male timber wolf. *American Zoologist*, **7**, 339–51.

Fox, M. W. (1968). Socialization, environmental factors, and abnormal behavioral development in animals. In *Abnormal Behavior in Animals*, ed. M. W. Fox, pp. 332–55. Philadelphia, PA: W. E. Saunders.

Fox, M. W. (1969). Behavioral effects of rearing dogs with cats during the 'critical period of socialization'. *Behaviour*, **35**, 273–80.

Fox, M. W. (1971). *Behavior of Wolves, Dogs and Related Canids*. New York: Harper and Row.

Fox, M. W. (1972). Socio-ecological implications of individual differences in wolf litters: a developmental and evolutionary perspective. *Behaviour*, **41**, 298–313.

Fox, M. W. (1978). *The Dog: Its Domestication and Behavior*. New York: Garland STPM Press.

Fox, M. W., Halperin, S., Wise, A. & Kohn, E. (1976). Species and hybrid differences in frequencies of play and agonistic actions in canids. *Zeitschrift für Tierpsychologie*, **40**, 194–209.

Fox, M. W. & Stelzner, D. (1966). Behavioural effects of differential early experience in the dog. *Animal Behaviour*, **14**, 273–81.

Frank, H. & Frank, M. G. (1982). On the effects of domestication on canine social development and behavior. *Applied Animal Ethology*, **8**, 507–25.

Frank, H. & Frank, M. G. (1985). Comparative manipulation-test performance in ten-week-old wolves (*Canis lupus*) and Alaskan malamutes (*Canis familiaris*): a Piagetian interpretation. *Journal of Comparative Psychology*, **99**, 266–74.

Freedman, D. G., King, J. A. & Elliot, O. (1961). Critical periods in the social development of dogs. *Science*, **133**, 1016–17.

Fuller, J. L. (1967). Experiential deprivation and later behavior. *Science*, **158**, 1645–52.

Goddard, M. E. & Beilharz, R. G. (1982). Genetic and environmental factors affecting the suitability of dogs as guide dogs for the blind. *Theoretical and Applied Genetics*, **62**, 97–102.

Goddard, M. E. & Beilharz, R. G. (1984). Factor analysis of fearfulness in potential guide dogs. *Applied Animal Behaviour Science*, **12**, 253–65.

Gottlieb, G. (1976). The roles of experience in the development of behaviour and the nervous system. In *Neural and Behavioural Specificity*, ed. J. P. Scott, pp. 25–54. New York: Academic Press.

Gray, D. R. (1993). The use of muskox kill sites as temporary rendezvous sites by arctic wolves with pups in early winter. *Arctic*, **46**, 324–30.

Harlow, H. F. & Harlow, M. K. (1966). Learning to love. *American Scientist*, **54**, 244–72.

Hart, B. L. & Hart, L. A. (1985). *Canine and Feline Behavioral Therapy*. Philadelphia, PA: Lea & Febiger.

Hinde, R. A. (1970). *Animal Behaviour*, 2nd edn. New York: McGraw-Hill.

Hinde, R. A. (1974). *Biological Bases of Human Social Behaviour*. New York: McGraw-Hill.

Immelmann, K. and Soumi, S. J. (1981). Sensitive phases in development. In *Behavioural Development*, ed. K. Immelmann, G. W. Barlow, L. Petrovich & M. Main, pp. 395–431. Cambridge: Cambridge University Press.

Jagoe, J. A. (1994). Behaviour Problems in the Domestic Dog: a Retrospective and Prospective Study to Identify Factors Influencing Their Development. Unpublished Ph.D. thesis, University of Cambridge.

Joffe, J. M. (1969). *Prenatal Determinants of Behaviour*. New York: Pergamon Press.

Levine, S. (1962). Plasma free corticosteroid response to electric shock in rats stimulated in infancy. *Science*, **135**, 795–6.

Levine, S. (1967). Maternal and environmental influences on the adrenocortical response to stress in weanling rats. *Science*, **156**, 258–60.

Line, S. & Voith, V. L. (1986). Dominance aggression of dogs towards people: behavior profile and response to treatment. *Applied Animal Behaviour Science*, **16**, 77–83.

Lockwood, R. (1979). Dominance in wolves: useful construct or bad habit? In *The Behavior and Ecology of Wolves*, ed. E. Klinghammer, pp. 225–44. New York: Garland STPM Press.

Lorenz, K. (1935). Der Kumpan in der Umwelt des Vogels. *Journal für Ornithologie*, **83**, 137–213, 289–413.

Lorenz, K. (1966). *On Aggression*. London: Methuen.

MacDonald, K. B. (1983). Stability of individual differences in behavior in a litter of wolf cubs (*Canis lupus*). *Journal of Comparative Psychology*, **97**, 99–106.

MacDonald, K. B. (1987). Development and stability of personality characteristics in pre-pubertal wolves: implications for pack organization and behavior. In *Man and Wolf*, ed. H. Frank, pp. 293–312. Dordrecht, The Netherlands: Dr W. Junk Publishers.

McCrave, E. A. (1991). Diagnostic criteria for separation anxiety in the dog. *Veterinary Clinics of North America: Small Animal Practice*, **21**, 247–55.

Mech, L. D. (1970). *The Wolf: the Ecology and Behavior of an Endangered Species*. New York: Natural History Press.

Meier, G. W. (1961). Infantile handling and development in Siamese kittens. *Journal of comparative and physiological Psychology*, **54**, 284–6.

Melzack, R. (1954). The genesis of emotional behavior: an experimental study of the dog. *Journal of comparative and physiological Psychology*, **47**, 166–8.

Melzack, R. & Scott, T. H. (1957). The effects of early experience on the response to pain. *Journal of comparative and physiological Psychology*, **50**, 155–61

Melzack, R. & Thompson, W. R. (1956). Effects of early experience on social behaviour. *Canadian Journal of Psychology*, **10**, 82–92.

Moyer, K. E. (1968). Kinds of aggression and their physiological basis. *Communications in Behavioral Biology. Part A*, **2**, 65–87.

Mugford, R. A. (1985). Attachment versus dominance: an alternative view of the man-dog relationship. In *The Human-Pet Relationship*, pp. 157–65. Vienna: IEMT (Institute for Interdisciplinary Research on the Human-Pet Relationship).

Murphree, O. D. (1973). Inheritance of human aversion and inactivity in two strains of the pointer dog. *Biological Psychiatry*, **7**, 23–9.

Murphree, O. D., Angel, C., DeLuca, D. C. & Newton, J. E. O. (1977). Longitudinal studies of genetically nervous dogs. *Biological Psychiatry*, **12**, 573–6.

Murphree, O. D. & Dykman, R. A. (1965). Litter patterns in the offspring of nervous and stable dogs. I. Behavioral tests. *Journal of Nervous and Mental Disease*, **141**, 321–32.

Newton, J. E. O. & Lucas, L. A. (1982). Differential heart-rate responses to person in nervous and normal pointer dogs. *Behavioural Genetics*, **12**, 379–93.

Niebuhr, B. R., Levinson, M., Nobbe, D. E. & Tiller, J. E. (1980). Treatment of an incompletely socialized dog. In *Canine Behavior*, ed. B. L. Hart, p. 83. Santa Barbara, CA: Veterinary Practice Publishing Co.

O'Farrell, V. (1986). *Manual of Canine Behavior*. Cheltenham, Glos.: BSAVA Publications.

Pedersen, V. & Jeppesen, L. L. (1990). Effects of early handling on later behaviour and stress responses in the Silver Fox (*Vulpes vulpes*). *Applied Animal Behaviour Science*, **26**, 383–93.

Pfaffenberger, C. J. & Scott, J. P. (1976). Early rearing and testing. In *Guide Dogs for the Blind: Their Selection, Development and Training*, ed. C. J. Pfaffenberger, J. P. Scott, J. L. Fuller, B. E. Ginsburg & S. W. Bielfelt, pp. 13–37. Amsterdam: Elsevier.

Pfaffenberger, C. J., Scott, J. P., Fuller, J. L., Ginsburg, B. E. & Bielfelt, S. W. (1976). *Guide Dogs for the Blind: Their Selection, Development and Training*. Amsterdam: Elsevier.

Plyusnina, I. Z., Oskina, I. N. & Trut, L. N. (1991). An analysis of fear and aggression during early development of behaviour in silver foxes (*Vulpes vulpes*). *Applied Animal Behaviour Science*, **32**, 253–68.

Rowell, T. E. (1966). Hierarchy in the organization of a captive baboon group. *Animal Behaviour*, **14**, 430–43.

Schenkel, R. (1967). Submission: its features and function in the wolf and dog. *American Zoologist*, **7**, 319–29.

Schjelderup-Ebbe, T. (1922). Beiträge zur Sozialpsychologie des Haushunds. *Zeitschrift für Tierpsychologie*, **88**, 225–52.

Scott, J. P. (1962). Critical periods in behavioral development. *Science*, **138**, 949–58.

Scott, J. P. (1963). The process of primary socialization in canine and human infants. *Monographs of the Society for Research in Child Development*, **28**, 1–49.

Scott, J. P. & Bielfelt, S. W. (1976). Analysis of the puppy testing program. In *Guide Dogs for the Blind: Their Selection, Development and Training*, ed. C. J. Pfaffenberger, J. P. Scott, J. L. Fuller, B. E. Ginsburg & S. W. Bielfelt, pp. 39–75. Amsterdam: Elsevier.

Scott, J. P. & Fuller, J. L. (1965). *Genetics and the Social Behaviour of the Dog*. Chicago: University of Chicago Press.

Scott, J. P. & Marston, M. V. (1950). Critical periods affecting the development of normal and mal-adjustive social behavior of puppies. *Journal of Genetic Psychology*, **77**, 25–60.

Scott, J. P. & Nagy, Z. M. (1980). Behavioral metamorphosis in mammalian development. In *Early Experiences and Early Behavior: Implications for Social Development*, ed. E. C. Simmel, pp. 15–37. New York: Academic Press.

Scott, J. P., Stewart, J. M. & DeGhett, V. J. (1974). Critical periods in the organization of systems. *Developmental Psychobiology*, **7**, 489–513.

Serpell, J. A. (1983). The personality of the dog and its influence on the pet-owner bond. In *New Perspectives on Our Lives with Companion Animals*, ed. A. H. Katcher & A. M. Beck, pp. 57–71. Philadelphia: University of Pennsylvania Press.

Serpell, J. A. (1987). The influence of inheritance and environment on canine behaviour: myth and fact. *Journal of small Animal Practice*, **28**: 949–56.

Shull-Selcer, E. A. & Stagg, W. (1991). Advances in the

understanding and treatment of noise phobias. *Veterinary Clinics of North America: Small Animal Practice*, **21**, 353–67.

Simmel, E. C. & Baker, E. (1980). The effects of early experiences on later behavior: a critical discussion. In *Early Experiences and Early Behavior: Implications for Social Development*, ed. E. C. Simmell, pp. 3–13. New York: Academic Press.

Slabbert, J. M. & Rasa, O. A. E. (1993). The effect of early separation from the mother on pups in bonding to humans and pup health. *Journal of the South African Veterinary Association*, **64**, 4–8.

Stanley, W. C. (1970). Feeding behavior and learning in neonatal dogs. In *Second Symposium on Oral Sensation and Perception*, ed. J. F. Bosma, pp. 242–90. Springfield, IL: Charles Thomas.

Stanley, W. C. (1972). Perspectives in behavior organization and development resulting from studies of feeding behavior in infant dogs. In *Third Symposium on Oral Sensation and Perception: The Mouth of the Infant*, ed. J. F. Bosma, pp. 188–257. Springfield, IL: Charles Thomas.

Stur, I. (1987). Genetic aspects of temperament and behaviour in dogs. *Journal of small Animal Practice*, **28**, 957–64.

Thompson, W. R. (1957). Influence of prenatal and maternal anxiety on emotionality in young rats. *Science*, **125**, 698–9.

Thompson, W. R., Watson, J. & Charlesworth, W. R. (1962). The effects of prenatal maternal stress on offspring behavior in rats. *Psychological Monographs*, **76**, 1–26.

Thorne, F. C. (1944). The inheritance of shyness in dogs. *Journal of Genetic Psychology*, **65**, 275–9.

Tuber, D. S., Hothersall, D. & Peters, M. F. (1982). Treatment of fears and phobias in dogs. *Veterinary Clinics of North America: Small Animal Practice*, **12**, 607–23.

van Hooff, J. A. R. A. M. & Wensing, J. A. B. (1987). Dominance and its behavioural measures in a captive wolf pack. In *Man and Wolf*, ed. H. Frank, pp. 219–51. Dordrecht, The Netherlands: Dr W. Junk Publishers.

Voith, V. L. & Borchelt, P. L. (1982). Diagnosis and treatment of dominance aggression in dogs. *Veterinary Clinics of North America: Small Animal Practice*, **12**, 655–63.

Voith, V. L. & Borchelt, P. L. (1985*a*). Fears and phobias in dogs. *Compendium of Continuing Education for the Practising Veterinarian*, **7**, 209–18.

Voith, V. L. & Borchelt, P. L. (1985*b*). Separation anxiety in dogs. *Compendium on Continuing Education for the Practising Veterinarian*, **7**, 42–52.

Whimbey, A. E. & Denenberg, V. H. (1967). Two independent behavioral dimensions in open-field performance. *Journal of comparative physiological Psychology*, **63**, 500–4.

Wilsson, E. (1984). The social interaction between mother and offspring during weaning in German shepherd dogs: individual differences between mothers and their effects on offspring. *Applied Animal Behaviour Science*, **13**, 101–12.

Wolfle, T. L. (1990). Policy, program and people: the three Ps to well-being. In *Canine Research Environment*, ed. J. A. Mench & L. Krulisch, pp. 41–7. Bethesda, MD: Scientists Center for Animal Welfare.

Woolpy, J. H. (1968). The social organization of wolves. *Natural History*, **77**, 46–55.

Woolpy, J. H. & Ginsburg, B. E. (1967). Wolf socialization; A study of temperament in a wild social species. *American Zoologist*, **7**, 357–63.

Wright, J. C. (1980). Early development of exploratory behavior and dominance in three litters of German shepherds. In *Early Experiences and Early Behavior: Implications for Social Development*, ed. E. C. Simmel, pp. 181–206. New York: Academic Press.

Wright, J. C. & Nesselrote, M. S. (1987). Classification of behavior problems in dogs: distributions of age, breed, sex and reproductive status. *Applied Animal Behaviour Science*, **19**, 169–78.

Zimen, E. (1981). *The Wolf: a Species in Danger* (transl. E. Mosbacher). New York: Delacorte Press.

Zimen, E. (1987). Ontogeny of approach and flight behavior towards humans in wolves, poodles and wolf-poodle hybrids. In *Man and Wolf*, ed. H. Frank, pp. 275–92. Dordrecht, The Netherlands: Dr W. Junk Publishers.

7 Feeding behaviour of domestic dogs and the role of experience

CHRIS THORNE

Introduction

Pets are an integral part of human society as is clear from the large number of animal-owning households worldwide. The domestic dog is one of the most popular companion animals with an estimated population of 90 million in Western Europe and the USA. One in every four households in Western Europe owns a dog, and the figure rises to two in every five households in the USA. The importance of dogs as pets has provided the impetus for intensive research into their nutritional needs (National Research Council, 1985; Burger & Rivers, 1989), and this in turn has helped to generate the great variety of commercially prepared, nutritionally complete foods that are available to dog owners today. These foods vary in their ingredients and format – compare, for example, a dry biscuit food to a canned meat food – and they also vary in their acceptability to both dog and owner. Responsibility for providing the pet with a nutritious diet resides with the owner and it is the owner who makes a personal evaluation of which foods are most suitable. In deciding which food to feed initially, owners will take the advice of others or use their own previous experience. However, the dog's response to the food is also important to the decision and, because every dog has individual likes and dislikes, owner's will often change to another food if it seems more acceptable to the pet. When presented with a choice between two equally nutritious foods a dog will invariably show a preference for one over the other. This raises several questions: how and when do these food preferences develop? Are these food preferences similar to those of the dog's wild relatives? Has the domestication process altered food selection behaviour, and why are some dogs more fussy or discriminating than others? In this chapter, I will explore the information relating to feeding behaviour in an effort to provide answers to these questions.

Palatability

Flavour preferences vary from species to species and we cannot assume that, because something tastes good to humans, it therefore tastes good to animals. Dogs prefer meat to vegetable protein and display preferences for one meat over another. These are, in order, beef, pork, lamb, chicken and horse-meat (Houpt, Hintz & Shepherd, 1978). Besides the flavour of the food, dogs also respond to the form in which the food is offered. They prefer canned or semi-moist food to dry food (Kitchell, 1978) and they prefer cooked to raw meat, and canned meat to the same meat freshly cooked (Lohse, 1974). Hence palatability, a concept that is based around the sensory properties of the food, its taste, odour and texture, is an important factor in food selection for the domestic dog. Palatability is a commonly used concept in discussions of companion animal feeding behaviour. It goes by many names, may be studied in various ways and is a keystone in the competitive market of prepared pet food.

The extent to which domestic dogs in the industrial West are able to select their own food, or decide the time at which they eat, or the quantity that they consume, is somewhat constrained due to their dependence on the owner. Palatability, however, plays a major role in their food preferences. The lifestyles of wild canids are very different from those of their domestic counterparts, and there is little evidence, one way or the other, for an important role of palatability in the natural environment. Wild canids are not presented with a supply of food each day, as are most domestic dogs, but have to expend considerable energy in locating and catching prey. In this situation, the palatability of the prey is probably of lesser importance than knowing that the food is safe and that it meets the animal's nutritional needs. Early domestic dogs would presumably have survived predominantly on what they could find or scavenge around human habitations, and this is still the situation for many dogs in the less industrialised nations (see Macdonald & Carr, Chapter 14; Boitani *et al.*, Chapter 15). Nevertheless, although the route by which domestic dogs obtain food has been, and still is, very different from that of their wild ancestors, the underlying behavioural mechanisms upon which food selection is based may still be intact, if modified somewhat by domestication.

Effects of domestication

Evidence for the domestication of the dog extends back at least 12 000 years (see Clutton-Brock, Chapter 2). By the end of the last glaciation, some 10 000 years ago, the partnership between man and dog was fully and irrevocably established (Turnbull & Reed, 1974; Davis & Valla, 1978). The timing of the development of this relationship between man and dog suggests that the dog was the first species to be domesticated. By about 3000 years ago many of the main types of dog that we recognize today, with their variety of shapes, sizes and colours, were already being depicted in works of art.

Domestication is an evolutionary process resulting from changes in the selection pressures on the species. The selection pressures of the natural environment would normally ensure that only those individuals that were most suited to that particular environment survived and bred. During the domestication process the animal is bred in an artificial environment, in the absence of competitors, and this may allow the survival of unusual genotypes, many of which would not have survived in the natural state. Zeuner (1963) proposed five stages of intensity of domestication:

1. Loose contacts with free breeding individuals.
2. Confinement to the human environment together with breeding in captivity.
3. Selective breeding organised by humans to obtain certain characteristics and occasional cross-breeding with wild forms.
4. Economic considerations of humans leading to planned development of breeds with desirable characteristics.
5. Wild ancestors persecuted and exterminated.

The domestic dog has clearly been through these stages, and the phase of planned development of desirable characteristics has been intensive, resulting in the current variety of different breeds. The effects of the domestication process are also apparent in the differences in behaviour patterns between breeds, as undesirable characteristics have been suppressed and desirable characteristics enhanced to match dogs to particular roles in society. For example, some sheepdogs show the normal sequence of hunting behaviour, that is stalking, staring (or showing 'eye') and chasing, but the final attack and kill have been partially suppressed (see Coppinger & Schneider, Chapter 3). Since changes have occurred during the domestication process to both the morphology and general behaviour of the dog, one would expect that the dog's feeding behaviour might also have been altered given the change in life-style associated with domestication and later selection for task performance.

The wolf is a highly social canid and groups will hunt co-operatively for large game. The digestive tract is specialized and is able to handle large quantities of food in a single meal – a distinct advantage in a competitive feeding situation. However, when associated with man, early domestic dogs would probably have survived by scavenging around human settlements. One would assume that human garbage would have been a primary source of food and hence domestic dogs would have needed to become more opportunistic and catholic in their feeding habits, and consume a much wider variety of food items than their wild ancestors. Whether the increased food variety would have resulted in early domestic dogs becoming more or less selective is anyone's guess, but it seems likely that any dogs as choosy or conservative as their wild ancestors would never have survived. The change in life-style associated with living with humans may also account for the domestic dog's preference for cooked over raw meats.

A brief review of the feeding characteristics of wild canids provides a useful basis for comparisons with the domestic form.

Ancestral feeding habits

The mammalian order Carnivora is characterized by great morphological, ecological and behavioural variation. Carnivores live in every habitat and vegetational zone from tropical rain forest to desert. In terms of behaviour, species range from those that lead relatively solitary lifestyles to those that live in large packs. Carnivora means literally 'eaters of flesh', which describes a characteristic of some, but not all, members of the order. Members of the cat family, the Felidae, are specialized for a carnivorous way of life to the extent that certain of their required nutrients are only found in meat; the domestic cat is an obligate carnivore and cannot survive on a vegetarian diet without supplementation. The lesser or red panda (*Ailurus fulgens*) is specialized for herbivory and includes young bamboo in its diet, although it also preys on small birds and mammals. The giant panda (*Ailuropoda melanoleuca*) is a specialist bamboo feeder but will include some meat in its diet when it is available. Thus, within the order Carnivora, there has been specialization for a wide variety of different feeding niches.

The dog family, Canidae, are able to obtain all of their nutrient requirements from vegetable material, although in nature the diet is predominantly meat. Studies of the feeding behaviour of the wolf (*Canis lupus*), the proposed ancestor of the domestic dog, have shown that its diet consists almost entirely of meat with a limited quantity of plant material ingested on occasions, apparently deliberately (Mech, 1970). Other members of the genus, however, take a wider variety of food types including a large proportion of insects and fruits. Jackals, for example, will raid cultivated fruit crops and also consume large quantities of grass (Ewer, 1973). The domestic dog's relatives display considerable flexibility in food selection, and the predominance of meat eating in the wolves studied by Mech (1970) may well have been a consequence of local environmental conditions and available food types (see e.g. Boitani *et al.*, Chapter 15).

Since the domestic dog may have arisen several times from different ancestral wolf stocks, it may have inherited a variety of feeding patterns and food preferences. Some breeds of domestic dog still demonstrate a remarkable ability to gorge, and will eat exceptionally large quantities of food whenever it is available. This behaviour is still a characteristic of some of the pack hounds, such as beagles and foxhounds, although this may be the result of the competitive group feeding that has been the usual husbandry practice for these breeds. The Labrador retriever also shows a tendency to over-eat when given the opportunity, while other breeds, particularly some of the toy and giant breeds, are so finicky that it is often difficult to maintain their optimum body weight. These differences in food selectivity strongly suggest that the domestication process has altered the feeding behaviour of the various breeds of dog. Some behaviour patterns associated with the selection of appropriate food items, which would have had adaptive significance for the dog's wild relatives, are still retained to some degree, and domestic dogs clearly indicate, when offered the opportunity, that they know what they like. Individuals within the same breed also show considerable variability in their food preferences, which suggests that experience also plays an important role in the development of food preferences.

Food selection in the dog is based on the perception of flavour as assessed by the chemical senses of odour and taste. There is no evidence to suggest that these senses have been significantly altered by the domestication process. We can obtain some idea of the types of food flavour that are important to the dog from the responses of these senses to chemical stimulation.

The chemical senses

Taste

Almost all of our knowledge of the sense of taste in the dog is based on neurophysiological studies, but these are poorly supported by behavioural studies and are predominantly based on the responses of the facial nerve. The facial nerve is only one of the neural paths involved in taste perception. The taste buds of the anterior two-thirds of the tongue are innervated by the chorda tympani branch of the facial nerve, those of the posterior third of the tongue are innervated by the lingual branch of the glossopharyngeal nerve, and those of the pharynx and larynx are innervated by the cranial laryngeal branch of the vagus nerve. The chorda tympani nerve in the dog is associated with the taste buds from the fungiform papillae located on the anterior two-thirds of the tongue (Olmsted, 1922). The lingual branch of the glossopharyngeal nerve is associated with taste buds from the vallate papillae, but no studies describe the role of the cranial laryngeal nerve. In addition there are the free nerve endings innervated by the trigeminal nerve.

From neurophysiological studies of the facial nerve four neural groups of taste bud have been identified in the dog. Each neural group has a resting firing rate that can be either increased (excited) or decreased (inhibited) by chemical stimulation. Those compounds that excite a neural group are perceived as a different taste experience to those that inhibit the same group, and this provides eight potential taste categories of which six have been identified (Boudreau *et al.*, 1985; Boudreau, 1989).

From Boudreau's studies, the most abundant taste buds in dogs are the Group A receptors that respond to sugars. Many different sugars, including some artificial sweeteners, stimulate these receptors, with fructose and sucrose being the most excitatory. The response to sugars is much weaker than the response shown to equivalent concentrations of amino acids. Most amino acids trigger positive responses identical to those produced by sugars and most taste sweet to humans. L-tryptophan and other bitter substances, such as quinine, are inhibitory. Thus the group A receptors, in terms of the types of chemicals that stimulate them, appear to be similar to the sweet/bitter dimension of human taste. In comparison, similar receptors in the cat respond to the sweet tasting amino acids, but appear not to respond to the simple sugars. The cat does not have a sweet-tooth and this lack of response to sugars is supported by behavioural evidence (Carpenter, 1956). This difference in the taste systems of these two domestic carnivores may be a reflection of their ancestral life-styles and diets. For an obligate carnivore such as the cat, the response to sugars would be of little use, whereas for the more omnivorous dog it would be useful in the selection of, say, ripe fruit, the sugars of which would be an excellent source of energy. In other words, the absence of the sweet response to sugars in the cat is probably an extreme adaptation to meat eating.

The second most abundant group of canine taste receptors are the acid units, Group B, that display low spontaneous activity rates and are responsive to distilled water and to inorganic acids. A few amino acids are also effective, including the sulphur compounds L-taurine and L-cysteine, while inosine monophosphate is a potent inhibitor of these receptors in the dog.

The other units are less well characterized but in general the nucleotide receptors, Group C, are characteristically found in carnivorous species in which meat, a taste characterized by the 'meaty' character of the nucleotides, is a major part of the diet. The furanol receptors, Group D, are triggered by compounds perceived by humans as fruity–sweet, such as furanol and methyl maltol. The greatest difference between the dog and most other mammals, except for the cat, is a total lack of salt-specific taste buds. For most herbivores and omnivores, detection of the salt content of food is of high priority since sodium is essential for normal bodily function. For wild carnivores, whose food consists primarily of prey, the diet will inevitably be salt-balanced and the lack of response to salt may represent another adaptation to meat-eating. The amino acid and nucleotide sensitive units would be used to distinguish meats of varying nutritional quality.

Not unexpectedly, the taste system of the dog appears adapted to the feeding requirements of a wild canid. The lack of salt-specific taste buds would tend to reflect a predominantly carnivorous diet in which the food is salt-balanced. The presence of both amino acid and nucleotide sensitive receptors – a characteristic of meat-eating species – also tends to indicate a specialized taste system adapted primarily to carnivory.

Olfaction

The world of the dog is perceived through the sense of smell to a much greater degree than in humans, and yet this sense is poorly understood. An appreciation of the importance of olfaction for the dog can be gained by comparison of the quantity of olfactory epithelium; about 75 cm^2 in the beagle compared to 3 cm^2 in man (Albone, 1984). Dodd & Squirrel (1980) found a range of 18–150 cm^2 of olfactory epithelium in various dog breeds. The dog has long been noted for its olfactory acuity and its ability to discriminate between complex mixtures of odours. Studies of comparative olfactory acuity have shown that the dog has a well-developed and acute sense of smell capable of detecting a variety of substances at concentrations ranging from one thousand to one hundred million times lower than humans can perceive (Neuhaus, 1953; Becker, Markee & King, 1957; Moulton, Ashton & Eayrs, 1960). The power of a dog's sense of smell can be appreciated more readily by behavioural tests rather than physiological experiments. In 1885, G. J. Romanes designed a test in which he set off at the head of a column of twelve men walking in single file. Each man carefully placed his feet in the footprints of the man in front. After an interval, the procession split into two halves, each continuing its own way. Romanes's dog, when released, unerringly followed the route taken by the six men headed by Romanes. Seventy years later, Kaimus (1955) carried out similar tests involving identical twins. If the dog was given the scent of one twin it would happily follow a trail laid by the other, but it could distinguish the twins if the scents were given together. Thus the power of the dog's sense of smell is similar to human powers of visual discrimination; the dog can discriminate identical twins when presented together in simultaneous tracking exercises but has more difficulty when confronted with one alone.

Olfaction is particularly important in food selection. Olfactorily intact dogs show a clear preference ranking of different meats, such as beef, pork, lamb, chicken or horse-meat, but anosmic dogs are unable to distinguish the meats, although they can still differentiate meat from cereal (Houpt *et al.*, 1978). Odour is also sufficient stimulus to initiate intake of a bland diet. Studies have shown that by suffusing a bland, dry food with a meat odour the feeding pattern of cats can be greatly altered, such that most feeding occurs in those periods when the odour is present (Mugford, 1977). Houpt (1978) has shown a similar effect with dogs that prefer a bland diet suffused with meat odour to one without meat odour. The effect, however, is short lived since the behavioural response habituates with repeated exposure to an odour that is not matched by the flavour of the food consumed. Thus odour is important for food selection and initiation of feeding, but there must be compatibility between the odour and taste characteristics to maintain preference or intake in the longer term.

Food selection – nature and nurture?

Adult food preferences are the result of an interaction between genetic predispositions, which make one flavour more palatable than another, and the effects of acquired environmental influences and experience. Both animals and humans appear to have innate food preferences that are directly associated with a food's nutritional properties. A sweet taste is often associated with a high concentration of carbohydrates and a bitter taste results from many toxic materials (LeMagnen, 1967). Most species, the felids being an exception, prefer a sweet taste (Soulairac, 1967), and many appear to have an innate preference for salt and an aversion to bitter-tasting foods (Denton, 1967; Rozin, 1967; Rozin & Kalat, 1971). Some flavour preferences appear early in life and appear to be relatively unaltered by subsequent learning experiences (Pfaffmann, 1936; Scott & Quint, 1946; Nachman, 1962; Jacobs, 1964; Young, 1966; Rozin & Kalat, 1971). Moreover, certain preferences can be selectively increased or decreased by inbreeding, suggesting a genetic component to some food preferences and aversions (Scott, 1946; Nachman, 1959).

In addition to genetically mediated preferences, however, a great variety of environmental influences are involved in the development of food likes and dislikes. Learning strongly influences the selection of food, since preferred odours, textures and tastes often bear no clear relationship to a food's nutritional value (LeMagnen, 1967; Rozin, 1967; Rozin & Kalat, 1971). Many animals, for example, rapidly decrease their food intake when a new, but otherwise nutritionally-adequate diet is abruptly substituted for a familiar one. After a period of time, however, the intake increases progressively and eventually becomes stabilized (Adolf, 1947; Garcia, Hankins & Rusiniak, 1974).

The behavioral processes involved in food selection are modified in response to the experience gained throughout life. The taste buds develop long before birth in some mammals (Bradley, 1972; Mistretta, 1972) and many mammalian species swallow *in utero* (Becker *et al.*, 1940; Pritchard, 1965; Lev & Ortic, 1972; Bradley & Mistretta, 1973). The foetus is surrounded by amniotic fluid rich in chemicals of changing concentration and the regular swallowing of these fluids during uterine development may provide stimuli that affect the maturing taste receptors. The neonate is suckled by the bitch, whose milk will vary in quality and flavour depending on what she has eaten. Before weaning young wild canids are offered regurgitated partly-digested food by the bitch, and the weaned young will tend to eat only from those food sources that are utilized by their parents or other adults. Thus, there are many opportunities for the young animal to experience flavours, to link flavours with safe food sources, and to develop a food selection strategy based around those foods that are locally available, nutritious and safe.

Development of the taste system

Ferrel (1984*a*) carried out an anatomical study of the development of the dog tongue and his findings indicated that the puppies peripheral gustatory system was functional at birth, but had not yet reached adult form. This would suggest that puppies may be responsive to chemical stimulation at or before birth. Ferrel (1984*b*) also investigated the responses of the gustatory nerves and found that chorda tympani nerve responses to chemical stimulation of the tongue were present at birth in the dog. Bradley (1972) has speculated that stimulation of the peripheral taste receptors by prolonged exposure to changing levels of chemicals in the amniotic fluid may be necessary to the normal development of central neural connections in the gustatory system. Taste experience *in utero* has also been suggested to play a role in the establishment of taste preferences and aversions expressed after birth (Mistretta, 1972; Mistretta & Bradley, 1977). Smotherman (1982) found that *in utero* exposure to apple solution in rats resulted in an increased preference for that flavour later in life. The preference was not the result of a general decrease in neophobic tendencies to all sapid solutions, but was specific to the flavour that had been experienced. Pedersen & Blass (1982) found that similar effects were produced by odours experienced *in utero*. Exposure to citral (a tasteless lemon scent) resulted in the pups preferentially attaching to nipples that had been scented with citral, and it was only those pups who had experienced the odour *in utero* who would attach to washed nipples to which citral was applied. It seems reasonable to assume that similar experiential effects occur in young dogs, and that they may represent the first stage in the development of flavour preferences.

The role of mother's milk

Females of all mammalian species nurse their offspring and are able to provide neonates with all necessary nutrients through the milk that they produce. During the nursing period, the young are provided with cues from the flavour of the milk which, in rat pups, causes them to seek out and preferentially ingest the food that their mother had been eating (Galef & Henderson, 1972). The clinical literature indicates that a wide variety of substances, including antibiotics, sulfonamides and most alkaloids, when ingested, pass intact into the nursing infant. Thus the complex long chain molecules associated with dietary taste and smell will also pass intact from the intestinal tract into the mother's milk to be passed intact to suckling young. Ling, Kan & Porter (1961) described changes in the flavour of cow's milk associated with their ingestion of certain natural foodstuffs, which indicates that even factors in the natural diet will produce a characteristic flavour to the milk. Commercial use has been made of food-additives that are designed to pass intact to nursing animal's milk. For example, the addition of a distinctive flavour to a sow's diet during lactation results in an increase in post-weaning food intake and growth rate in her piglets when they are subsequently offered food with the same flavour added (Campbell, 1976). Although such studies have not been carried out on the domestic dog, it is likely that a bitch's milk will contain food flavours which provide her puppies with flavour cues that may be involved in the development of early food preferences.

Early experience of solid food

In many species, exposure to a specific dietary flavour early in life has been shown to enhance subsequent preference for that flavour. The strength and persistence of such a preference is influenced by such factors as the species studied, the developmental age of the animal, the attractiveness of the flavour employed and the duration of exposure. Turtles and snakes, for instance, exhibit long lasting preferences for the first fed food over subsequently presented foods, even if it is not a natural food prey item (Burghart & Hess, 1966; Fuchs & Burghart, 1971). Similarly, young rats restricted to a single flavour choose that flavour subsequently, whereas those given a variety of flavours will more readily consume a novel one.

Several studies have investigated the role of early flavour experience on food choice in the dog. In the most extreme study, chow chow puppies were hand-reared from birth to six months of age on one of three diets (Kuo, 1967). Those puppies fed on a soya-bean diet would eat no novel food, those reared on a mixed vegetarian diet would eat no animal protein, but those reared on a mixed diet would eat any new food except those with bitter, sour or stale tastes. Thus, limited flavour experience in these pups led to a fixation of food preferences, while the provision of some flavour or textural variety within the diet enhanced the acceptance of novel foods.

In studies using natural weaning and single food flavour experience (Mugford, 1977), basenji and terrier pups fed a single canned food from weaning for the next 16 weeks preferred a novel food to their accustomed diet. The persistence of this preference, however, was dependent on the relative palatability of the diets. Novelty of the food combined with low relative palatability produced only a short-lived preference for the novel diet, whereas when the diets were of similar palatability the effect was more persistent. Ferrel (1984*c*) carried out a similar study in which she assessed eventual food preferences in beagle pups offered single semi-moist foods, each with a characteristic flavour, for three weeks from three-and-a-half weeks of age. She noted considerable individual variability of response, but again concluded that palatability and novelty were the important factors in diet selection. One diet group did in fact show a tendency to prefer the food with the accustomed flavour, but this was confounded by the higher palatability of that diet.

We can conclude from these findings that, if there is a sensitive period for the development of flavour fixation, it either occurs before three-and-a-half weeks of age, or that it extends into adulthood. It should be emphasized, however, that a fixation of food preferences is only observed when the animal receives limited early flavour and texture experience. Providing the pup with variety in its diet always reduces, and usually eliminates, any tendency for food preferences to become fixed, resulting in a general preference for novel foods. Novelty and palatability appear to be the most important factors in the control of food choice, and this highlights the labile nature of the food selection process in the dog.

Flavour experience in the adult

Positive nutrition

Dogs appear to have little, if any, direct perception of the nutritional status of a food, but they will learn rapidly to associate a food's flavour with its physiological consequences. Learning occurs in situations where the flavour experience and the physiological effect are well separated in time, a situation that is not easily reconciled with the usual necessity for a close temporal link between the stimulus and reward or punishment in classical models of conditioning (McFarland, 1978). This type of learning will result in a preference for flavours associated with nutritional benefit and an avoidance of flavours associated with foods which are deficient, imbalanced or toxic. Baker *et al.* (1987) demonstrated in rats that these flavour/nutrient associations result in the development of nutrient specific hungers in which the preference is only observed when the rat is deficient in that specific nutrient. If the rat is given a food preloaded with the specific nutrient, the preference for a flavour associated with that nutrient is not observed. Although preference for flavours associated with a particular nutrient have been demonstrated in the dog, nutrient specific hungers have not.

Neophobia

Neophobia is a fear of new objects or situations, or a rejection of new foods. Food neophobia in dogs is not common, but has been demonstrated in one of our studies. A variety of small breeds – poodles, dachshunds, Yorkshire terriers and Cavalier King Charles spaniels – were reared on specific dietary regimens from weaning to two years old, one of which provided limited flavour experience in the form of a nutritionally complete puppy food, and two of which provided a variety of prepared and fresh foods, respectively. These dogs were subsequently offered a novel food of very unusual texture and the two groups compared. Those fed a variety of flavours showed an immediate preference for the novel food, but the flavour restricted groups preferred their usual food. Hence, the degree of previous experience of different flavours and textures is fundamental to the food selection behaviour of the individual.

Aversion

Aversion develops rapidly to any food whose ingestion produces negative physiological responses. Lithium chloride administration during or following food intake results in nausea and an immediate reduction in intake of the food just eaten. The development of aversion is very rapid and is clearly an adaptation for the avoidance of toxic foods.

The phenomenon of aversion has been exploited with variable success to eliminate predation on sheep by coyotes. One reported successful study offered sheep carcasses, laced with lithium chloride, to wild coyotes. The resulting conditioned aversion subsequently blocked attack behaviour on live sheep and the effect had not extinguished after nine weeks (Ellis, Catalano & Schechinger, 1977). In the domestic dog, lithium chloride treatment is less effective, some dogs will eat the laced vomitus and the effect is often short lived (Rathore, 1984). This lack of response to lithium chloride in the domestic dog may be an effect of domestication. As already suggested, it would have been important for early domestic dogs to have had highly opportunistic feeding habits. This would be aided by an enhanced ability to experiment with novel foods and to tolerate foods of low quality.

Neophilia

The preference for new foods over a usual food is common in dogs (Mugford, 1977; Griffin, Scott & Cante, 1984). Neophilia and neophobia provide feeding strategies that are important to the survival of the wild animal – neophobia provides the means of avoiding potentially toxic foods, whereas neophilia provides a means for evaluating potential new food sources for nutritional quality and providing an alternative to the usual food, if it becomes scarce. Which particular behaviour is shown appears to be dependent on context. A novel food whose characteristics are outside the animal's feeding experience will tend to be rejected, whereas those that are similar to the usual food are accepted and tried. Similarly, if the animal is placed in a novel environment, presumably increasing its stress level, it will become more neophobic and prefer its familiar food over a novel one (Thorne, 1982).

These behavioural strategies, and there are probably others, provide a basis for learning about foods and developing a food selection strategy. They are adaptive in that they achieve avoidance of potentially dangerous foods but, wherever possible, result in trial of potential new foods so that nutritional quality can be assessed. Aversion provides bottom line safety by ensuring that foods which produce negative physiological consequences do not become part of the diet. Overlaying this behavioural basis of food selection are the differences associated with breed and the individual. There is evidence that the various breeds respond differently to food. For example, Cavalier King Charles spaniels and Labrador retrievers are particularly non-selective about food. In addition, every domestic dog is an individual with its own particular food likes and dislikes. Such individual differences may vary from long-term acceptance of one food to a demand for extensive variety.

Preferences or aversions associated with food, whether genetic or learned, and the particular palatability of the diet consumed, profoundly affect the abilities of all animals to meet all of their daily requirements. These factors are particularly important in the natural habitat (Ruiter, 1967), since no single food source will meet the animal's needs exactly, and the combination of preferences for, and aversions to, particular foodstuffs should lead the animal to eat a safe and nutritious diet. The lives of dogs may be very different from those of their wild ancestors, yet essentially similar developmental and learning processes are involved in the selection of food by the domestic species.

Conclusions

The domestic dog's canid ancestors developed as specialists to meet the demands of their ecological niche. This is evident from their anatomy that is particularly suited for endurance running, enabling them to tire and capture large prey; from their dentition that is characteristic of the carnivores, incorporating enlarged canine teeth for gripping and puncturing, and large carnassial teeth for cutting flesh; and from their vision that has good binocular function for accurate estimation of a prey animal's position, and high sensitivity to movement over long distances. Likewise, their senses of smell and taste are particularly sensitive to those chemicals that are relevant to the assessment of meat quality. Despite these inherited specializations, the processes involved in food selection show considerable flexibility with the repertoire of acceptable food items changing in response to food availability and quality. (The learning processes involved in the development of food selectivity may themselves be genetically mediated, since canids appear to know what to learn and which sensory cues are more relevant.) Changes in an individual's responses to food items occur throughout its life and provide an adaptive behavioural strategy which results in the continual trial of new food items that may then be added to the individual's repertoire of acceptable foods. The greater the range of safe and nutritious foods the individual knows, the more likely it is to survive in times of scarcity.

The first safe solid food that the young wild canid experiences is provided by its parents. Some flavour cues from the food that the mother has eaten may reach the neonate through its mother's milk, although the evidence for this is still uncertain. When puppies first start self-feeding they will tend to eat those foods that are also eaten by other adults. In this way they will begin to recognize foods that are safe and available locally. As the pups mature and start to hunt for their own food, two contrasting behavioural strategies, neophobia and neophilia, will help them to avoid unusual foods that may be toxic and to sample new foods which result in dietary variety.

In the domestic dog these strategies still operate to a greater or lesser extent. Both variety and palatability are important factors in the dog's food selection behaviour, and most domestic dogs respond favourably to variety in their diet. Overall, the domestic dog still shows a strategy of food selection which is similar to that shown in wild canids and, although some aspects of feeding behaviour may have been altered by domestication, much of the food selection behaviour that would have had adaptive advantages for wild canids is still seen in the domestic dog.

Many aspects of feeding behaviour in the domestic dog are still not fully understood. For example, most dogs will ingest unusual foods and even non-food items. Grass is the most commonly reported unusual item that is consumed by dogs, and many suggestions have been made to explain this ingestive peculiarity. Explanations for grass-eating include: providing a source of nutrients that are lacking in the diet; a means of dislodging intestinal parasites, and a natural emetic to expel food that is causing digestive upset. There is no good evidence to support any of these hypotheses, although it is unlikely that grass is consumed to obtain nutrients since dogs that are fed a nutritionally complete diet will still eat grass. Grass-eating is, however, usually associated with other behaviour patterns suggestive of some form of mild digestive upset. Coprophagy, the eating of faeces, is a familiar behaviour among dogs and is frequently a cause of disgust and disappointment for the owner. For many species, coprophagy is a natural and essential ingestive behaviour. The rabbit, for example, produces two forms of faeces, one soft and one hard, the former being partially digested material rich in nutrients, which is eaten again to ensure extraction of maximum nutrients from the vegetarian diet. A similar phenomenon may help to explain the behaviour of dogs when they eat the faeces of herbivores that are still rich in nutrients from vegetation, which the dog's digestive system is not capable of processing. However, this would not explain the great liking that many dogs show for the faeces of foxes, other dogs, or even for their own faeces. This form of pica, the ingestion of non-food items, may be a consequence of domestication and the need for the domestic dog to make use of any potential food item that could be found around human habitations. Such selection pressures would have resulted in canids that were far more cosmopolitan in their feeding habits than their wild relatives.

Despite these unusual feeding habits, most dogs prefer their food to be of high quality, and the challenge for the commercial manufacturer is to produce a food that is both nutritious and acceptable to the majority of animals. Although modern domestic dogs, when needs must, will eat almost anything, they still show strong preferences for specific foods. A commercial pet food is the culmination of extensive nutritional, behavioural and market research. Ideally, it provides a single nutritionally complete food that meets the requirements of both owners and dogs, and is capable of matching the nutritional needs of a dog over its entire lifespan – probably the most demanding specification for any food. Studies of the nutrition and food selection behaviour of the dog are still generating new data which help to ensure that prepared pet foods are optimally designed. New discoveries continue to be made and therefore nutritional and behavioural studies of the domestic dog will continue for the foreseeable future.

References

Adolf, E. F. (1947). Urges to eat and drink in rats. *American Journal of Physiology*, **151**, 110–25.

Albone, E. S. (1984). *Mammalian Semiochemistry; Investigation of Chemical Signals Between Mammals*. Chichester, West Sussex: John Wiley & Sons Ltd.

Baker, B. J., Booth, D. A., Duggan, J. P. & Gibson, E. L. (1987). Protein appetite demonstrated: learned specificity of protein-cue preference to protein need in adult rats. *Nutrition Research*, **7**, 481–7.

Becker, F., Markee, J. E. & King, J. E. (1957). Studies on olfactory acuity in dogs. 1. Discriminatory behaviour in problem box situations. *Animal Behaviour*, **5**, 94–103.

Becker, R. F., Windle, M. F., Barth, E. E. & Schulz, M. D. (1940). Fetal swallowing, gastrointestinal activity and defecation in amnio. *Surgical Gynecology and Obstetrics*, **70**, 603–14.

Boudreau, J. C., Sivakumar, L., Do, L. T., White, T. D., Orovec, J. & Hoang, N. K. (1985). Neurophysiology of geniculate ganglion (facial nerve) taste systems: species comparisons. *Chemical Senses*, **10**, 89–127.

Boudreau, J. C. (1989). Neurophysiology and stimulus chemistry of mammalian taste systems. In *Flavour*

Chemistry: Trends and Developments, ed. R. Teranishi, R. G. Buttery & F. Shahidi, pp. 122–37. American Chemical Society Symposium Series, 388.

Bradley, R. M. (1972). Development of the taste bud and gustatory papillae in human fetuses. In *The Third Symposium on Oral Sensation and Perception: The Mouth of the Infant*, ed. J. F. Bosma, pp. 137–62. Springfield, IL: Charles C. Thomas.

Bradley, R. M. & Mistretta, C. M. (1973). Swallowing in fetal sheep. *Science*, **179**, 1016–17.

Burger, I. H. & Rivers, J. P. W. (1989). *Nutrition of the Dog and Cat.* Waltham Symposium No. 7. Cambridge: Cambridge University Press.

Burghart, G. M. & Hess, E. H. (1966). Food imprinting in the snapping turtle (*Chelydra serpentia*). *Science*, **151**, 108–9.

Campbell, R. G. (1976). A note on the use of a feed flavour to stimulate the feed intake of weaner pigs. *Animal Production*, **23**, 417–19.

Carpenter, J. A. (1956). Species differences in taste preferences. *Journal of comparative and physiological Psychology*, **49**, 139–44.

Davis, S. J. M. & Valla, F. R. (1978). Evidence for domestication of the dog 12 000 years ago in the Natufian of Israel. *Nature*, **276**, 608–10.

Denton, D. A. (1967). Salt appetite. In *Handbook of Physiology*, Vol. 1, ed. C. F. Code, pp. 433–59. Washington, DC: American Physiological Society.

Dodd, G. H. & Squirrel, D. J. (1980). Structure and mechanism in the mammalian olfactory system. *Symposia of the Zoological Society of London*, **45**, 35–6.

Ellis, S. R., Catalano, S. M. & Schechinger, S. A. (1977). Conditioned taste aversion: a field application to coyote predation on sheep. *Behavioral Biology*, **20**, 91–5.

Ewer, R. F. (1973). *The Carnivores*. London: Weidenfield & Nicolson.

Ferrel, F. (1984*a*). Taste bud morphology in the fetal and neonatal dog. *Neuroscience and Biobehavioral Reviews*, **8**, 175–83.

Ferrel, F. (1984*b*). Gustatory nerve response to sugars in neonatal puppies. *Neuroscience and Biobehavioural Reviews*, **8**, 185–90.

Ferrel, F. (1984*c*). Effects of restricted dietary flavour experience before weaning on post-weaning food preferences in puppies. *Neuroscience and Biobehavioural Reviews*, **8**, 191–8.

Fuchs, J. L. & Bughart, G. M. (1971). Effects of early feeding experience on the responses of garter snakes to food chemicals. *Learning and Motivation*, **2**, 271–9.

Galef, B. G. & Henderson, P. W. (1972). Mother's milk: a determinant of the feeding preferences of weanling rat pups. *Journal of Comparative and Physiological Psychology*, **78**, 213–19.

Garcia, J., Hankins, W. & Rusiniak, K. (1974). Behavioral regulation of the milieu interne in man and rat. *Science*, **185**, 824–31.

Griffin, R. W., Scott, G. C. & Cante, C. J. (1984). Food preferences of dogs housed in testing-kennels and in consumers' homes: some comparisons. *Neuroscience & Biobehavioural Reviews*, **8**, 253–9.

Houpt, K. A. (1978). Palatability and canine food preferences. *Canine Practice*, **5**(6), 29–35.

Houpt, K. A., Hintz, H. F. & Shepherd, P. (1978). The role of olfaction in canine food preferences. *Chemical Senses and Flavour*, **3**, 281–90.

Jacobs, H. L. (1964). Observations on the ontogeny of saccharine preference in the neonate rat. *Psychonomic Science*, **1**, 105–6.

Kaimus, H. (1955). The discrimination by the nose of the dog of individual human odours and in particular of the odours of twins. *Animal Behaviour*, **3**, 25–31.

Kitchell, R. L. (1978). Taste perception and discrimination by the dog. *Advances in Veterinary Science and Comparative Medicine*, **22**, 287–314.

Kuo, Z. Y. (1967). *The Dynamics of Behaviour Development: An Epigenetic View*. New York: Random House.

LeMagnen, J. (1967). Habits and food intake. In *Handbook of Physiology*, Vol. 1, ed. C. F. Code, pp. 11–30. Washington, DC: American Physiological Society.

Lev, R. & Orlic, D. (1972). Protein absorption by the intestine of the fetal rat *in utero*. *Science*, **177**, 522–4.

Ling, E. R., Kan, S. K. & Porter, J. W. G. (1961). The composition of milk and the nutritive value of its components. In *Milk: The Mammary Gland and its Secretion*. Vol. II, ed. S. K. Kan & A. T. Cowrie, pp. 195–263. New York: Academic Press.

Lohse, C. L. (1974). Preferences of dogs for various meats. *Journal of the American Hospital Association*, **10**, 187–92.

McFarland, D. J. (1978). Hunger in interaction with other aspects of motivation. In *Hunger Models: Computable Theory of Feeding Control*. ed. D. A. Booth, pp. 375–405. London: Academic Press.

Mech, L. D. (1970). *The Wolf: Ecology of an Endangered Species*. New York: Natural History Press.

Mistretta, C. M. (1972). Topographical and histological study of developing rat tongue, palate and taste buds. In *The Third Symposium on Oral Sensation and Perception: The Mouth of the Infant*, ed. J. F. Bosma, pp. 137–62. Springfield, IL: Charles C. Thomas.

Mistretta, C. M. & Bradley, R. M. (1977). Taste *in utero*: theoretical considerations. In *Taste and Development – The Genesis of Sweet Preference*, ed. J. M. Weiffenbach, pp. 51–69. Bethesda, MD: DHEW Publications.

Moulton, D. G., Ashton, E. H. & Eayrs, J. T. (1960). Studies in olfactory acuity. 4. Relative detectability of n-aliphatic acids by the dog. *Animal Behaviour*, **8**, 117–28.

Mugford, R. A. (1977). External influences on the feeding

of carnivores. In *The Chemical Senses and Nutrition*, ed. M. R. Kare & O. Maller, pp. 25–50. New York: Academic Press.

Nachman, M. (1959). The inheritance of saccharine preference. *Journal of comparative and physiological Psychology*, **52**, 451–7.

Nachman, M. (1962). Taste preferences for sodium salts by adrenalectomized rats. *Journal of comparative and physiological Psychology*, **55**, 1124–9.

National Research Council (1985). *Nutrient Requirements of Dogs*. Washington, DC: National Academy of Sciences.

Neuhaus, W. (1953). Über die Riechschärfe des Hundes für Fettsäuren. *Zeitschrift für vergleichende Physiologie*, **35**, 527–52.

Olmsted, J. M. D. (1922). Taste fibres and the chorda tympani. *Journal of Comparative Neurology*, **34**, 337–41.

Pedersen, P. E. & Blass, E. M. (1982). Prenatal and postnatal determinants of the 1st suckling episode in albino rats. *Developmental Psychobiology*, **15**, 349–55.

Pfaffmann, C. (1936). Differential responses of the new-born cat to gustatory stimuli. *Journal of Genetical Psychology*, **49**, 61–7.

Pritchard, J. A. (1965). Deglutition by normal and anencephalic fetuses. *Journal of Obstetrics and Gynecology*, **25**, 289–97.

Rathore, A. K. (1984). Evaluation of lithium chloride taste aversion in penned domestic dogs. *Journal of Wildlife Management*, **48**, 1424.

Rozin, P. (1967). Thiamine specific hunger. In *Handbook of Physiology*, Vol. 1, ed. C. F. Code, pp. 411–31. Washington, DC: American Physiological Society.

Rozin, P. & Kalat, J. W. (1971). Specific hungers and poison avoidance as adaptive specialisations of learning. *Psychological Reviews*, **78**, 459–86.

Ruiter, L. de (1967). Feeding behaviour of vertebrates in the natural environment. In *Handbook of Physiology*, Vol. 1, ed. C. F. Code, pp. 97–116. Washington, DC: American Physiological Society.

Scott, E. M. (1946). Self selection of diet. I. Selection of purified components. *Journal of Nutrition*, **31**, 397–406.

Scott, E. M. & Quint, E. (1946). Self selection of diet. III. Appetites for B vitamins. *Journal of Nutrition*, **32**, 285–91.

Smotherman, W. P. (1982). In utero chemosensory experience alters taste preferences and corticosterone responsiveness. *Behavioural and Neural Biology*, **36**, 61–8.

Soulairac, A. (1967). Control of carbohydrate intake. In: *Handbook of Physiology*, Vol. 1, ed. C. F. Code, pp. 387–98. Washington, DC: American Physiological Society.

Thorne, C. J. (1982). Feeding behaviour in the cat – recent advances. *Journal of small Animal Practice*, **23**, 555–62.

Turnbull, P. F. & Reed, C. A. (1974). The fauna from the terminal Pleistocene of Palegawra Cave. *Fieldiana Anthopology*, **64**, 99–101.

Young, P. T. (1966). Hedonic organization and regulation of behaviour. *Psychological Reviews*, **73**, 59–86.

Zeuner, F. E. (1963). *A History of Domesticated Animals*. New York: Harper and Row.

8 Social and communication behaviour of companion dogs

JOHN W. S. BRADSHAW AND HELEN M. R. NOTT

Introduction

During the process of domestication, people have taken advantage of the social system of the dog, and have exploited and enhanced the tendency for dogs to behave in a subordinate way towards humans. Many of the behaviour patterns of dogs bear a close resemblance to those seen in wolves, and so it has become customary to assume that the social behaviour of dogs is simply a corrupted version of that seen in wolves. However, it has become clear in recent years that the social systems of carnivores are highly flexible, even within species, and that they often depend upon ecological factors, such as the availability and distribution of food (Moehlman, 1989; Macdonald & Carr, Chapter 14). It is self-evident that the ecology of wolves differs in almost every respect from that of those breeds of dog, such as the Pekinese, that have lived in close association with man for many hundreds of years. There are, of course, intermediate stages, such as the dingo, the pariah dogs and feral dogs, whose origins and/or ecology lie somewhere between the wolf and the companion breeds of dog, and these may provide some clues as to the form that the social structure would take in the domesticated breeds if the influence of man were removed (see Macdonald & Carr, Chapter 14; Boitani *et al.*, Chapter 15). However, during the derivation of the more modern breeds, humans have taken over one of the most important functions of the wolf's social system, that is, the almost complete suppression of breeding for most members of the group by the *alpha* male and female. Rather than the social structure itself determining which individuals will breed each season, and which will not, the human breeder selects the partners for each mating, based on the particular morphological or behavioural characteristics that he or she desires to see in the offspring (although bitches can sometimes nullify this choice by refusing to mate with the selected male). It would be surprising if this profound change had not produced a corresponding alteration in the social repertoire of the more refined breeds (e.g. those showing the greatest alteration in appearance and/or behaviour from that of the wolf), but this possibility seems to have received little attention from researchers. This chapter is an attempt to draw together studies of dogs that enjoy a close relationship with humans, which we therefore refer to as companion dogs, to distinguish them from those, such as free-ranging and feral dogs, which have a looser relationship with society.

Because of the ecological and genetic differences between dog and wolf, the social behaviour of companion dogs is worth studying in its own right. It must always be borne in mind that the degree of dependence on man is likely to have a major effect on any inherited aspects of social behaviour, and that any inherited tendencies will be further modified by the circumstances under which each individual dog lives. Therefore, there is probably as much to be learned from comparisons between breeds with different histories of domestication, as from a search for those underlying facets of social behaviour that are characteristic of domestic dogs as a whole.

Canid sociality

Examination of the social systems employed by wild canids suggests some possibilities that could have been selected for during domestication. Fox (1978) divided the different types of social organization in the canids into three major categories.

In Type I canids, a temporary pair bond is formed between male and female during the breeding season. The male usually stays with the female once the cubs are born, and assists in their rearing by bringing food and defending the den. For the rest of the year all individuals live as solitary hunters.
Permanent pair-bonds are a characteristic of the Type II system, and in addition the young often stay with their parents until the following breeding season. If there is sufficient food available they may remain in the family group, and assist in the rearing of the next litter, but it is more likely that they will disperse and establish their own territories.
The most complex form of social organisation, classed as Type III, is the pack, usually consisting of related individuals. Normally, only one male and one female breed at a time, the remaining members of the pack assisting in the rearing of young. Co-operation between pack members is not restricted to breeding, but also extends to group hunting, which increases the range of prey available, by the inclusion of species too large to be caught by an individual pack member.

It would be misleading to describe Types II and III in terms of a single typical species, because of the flexibility that the more social canids exhibit in relationships with their conspecifics. The species most relevant as far as the domestic dog is concerned, the wolf, can be put into all three categories, depending upon geographical race and the circumstances under which each individual or group lives. Since it is possible that the domestic dog has arisen from multiple domestications of several races of wolf (see Clutton-Brock, Chapter 2), there is no reason to assume that all domestic dogs should exhibit the same social repertoire. It may not even be possible to put each breed of dog on a continuum from the most wolf-like to the most domesticated, since our artificial selection, starting from races of wolves that are themselves behaviourally distinct, is likely to have taken different directions as each breed was moulded for different purposes. When sufficient data has been gathered, we should not be surprised to find that there is as much diversity in social behaviour within the species *Canis familiaris* as might normally be expected within a whole genus or even a whole family of undomesticated species.

Unfortunately, the close association between dogs and humans also makes the social behaviour of dogs difficult to study in isolation. Some progress has been made through observations of feral dogs (see Boitani *et al.*, Chapter 15), but the activities of free-roaming dogs almost inevitably bring them into conflict with people, and the resultant persecution militates against the establishment of permanent groups. Companion dogs are limited to an even greater extent; many pass their lives entirely isolated from conspecifics apart from short daily exercise periods, and even where dogs are kept in groups, the sex-ratio is almost always determined by the owner, rather than by natural immigration and emigration. It seems unlikely that the social systems of the domestic dog ever operate in an entirely unrestricted way for long enough to exhibit all the complexity of which the animals are capable, and so it is necessary to piece together the whole picture from studies that are aimed at particular aspects of social behaviour. Some of these, including the socialization process itself, are covered elsewhere in this book. Others, such as the reasons why male dogs rarely take care of puppies, in contrast to the high level of care-giving by male wolves, remain unstudied. In this chapter we consider just three aspects of social behaviour: communication, intragroup interactions that indicate social hierarchies, and intergroup interactions. Because there is as yet no other framework on which to base the sociality of the domestic dog, analogies with the wolf will be used throughout.

Communication

Effective communication is essential for the formation and maintenance of social relationships. Dogs have three main methods of communication: auditory, visual and olfactory. In discussion of each of these methods, reference will also be made to their use by wolves since this can often aid in our interpretation of the probable function of certain behaviour patterns observed in domestic dogs.

Auditory communication

Auditory communication may be employed over a range of distances and is particularly useful when vision is impaired, for example in thick vegetation. Studies by Fox (1978) described a range of signals given by dogs, which can also be recorded in wolves. These include a diverse range of sounds from grunts, whines, yelps and screams to tooth snapping, coughing, growling and, of course, barking. The contexts in which each of these sounds are used are summarized in Table 8.1.

Table 8.1. *Contexts in which different methods of auditory communication are used by domestic dogs*

Sound	Behavior
Bark	Defence
	Play
	Greeting
	Lone call
	Call for attention
	Warning
Grunt	Greeting
	Sign of contentment
Growl	Defence warning
	Threat signal
	Play
Whimper/Whine	Submission
	Defence
	Greeting
	Pain
	Attention seeking

Despite having a wide range of potential signals that are used in a variety of different situations, dogs tend to rely more on barking than do other species of canid. It therefore seems possible that there has been selection pressure for the tendency to bark in dogs. Joslin (cited in Mech, 1970) observed two types of bark produced by wolves. The first, the alarm bark, is short and usually followed by silence. The other is more of a threatening or challenging bark, which is used at the approach of intruders. It is possible that humans have selected dogs to bark more readily in order to call attention to potential hazards or problems ('watchdog barking') and also during the pursuit of prey, directing human hunters towards the kill. It does, however, seem unlikely that this was due to conscious selection, since dogs that have evolved in different parts of the world generally all show the same propensity to bark, although experimental studies suggest that the latency to bark is a strongly heritable characteristic (Scott & Fuller, 1965). Coppinger & Feinstein (1991) point out that young animals of many species of canid tend to bark more frequently than adults. It is therefore possible that, during selection for tameness, juvenile characteristics were also selected for, including the propensity to bark. Studies on foxes selected over 20 generations for tameness by a group of Soviet biologists showed that over successive generations the foxes gradually began to sound more and more like dogs (cited in Coppinger & Feinstein, 1991).

The exact message in a dog's bark therefore remains unclear. In some cases dogs do seem to bark in similar context to those that elicit barking in wolves, in other words to attract attention in an alarm or to signal the presence of intruders. In many cases, however, the bark seems to convey no single clear signal and it is debateable whether such noises are actually used as a method of communication in themselves, or simply serve to attract attention to visual signals that the sender may also be giving.

Another method of dog vocal communication, which warrants further discussion, is the howl. The literature on wolf ecology suggests that wolves howl for one of two reasons. The first and most frequent occurrence of howling is to aid in the assembly of the pack, particularly before a hunt. In addition, individuals will howl when alone either to seek contact with other pack members (Mech, 1970) or to attract other wolves during the breeding season (Klinghammer & Laidlaw, 1979). Some dogs howl when alone, and it would probably be fair to say that such individuals are seeking social contact either with other dogs or with humans. However, some dogs also howl at objects, or in response to the sound of singing or the violin, or seemingly just at the sky or the moon. The exact reason for this behaviour, if any, remains unclear and would be difficult to study because of the high degree of individuality in its occurrence in the more refined breeds.

Some other methods of communication also seem to be individualistic in their occurrence. One example of this is teeth chattering or snapping. Some dogs use this in situations of play, others in warning or defence, and others still in anticipation or when generally excited. Again whether this signal actually conveys a relevant message to the recipient is not clear and, as with howling, would be hard to study objectively.

Visual communication

Dogs of the less physically modified breeds have many visual communication methods in common with wolves, including those postures that indicate dominance status, aggression and fear. Abrantes (1987) has classified these postures shown by wolves using two primary dimensions: aggressive/fearful and dominant/submissive. A dominant wolf is characterized by an upright body posture with the head and tail held high and the ears pricked. An aggressive dominant wolf will couple this body posture with raised hackles, curled lips and bared teeth. In contrast, subordinate wolves hold their bodies low, the ears flat, and the tail held low and close to the body, creating the general impression of a smaller animal. Subordinate fearful behaviour exaggerates these postures with wolves cringing, tucking their tails between their legs and generally reducing the overall apparent body size. Subordinate wolves often approach more dominant individuals in an enthusiastic greeting with extreme wagging of the tail whilst maintaining a low general body posture (Fox & Bekoff, 1975). Such behaviour may also be associated with nuzzling and licking the face of the more dominant animal, much as wolf pups do to other pack members in order to encourage them to regurgitate food. Submission may also include more extreme visual signals including rolling over and displaying the inguinal region (Fig. 8.1); submissive urination

Fig. 8.1. The posture for passive submission is very similar in most breeds of dog, and also in wolves. In order to submit, the dog rolls on its back and exposes its inguinal region which is usually sniffed by the other dog. Photograph: Steve Wickens.

may also occur (Schenkel, 1967). Some breeds of domestic dogs show very similar behaviour patterns, often seen in interactions between dogs and their owners, for example the enthusiastic greeting ritual when an owner returns home after a period of absence.

Visual communication between wolves or dogs is particularly apparent in their facial expressions. One of the most effective signals used by more dominant dogs is the direct stare. When two dogs first meet, subordinate individuals break eye contact earlier than dominant ones. It has been suggested that this brief exchange sets up social priorities (Beaver, 1982), although Bradshaw and Lea (1993) could find no direct link between the stare and the subsequent outcome of an interaction (also see Fig. 8.9). Confusion can often arise if a dog continues to be stared at despite having already broken eye contact. Such a dog may attack without any of the usual threat signals purely through fear at such a situation. Similarly, if a dog continues to stare at a more dominant dog, the dominant animal's intentions may be reinforced by more direct threatening signals, including baring of the teeth and snarling.

Similar staring may also occur in play between dogs. In play, the dogs' general body postures convey no threat, and so staring can be used without the risk of confusion. The variety of visual communication methods shown in play are difficult to describe as they are often characteristic of specific individuals. The more universal signals include the play bow, pawing with a front foot, twisting jumps and open mouthed panting (Bekoff, 1977). Associated with these behaviour patterns is the one that is probably the most characteristic of dogs – tail-wagging. Tail movements are used in a variety of contexts related to a variety of moods (Fox & Bekoff, 1975). Loose, free tail-wagging indicates general friendliness, and often extends to incorporate the entire rump in subordinate animals. More anxious or nervous dogs tend to wag their drooping tails more stiffly, seemingly as an appeasement signal. Rapid, stiff, upright 'flagging' of the tail indicates threat and the possibility of aggression.

Despite using a wide variety of visual communication methods it has been suggested that dogs may be less reliant on them than their wild ancestors. Selection by humans for certain morphological characters has reduced some dogs' abilities to use certain structures for visual communication (Beaver, 1981, 1982; Blackshaw, 1985; Fig. 8.2). For example, dogs with drooping ears and/or docked tails may be less able to signal their status than those with more wolf-like body conformation. Similarly, long-haired breeds will be unable to raise their hackles effectively and some may be unable to communicate through eye contact or staring. Due to this reduction in reliance on visual communication it has been suggested that dogs rely more on other forms of communication, particularly olfactory signals (Bradshaw & Brown, 1990)

Fig. 8.2. Breeds with a highly neotenised appearance, like this French bulldog, are incapable of performing more than a small fraction of the visual signals produced by wolves. Photograph: Steve Wickens.

Olfactory communication

The sense of smell in dogs is extremely acute, due to a huge area of olfactory epithelium (Bradshaw, 1992). In addition, dogs possess a vomeronasal organ (VNO) that consists of two fluid filled sacs connected to the nasopalatine canal by fine ducts. In other carnivores that possess a VNO, scents are pumped into the organ accompanied by flehmen behaviour. Flehmen is not seen in domestic dogs, but morphological examination of the innervation of the VNO suggests that it is functional in sexual behaviour (Hart, 1983).

Smells have the advantage of remaining in the environment for a long time and are an advantageous method of communication in dense vegetation where visual communication is impaired. In addition, due to their longevity, scent marks function when the 'owner' is not present.

Olfactory communication in the dog is conducted via two main methods; the deposition of scents in the environment such as faeces, urine and anal sac secretions, and the distinctive body odours of individual dogs. These latter odours are produced via a variety of glandular secretions. Faeces are frequently used as scent markers in communication between wolves. Packs deposit faeces more frequently on trails within their territory whereas lone individuals leave the trails in order to defecate (Peters & Mech, 1975). Similarly scent marks are deposited at greater densities along the edge of territories and especially if a foreign scent mark is encountered. Thus wolves can communicate their residency without having to continually patrol their territory edge.

Domestic dogs tend to defecate more when not on a lead and if the owner is not present, however this could be an effect of conditioning by the owner. Dogs do show some investigatory behaviour towards the faeces of other dogs but there is no evidence for olfactory communication. Further studies are needed on free-living dogs whose behaviour has been less modified by humans in order to confirm this.

In contrast, there is evidence that domestic dogs use urine deposits as a method of communication. One of the most obvious behaviour patterns shown by male dogs is the raised-leg urination (RLU). Studies on dogs, wolves and other canids have shown that this is used in olfactory communication. Wolves tend to urine mark mainly at elevated sites during RLU, and this is carried out primarily by the dominant male and female wolves in a pack. When changes in the dominance order occur the new *alpha* individuals show similarly high frequencies of urination. All wolves show a high rate of investigation of such scent marks, and they may also help intruders to avoid encounters with resident animals. Rothman & Mech (1979) observed very few RLUs in lone wolves. In the domestic dog, raised-leg urination results in small quantities of urine being left at numerous locations. Dogs can adopt a wide variety of postures when urinating and the majority of these can be considered as scent marking postures. This, however, makes it difficult to distinguish between scent marking and elimination. Macdonald (1985) suggested that the volume of urine voided should be used to distinguish between these behaviour patterns in canids and, although there are problems with this criterion, it is a useful general rule.

The most extensive studies on the urination behaviour of domestic dogs has been carried out on free-ranging dogs in the USA by Bekoff (1979, 1980). Of all the urination observed, males performed RLUs 97.5% of the time and females squatted to urinate 67.6% of the time. Both sexes did, however, perform each posture on occasions. Males tended to urinate more than the females and a greater proportion of their urination could be classified as scent marking.

Some dogs will perform RLUs without the production of any urine; a behaviour termed 'raised-leg display' (RLD). It is not known whether the RLD is distinct from RLUs or simply reflects an empty bladder, although Bekoff showed that a dog was more likely to perform a RLD than an RLU when other dogs were visible.

Domestic dogs will also overmark the urine of other dogs, a behaviour commonly observed in canids. This behaviour occurs in both males and females, and dogs can often be observed 'queuing' behind another dog urinating in order to overmark the same site. Similar behaviour has been reported in wolves where packs tend to over-mark the urine of lone wolves, but not vice versa. This again suggests that urine marks are used to denote territories or home ranges and that by over-marking, 'strange' odours are masked.

Many male dogs, and to a lesser extent females, scratch the ground with the hind legs after urinating or defecating. This behaviour, called ground scratching, could serve a number of different functions. It is possible that such behaviour is designed to spread the scent, although in practice the mark is not often hit. Alternatively, it has been suggested that the scratching action itself may leave scent in the environment produced by either interdigital glands, sweat glands on the foot pads, or sebaceous glands in the fur between the toes. Interestingly, northern wolves do not appear to possess sweat glands on their pads (Sands, Coppinger & Phillips, 1977), although they also exhibit ground scratching, suggesting that this is a less probable explanation for this behaviour, although interdigital glands may still be involved. Bekoff (1979) has suggested that ground scratching may be a visual method of communication. This could either function directly by the behaviour itself or via scratch marks in the ground. Interestingly, wolves have not been observed scratching after a squat urination, only after RLU or defaecation. In addition RLU is performed predominantly by high ranking males, supporting the hypothesis that it is used in the communication of rank and/or territory (Peters & Mech, 1975). There is, however, no information on whether ground scratching is performed by lone wolves, which might help to disentangle these two factors.

Urine is not only used to indicate residency or status. Male dogs are able to detect a bitch in oestrous over large distances simply by the smell of her urine (Doty & Dunbar, 1974). The female's urine contains pheromones, probable metabolites of oestrogen, and since bitches in oestrus seem to urinate more frequently than dioestrous bitches, the pheromone is spread over a wider area.

Anal sac secretions

All species of canid, including the domestic dog, possess anal sacs. Anal sacs are paired reservoirs, one either side of the anus, that lead into ducts opening near to the anal orifice. The sacs store secretions from apocrine glands and a few sebaceous glands that are confined to the wall of the duct. The contents of anal sacs are discharged during defaecation. Natynczuk, Bradshaw & Macdonald (1989) showed that there were differences in the volatiles and the constituent compounds between different groups of individuals. This suggests possible sex and/or genetic differences that individual dogs could use in their assessment of others. The secretions also differed considerably between dogs and with variations from day to day in rate of secretion, colour and general odour (Doty & Dunbar, 1974; Bradshaw, Natynczuk & Macdonald, 1990). These findings suggest that anal sac secretions are highly individual-specific and may be important in individual and territorial recognition. They may function either via a learned association between the smell of the secretion and the dog's presence, or via odour matching between the smell of the secretion and the smell of the dog. Since an individual's secretions change subtly over time, any such association or matching would need to be continually updated by the recipient to be effective.

In captive wolves, anal sac secretions are left on faeces predominantly by the dominant male, suggesting a territory or rank-advertising role, although there is no difference between females of varying rank. Because of problems in their detection, no thorough studies have been carried out into the use of anal sac secretions by wild wolves.

General odours

General body odours are produced by a variety of skin glands. Sebaceous glands produce oily secretions that are more long lasting, whereas sudoriferous glands produce shorter living watery secretions. Sudoriferous glands can be further divided into eccrine glands, the sweat glands found on the feet of dogs, and apocrine glands that are more widely distributed over the body.

Apocrine glands tend to be more dense around the head, the anal region, the upper surface of the base of the tail (supracaudal gland) and the perineum. When dogs are shown life-size paintings of dogs they will approach these areas of higher gland density and sniff them (Fox, 1971). This suggests that these areas play a role in olfactory communications (Fig. 8.3).

Fig. 8.3. Even dogs that live as a group will frequently attempt to sniff one another, although it is unclear how they benefit from the olfactory information they gain. Photograph: Steve Wickens.

Intragroup interactions

With the possible exception of feral dogs and dingoes, humans essentially control the age/sex class composition of groups of domestic dogs. The access of males to females is often restricted so that pairings can be controlled, and while several females can often coexist peacefully, expression of the aggression that often occurs between males is rarely allowed, at least not to the extent that it might be resolved and a male dominance hierarchy be established. Thus the groups of dogs that can be readily studied at close quarters are precisely those in which the greatest degree of artificiality occurs. How important this artificiality might be, can be judged by comparing the structure of such groups to that of the typical wolf pack.

There is some disagreement about the key features of wolf sociality; the description that follows is drawn from those of Zimen (1982) and Ginsburg (1987). The pack often consists of a breeding pair, the *alpha* individuals and their offspring. The *alpha* pair more or less successfully suppress breeding in the rest of the pack by agonistic behaviour (Packard *et al.*, 1985). Two parallel hierarchies can be detected in the pack, one male and one female. Both are essentially pyramidal in structure, since rank differences are most obvious between high-ranking individuals, and are less distinct between middle-ranking adults and between the pups. There is generally a close relationship between age and rank, the oldest animals occupying the top of each hierarchy. Cross-sex dominance relationships between males and females of similar rank are weak or non-existent. The *alpha* female is highly aggressive towards other females in her pack before and during the mating season, apparently in order to prevent them from breeding. The *alpha* male tends to be highly aggressive towards intruders, but not to other pack members. A *beta* male can sometimes be distinguished, and an individual with this rank will often be the most aggressive male in the pack, but will reserve aggression towards the *alpha* male for direct challenges to his leadership. Low-ranking wolves tend to be sociable both inside and outside the pack.

Given that wolves appear to have single-sex social hierarchies, the biased sex-ratios of most groups kept by dog breeders should not prove a barrier to the establishment of hierarchies in these groups, with the obvious proviso that only one (generally the female) hierarchy will be detectable. It would be difficult to identify an *alpha* female in a dog group by her suppression of the mating of other females, since sexual activity is generally controlled by the owner. However, the status of high- and low-ranking wolves can be readily identified by behaviour patterns that are characteristic of rank, and are not retained by individual wolves when they change position in the pack (Table 8.2). Expression of similar behaviour patterns by dogs might indicate a similar dominance structure to that of wolves, even in the absence of functional correlates, such as breeding success. The morphological differences between wolves and dogs of different breeds evidently contribute to a reduction in the effectiveness of some aspects of these displays, for example the raising of the hackles. Others may have disappeared through artificial selection at the CNS level of organization, so that although the necessary external structures are present, the corresponding reflexes are absent.

It is usually assumed that groups of dogs set up dominance hierarchies that reduce aggression between individual members (e.g. Scott & Fuller, 1965), and may conceivably influence reproductive success. In studies of canine social behaviour, the status of each individual is often established experimentally by means of pairwise competitions over an indivisible resource, such as a bone. Such tests do not, however, necessarily establish that one individual is socially dominant over another, since it is possible that escalation of the conflict is avoided by each dog assessing the other's resource holding potential (*sensu* Parker & Rubenstein, 1974) and thereby the most likely outcome of a fight (see Bernstein, 1981; Parker, 1984). Such assessments could (theoretically) take place on the first occasion that two individuals meet, and need not depend on, and certainly would not prove the existence of, any underlying social structure. Competitions between male dogs often result in fights, the winner of which is usually the heavier animal (Scott & Fuller, 1965). Although not conclusive, this is exactly the outcome predicted if there were no male hierarchy, and all conflicts were resolved by direct assessment of resource holding potential. Females establish pairwise relationships on the basis of vocalizations and threats, which could again indicate either an underlying social structure, or simply assessments of each others' abilities to obtain and defend resources. Hierarchies constructed

Table 8.2. *Dominant and submissive behaviour patterns in wolves*

Dominant behaviours	Submissive behaviours
Dominant pose: stiff, tall stance; ears up or forward, tail out or up	*Submissive pose*: crouched posture, ears flat, tail tucked, forehead smooth, pulling corners of the mouth back ('grinning'), licking or extruding the tongue, lowering and averting the gaze and/or head
Feet on: dominant places its forelegs across the shoulders of a subordinate	*Submissive arched posture*: back very arched and neck curved down and to the side, head low, muzzle extended up, tail tucked, ears flat; often leads to dropping to the ground and lifting hind leg to expose inguinal region with tail tucked and ears flat
Muzzle pin: dominant either bites or grabs subordinate's muzzle, forcing it to the ground and keeping it there	*Submissive sit*: sitting back, tucking chin into chest and sometimes pawing at the dominant and averting gaze and/or head.
Stand across: dominant stands stiffly across the forequarters of a lying subordinate	

Source: Adapted from Mech (1970); Jenks & Ginsburg (1987).

on the basis of the outcome of these contests tend to be linear in the more aggressive breeds but, in less aggressive breeds, several individuals can appear to hold the same rank. Breed differences are also apparent in the contexts over which dogs will compete. For example, Shetland sheepdogs seem to possess a well-defined dominance structure when competing for space, but not for food, while basenjis show the opposite tendency (Scott & Fuller, 1965).

The term 'dominance', when applied to domestic dogs, has been used repeatedly to describe the outcome of all contests, without specifying whether these are determined by an underlying social structure, or temporary asymmetries between pairs of individuals. Where social dominance is fully expressed, status itself becomes the focus of much competition (Bernstein, 1981), as can be seen in the dominance and submissive displays of wolves. Because social dominance is often confused with aggression, there is little published information that relates directly to the idea of status within groups of domestic dogs. Steve Wickens, a graduate student working with the first author, has been investigating possible indicators of status in single-breed groups of dogs. With the exception of breeds such as huskies, which regularly use many of the signals seen in wolf packs (see e.g. Nott & Bradshaw, 1994), signalling within well-established groups of domestic dogs seems to occur infrequently. Our studies of interactions between unfamiliar dogs (see below), as well as everyday experience of dogs, tells us that many breeds are capable of making most of the signals in Table 8.2, morphology permitting. The reasons for the paucity of signalling within many established groups are therefore unclear.

In one group of female Cavalier King Charles spaniels, overt visual signals and aggressive behaviour were rarely observed, and the only common indicator of competition was the extent to which each dog deferred to another when one or both were moving towards a resource or goal. We have termed this behaviour pattern 'displacement'. Within dyads, there was nearly always a significant asymmetry in this measure, such that a hierarchy could be constructed (Fig. 8.4). The structure that emerged consisted of *alpha* and *omega* individuals (BW1 and TW1 respectively), with a less clearly-defined set of relationships in between, in which some individuals (e.g. BW2 and TW2) held similar rank. This hierarchy was independent of the context under which it was measured, suggesting that it does indeed reflect an underlying social structure (no significant deviations emerged whether the dogs were competing over food, their owner, space, or even when there was no obvious reason for competition to take place). However, there was little indication that any of the

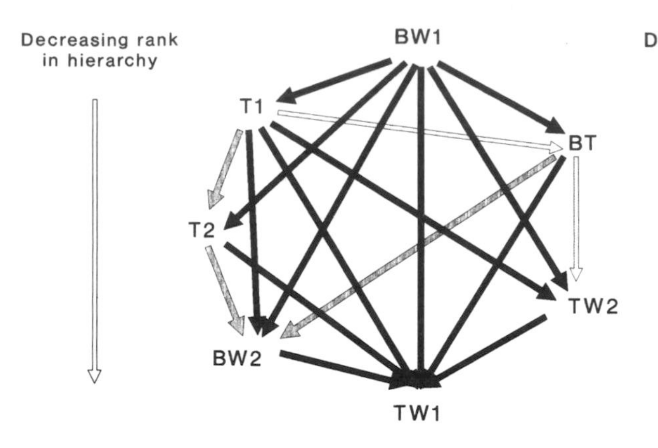

Fig. 8.4. Hierarchy derived from ratios of displacements (N = 2607) recorded between seven Cavalier King Charles bitches. Arrows point towards the dog that made the smaller number of displacements within each pair. Solid arrows indicate a significant deviation from a 1:1 ratio by chi-square goodness-of-fit statistic at $P < 0.01$, heavily shaded arrows at $P < 0.025$, and lightly shaded at $P < 0.05$.

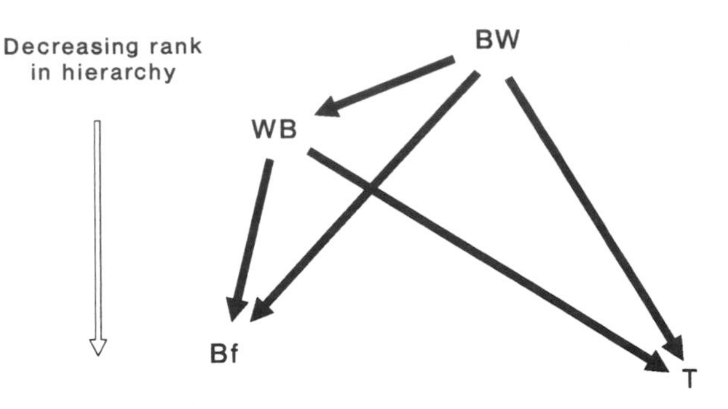

Fig. 8.5. Hierarchy derived from ratios of displacements (N = 901) recorded between four French bulldog bitches. Arrows point towards the dog that made the smaller number of displacements within each pair; all are significantly different from a 1:1 ratio by chi-square goodness-of-fit statistic.

dogs distributed their displacements disproportionately among other members of the group. It might be expected that competition for status would result in high rates of interaction between pairs of individuals of similar rank, and lower rates between those of disparate rank (Jenks & Ginsburg, 1987), and yet, BW1 interacted with BW2 as often as with T1.

The position of each dog in the hierarchy appeared to be affected by age. The oldest dog, TW1 (10 years), occupied the *omega* position, and the next oldest, BW2 and TW2 (7 years), the rank immediately above. The *alpha* dog (BW1) was five years old, and the two second-ranking dogs T1 and BT were two and three years old respectively. Several changes of rank must have occurred within this group, since BW1 was the daughter of BW2, who was a littermate of TW2, and daughter of TW1.

The importance of distinguishing aggressive from status-indicating interactions can be illustrated using Wickens' study of a group of French bulldogs. The group consisted of four females and a young male (Bm), born to one of the females (Bf). Again, based on displacements, an *alpha* female (BW) could be detected (Fig. 8.5), which, at four years old, dominated both older (Bf and T) and younger (WB) females. Since only T produced young during the study, status as revealed by displacements was unrelated to breeding success, although Bf attempted to take over and exclude T from this litter.

While the relative rankings of the females were unaffected by the context in which the displacements occurred, the position of the male Bm in relation to the females did change from one situation to another. Overall, while he was dominant over Bf and T, his rank compared to BW and WB was unclear. In the contexts of competition for access to a familiar male dog, and a novel object, Bm was dominant to all the females, but BW was dominant over Bm for some food-related contexts. Given that male and female hierarchies are only loosely interconnected in wolves, these inconsistencies do not contraindicate the existence of a (female) hierarchy in this group.

Out of the 45 aggressive encounters observed during social interactions within this group, 36 occurred between BW , WB and Bm, the individuals with the highest rank. Bm, in particular, was also aggressive towards outsiders, both male and female. Aggression within the group was observed when pairwise competitions for food items were staged, although the aggressor was often the eventual loser of the contest. As is consistent with the idea of an underlying social organization based upon status, aggression does not appear when the outcome of an encounter can be predicted from rank.

Whether or not the measures described above eventually prove to be useful for determining social structure in domestic dogs in general, they have already highlighted the futility of applying criteria used to detect dominance in wolves (the necessary behaviour patterns simply do not occur in some breeds), and the importance of treating the domestic dog as a distinct species (or set of 'species').

Development of dominance

During the socialization period of domestic dog puppies it is likely that a great deal more is learned than simple species identity. If the members of a litter remain in the same group for their entire lives, then it may be important to establish the basis of a dominance hierarchy at this early stage, as it appears to be in wolves (Fox, 1972). Even if this is not the case behaviour patterns learned from competition with siblings are likely to be useful in the process of immigration into other groups.

Behaviour that is recognisably competitive, and which might be involved in the establishment of dominance, first appears within litters when the pups are about three to four weeks old. If the pups remain together, each pairwise relationship becomes stable by about the eleventh week (Scott & Fuller, 1965). During the intervening period, our own research, and that of Wright (1980) has demonstrated considerable instability in the relationships between individual littermates; far more than might be expected to underlie a straightforward progression towards a stable hierarchy.

Nightingale (1991) and Hoskin (1991), studying litters of Border collies and French bulldogs respectively, found that some individual puppies would move from the top to the bottom of the competitive hierarchy, and back again, within a matter of days. For example, in five sets of pairwise competitions for a toy, held among the collies over a two-and-a-half week period, winners on one occasion were as likely to lose on the next as they were to win (Fig. 8.6). Such reversals were also apparent from observations of social play within both groups, although no hierarchy could be detected until the pups were about six weeks old (see e.g. Fig. 8.7(*a*)). For example, although Mb was clearly dominant in social play at 63 days (Fig. 8.7(*b*)), he had occupied a subordinate position to all his littermates except Fm just nine days previously. This represented a statistically significant change in the ratio of dominant postures given: received by Mb ($\chi^2 = 18.4$, d.f. = 1, $P < 0.0001$).

Social play and pairwise competitions also produced contradictory hierarchies, even when recorded on the same day (Fig. 8.8). None of these hierarchies, however, was clear-cut; the male Ma was unable to monopolize possession of a toy in any of his three contests, whereas during play none of the other pups dominated him. Similar, if less extreme, discrepancies between these two methods were noted by Wright (1980), who maintained that such tests measured two types of dominance, which he termed 'social dominance' and 'competitive dominance'. Certainly, during the socialization period from 3–10 weeks (Scott, 1962) there is little evidence for any underlying hierarchy based upon status, which could later emerge and stabilize during the juvenile period. Rather, these social tussles are perhaps better regarded as rehearsals for roles to be played later in life (Martin, 1984; see also Serpell & Jagoe, Chapter 6).

These rapid changes in the expression of dominance behaviour by individuals within a litter may help to explain the lack of success of so-called 'puppy tests' at predicting aggressive or owner-directed dominance behaviour later in life (Beaudet & Dallaire, 1993; M. S. Young, personal communication). If such tests are to have any value, they should be carried out later, when the 'personalities' of the pups have stabilized. Research is urgently needed to pinpoint the age when this occurs, and whether there is a link between this type of maturation and the end of physical growth, which varies greatly between breeds.

Interactions between dogs from different groups

Dogs vary in the extent to which they behave aggressively towards other dogs, and some of this variation can be attributed to breed differences (Hart & Hart, 1985; Hart, Chapter 5; Lockwood, Chapter 9). Although some dogs are kept isolated from their conspecifics, either because of aggressive tendencies, or through their owners' choice, many are allowed to encounter other dogs freely during exercise periods. The behaviour patterns that make up such encounters suggest further changes that have resulted from domestication. Visual communication appears to play little part in many of these interactions, possibly because dogs of different breeds have incompatible visual signals due to the modification of their signalling structures.

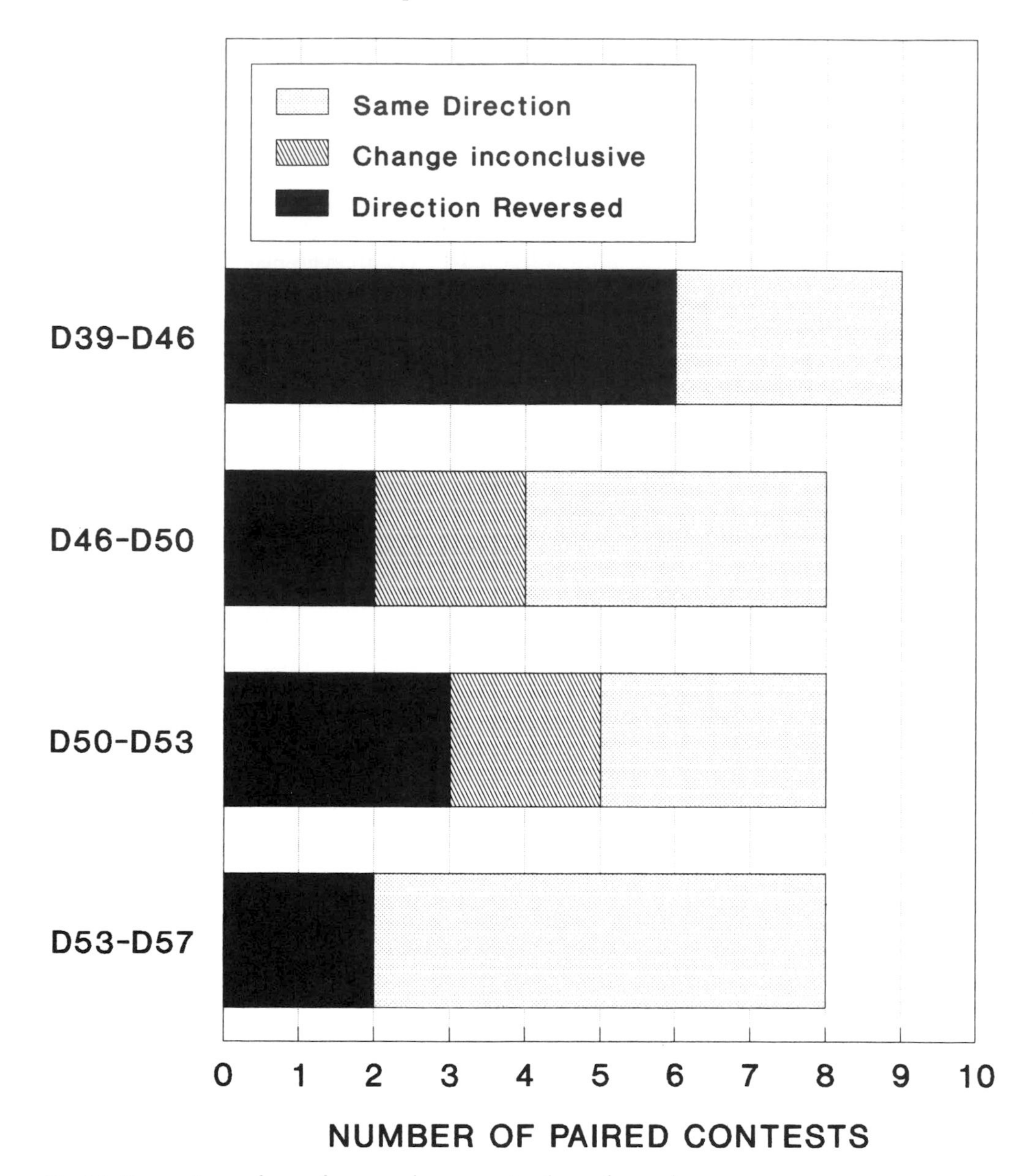

Fig. 8.6. Comparisons of sets of contests between pairs drawn from a litter of five Border collie puppies. On days 29, 46, 50, 53 and 57 postpartum, all possible pairs of the pups were presented with a toy, initially placed equidistant from each, and their behaviour was recorded for the next minute (on each day except day 39, one result had to be discarded due to the inactivity of a pup). A 'win' was recorded if one pup retained possession of the toy for twice as long as the other; usually (33/46 tests) one of each pair monopolized the toy, and only two inconclusive results were recorded, both on day 50. The graph shows the number of pairs for which a win on one test day was followed by the same, or the reverse, result on the next test day.

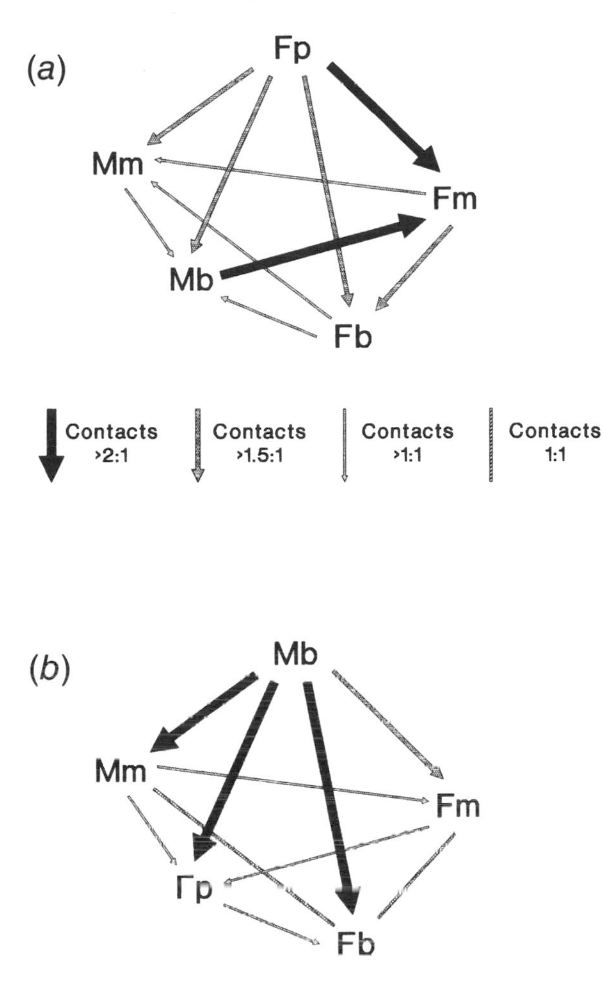

Fig. 8.7. Hierarchies derived from social play between five French bulldog puppies (3 females, 2 males, denoted F and M), at 28 days (*a*) and 63 days postpartum (*b*). For each pup, the total number of dominant postures (bite, paw, kick, head over back, stand over), and their target, were recorded during several bouts of social play between all five puppies [N(*a*) = 355; N(*b*) = 536]. The arrows are in the direction dominant to subordinate, and their thickness indicates the degree of asymmetry within each pairwise relationship (see key). Note that at 28 days it is impossible to arrange the pups in a linear hierarchy; the apparently dominant position of Fp may be due to chance.

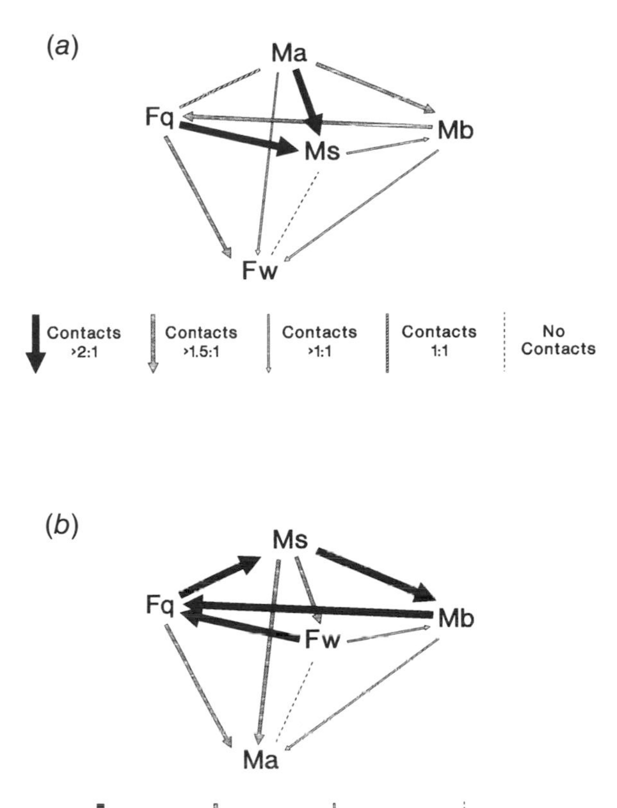

Fig. 8.8. Hierarchies derived for five Border collie puppies (3 males, 2 females, denoted M and F) at 50 days postpartum. (*a*) Partly transitive hierarchy derived from social play (scored and illustrated as indicated in the caption to Fig. 8.7). Kendall rank correlation coefficient between pairwise scores in (*a*) and (*b*) = 0.217; $P < 0.395$. (*b*) Circular hierarchy from outcomes of pairwise competitions over a toy. The arrows point from the pup that took most possession of the toy, and their thickness indicates the degree of monopolization.

The majority of interactions seem to focus on exchanges of olfactory information, in which each dog sniffs the head and anogenital area of the other. Although this is in itself unremarkable, the precise sequence in which the patterns occur is more revealing. Sniffing tends to progress from the head towards the tail, but whether or not this progression is completed, the dog that is being sniffed is the one that is most likely to terminate the interaction (see Fig. 8.9). The dog that makes the initial approach (the Initiator) is the one most likely to sniff the other (the Recipient); only rarely does the Recipient sniff the Initiator while avoiding being sniffed itself (Bradshaw & Lea, 1993). Thus each dog appears to be trying to gain olfactory information about other dogs, while avoiding giving away olfactory information about itself.

Male dogs are more likely to sniff the anogenital area than are females, irrespective of the sex of the partner. This is also true of olfactory communication

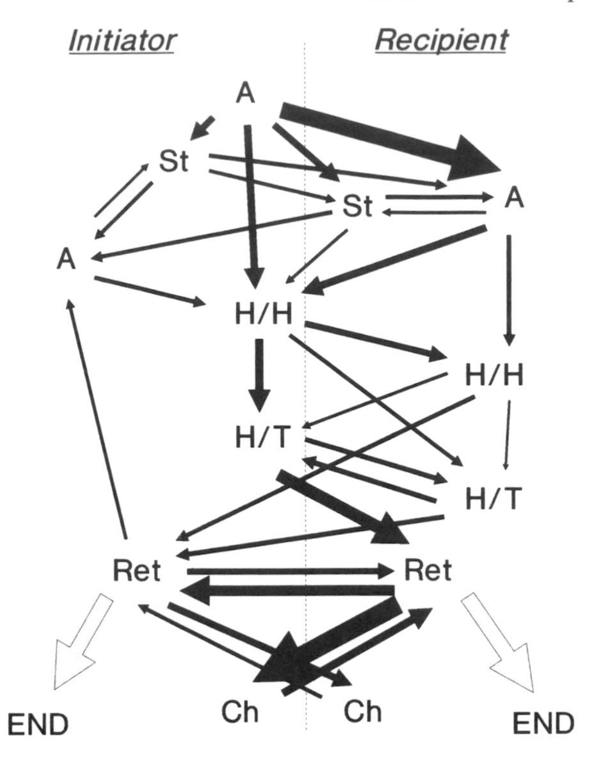

Fig. 8.9. Sequences of behaviour patterns observed between pairs of dogs interacting spontaneously during exercise, constructed from significant ($P < 0.05$) first-order transitions from one pattern to another. Only sequences containing five or more components are included ($N = 218$). The dog making the first approach is classified as the initiator. A, Approach; St, Stare; H/H, sniff head; H/T, sniff anogenital area; Ret, retreat; Ch, chase. The width of the arrows is proportional to the frequency of the transitions. Three additional behaviour patterns with significant transitions are omitted from the diagram for clarity (see Bradshaw & Lea, 1993).

between wolves (Mech, 1970). However, there are also differences between the sequences described for dogs, and those typical of wolves. First of all, only low-ranking wolves are sufficiently unaggressive to strangers to allow olfactory inspection to take place (Fox, 1973). Second, high-ranking wolves regularly stand and present their anal regions for sniffing by subordinate members of their pack; highly subordinate wolves often attempt to thwart being sniffed by covering their anal regions with their tails, a posture also seen in dogs, particularly females. Since, in the encounters between dogs (as recorded in Fig. 8.9) both animals try to avoid anogenital inspection by the other, they appear to be behaving like subordinate wolves.

Summary

Studies of the social systems of companion dogs per se are not yet sufficiently numerous for any conclusions to be more than tentative. In particular, the diversity of the different breeds of dog suggests that there may be several types of social system, the complexity of which may vary both quantitatively and qualitatively. The extent to which wolf-like characteristics are exhibited varies not only between breeds, but also from one situation to another. In intergroup encounters, dogs behave somewhat like subordinate wolves. They seem to place more emphasis upon chemical than visual signalling; indeed the mere fact that giant breeds appear to recognize even toy breeds as conspecifics, and vice versa, suggests that species identity may be more encoded in smells than in appearance. Visual signals between members of different breeds may be unreliable due to modification of the signalling structures by selective breeding. Sniffing appears to be the goal of many dog–dog encounters, and the diligence with which males urine-mark also suggests that both personal and deposited odours are an important mode of communication.

Within permanent groups of the more refined breeds, wolf-type signalling seems to be rare. Female hierarchies can be postulated on the basis of displacements observed between bitches that live together, and the lack of dependence of these hierarchies on the context under which they are measured argues that they are genuinely based upon each individual's status within the group. This status appears to be age-related, but the *alpha* position is not always held by the oldest individual in the group. Comparisons with wolves in this respect may be misleading however, since human intervention prolongs the lifespan of individual dogs compared to that likely in the wild. The means by which bitches progress from puppyhood to the *alpha* position has yet to be documented, but studies of dominance interactions during the socialization period indicate that at this stage no consistent hierarchy exists; the critical events therefore presumably take place after the end of the second month of life.

Acknowledgements

Our outstanding gratitude goes to Betty Slightam, Sandy Hawkins, Sue Hull and Richard Davies for their assistance and patience during studies of their dogs. We also thank Steve Wickens, Sarah Brown, Rory Putman, Ray Coppinger, David Macdonald and Ian Robinson for valuable discussions. JWSB gratefully acknowledges the financial support of the Waltham Centre for Pet Nutrition and the Science and Engineering Research Council for parts of the research described.

References

Abrantes, R. A. B. (1987). The expression of emotions in man and canid. In *Canine Development throughout Life*, Waltham Symposium No. 8, ed. A. T. B. Edney. *Journal of small Animal Practice*, **28**, 1030–6.

Beaudet, R. & Dallaire, A. (1993). Social dominance evaluation: observations on Campbell's test. *Bulletin on Veterinary Clinical Ethology*, **1**, 23–9.

Beaver, B. V. (1981). Friendly communications by the dog. *Veterinary Medicine: Small Animal Clinician*, **76**, 647–9.

Beaver, B. V. (1982). Distance-increasing postures of dogs. *Veterinary Medicine: Small Animal Clinician*, **77**, 1023–4.

Bekoff, M. (1977). Social communication in canids: evidence for the evolution of a stereotyped mammalian display. *Science*, **197**, 1087–9.

Bekoff, M. (1979). Scent-marking by free ranging domestic dogs. Olfactory and visual components. *Biology of Behaviour*, **4**, 123–39.

Bekoff, M. (1980). Accuracy of scent mark identification for free-ranging dogs. *Journal of Mammalogy*, **57**, 372–5.

Bernstein, I. S. (1981). Dominance: the baby and the bathwater. *The Behavioral and Brain Sciences*, **4**, 419–57.

Blackshaw, J. K. (1985). Human and animal inter-relationships. Review series 3: Normal behaviour patterns of dogs. Part 1. *Australian Veterinary Practitioner*, **15**, 110–12.

Bradshaw, J. W. S. (1992). Behavioural biology. In *The Waltham Book of Dog and Cat Behaviour*, ed. C. J. Thorne. pp. 31–52. Oxford: Pergamon Press.

Bradshaw, J. W. S. & Brown, S. L. (1990). Behavioural adaptations of dogs to domestication. In *Pets, Benefits and Practice*, Waltham Symposium No. 20, ed. I. H. Burger, pp. 18–24. *Journal of small Animal Practice*, **31**(12) suppl.

Bradshaw, J. W. S. & Lea, A. M. (1993). Dyadic interactions between domestic dogs during exercise. *Anthrozoös*, **5**, 234–53.

Bradshaw, J. W. S., Natynczuk, S. E. & Macdonald, D. W. (1990). Potential for applications of anal sac volatiles from domestic dogs. In *Chemical Signals in Vertebrates*, 5, ed. D. W. Macdonald, D. Müller-Schwarze & S. E. Natynczuk, pp. 640–4. Oxford: Oxford University Press.

Coppinger, R. P. & Feinstein, M. (1991). Why dogs bark. *Smithsonian Magazine*, January 1991, 119–29.

Doty, R. L. & Dunbar, I. F. (1974). Attraction of beagles to conspecific urine, vaginal and anal sac secretion odours. *Physiology and Behaviour*, **12**, 325–833.

Fox, M. W. (1971). *Behaviour of Wolves, Dogs and Related Canids*. London: Jonathan Cape.

Fox, M. W. (1972). Socio-ecological implications of individual differences in wolf litters: a developmental and evolutionary perspective. *Behaviour*, **41**, 298–313.

Fox, M. W. (1973). Social dynamics of three captive wolf packs. *Behaviour*, **47**, 290–301.

Fox, M. W. (1978). *The Dog: Its Domestication and Behaviour*. New York: Garland STPM Press.

Fox, M. W. & Bekoff, M. (1975). The behaviour of dogs. In *The Behaviour of Domestic Animals*, 3rd edn. ed. E. S. E. Hafez, pp. 370–409. London: Baillière Tindall.

Ginsburg, B. E. (1987). The wolf pack as a socio-genetic unit. In *Man and Wolf*, ed. H. Frank, pp. 401–13. Dordrecht, The Netherlands: Dr W. Junk.

Hart, B. L. (1983). Flehmen behaviour and vomeronasal organ function. In *Chemical Signals in Vertebrates*, 3, ed. D. Müller-Schwarze & R. M. Silverstein, pp. 87–103. New York: Plenum Press.

Hart, B. L. & Hart, L. A. (1985). Selecting pet dogs on the basis of cluster analysis of breed behaviour profiles and gender. *Journal of the American Veterinary Medical Association*, **186**, 1181–95.

Hoskin, C. N. (1991). Development of the Dominance Hierarchy Amongst a Litter of French Bulldog Pups. Unpublished B.Sc. thesis, University of Southampton.

Jenks, S. M. & Ginsburg, B. E. (1987). Socio sexual dynamics in a captive wolf pack. In *Man and Wolf*, ed. H. Frank, pp. 375–99. Dordrecht, The Netherlands. Dr W. Junk.

Klinghammer, E. & Laidlaw, L. (1979). Analysis of 23 months of daily howl records in a captive grey wolf pack (*Canis lupus*). In *The Behaviour and Ecology of Wolves*, ed. E. Klinghammer, pp. 153–81. New York: Garland STPM Press.

Macdonald, D. W. (1985). The carnivores: Order Carnivora. In *Social Odours in Mammals*, Vol. 2, ed. R. E. Brown & D. W. Macdonald, pp. 619–722. Oxford: Clarendon Press.

Martin, P. (1984). The (four) whys and wherefores of play in cats: a review of functional, evolutionary, developmental and causal issues. In *Play in Animals and Humans*, ed. P. K. Smith, pp. 71–94. Oxford: Basil Blackwell.

Mech, L. D. (1970). *The Wolf: The Ecology and Behaviour of an Endangered Species*. New York: Natural History Press.

Moehlman, P. D. (1989). Intraspecific variation in canid social systems. In *Carnivore Behaviour, Ecology and Evolution*, ed. J. L. Gittleman, pp. 143–63. New York/London: Chapman & Hall.

Natynczuk, S., Bradshaw, J. W. S. & Macdonald, D. W. (1989). Chemical constituents of the anal sacs of domestic dogs. *Biochemical Systematics and Ecology*, **17**, 83–7.

Nightingale, A. (1991). The Development of Social Structure During the Primary Socialisation Period in Border Collies. Unpublished B.Sc. thesis, University of Southampton.

Nott, H. M. R. & Bradshaw, J. W. S. (1994). Companion animals. In *Video Techniques in Animal Ecology and Behaviour*, ed. S. D. Wratten, pp. 145–61. London: Chapman & Hall.

Packard, J. M., Seal, U. S., Mech, L. D. & Plotka, E. D. (1985). Causes of reproductive failure in two family groups of wolves (*Canis lupus*). *Zeitschrift für Tierpsychologie*, **68**, 24–40.

Parker, G. A. (1984). Evolutionarily stable strategies. In *Behavioural Ecology: an Evolutionary Approach*, 2nd edn, ed. J. R. Krebs & N. B. Davies, pp. 30–61. Oxford: Blackwell Scientific Publications.

Parker, G. A. & Rubenstein, D. I. (1974). Role assessment, reserve strategy, and acquisition of information in asymmetrical animal conflicts. *Animal Behaviour*, **29**, 221–40.

Peters, R. P. & Mech, L. D. (1975). Scent marking in wolves. *American Scientist*, **63**, 628–37.

Rothman, J. and Mech, L. D. (1979). Scent marking in lone wolves and newly formed pairs. *Animal Behaviour*, **27**, 750–60.

Sands, M. W., Coppinger, R. P. & Phillips, C. J. (1977). Comparisons of thermal sweating and histology of sweat glands of selected Canids. *Journal of Mammalogy*, **58**, 74–8.

Schenkel, R. (1967). Submission: its features and function in the wolf and dog. *American Zoologist*, **7**, 319–29.

Scott, J. P. (1962). Critical periods in behavioural development. *Science*, **138**, 949–58.

Scott, J. P. & Fuller, J. L. (1965). *Genetics and the Social Behaviour of the Dog*. Chicago: University of Chicago Press.

Wright, J. C. (1980). The development of social structure during the primary socialization period in German Shepherds. *Developmental Psychobiology*, **13**, 17–24.

Zimen, E. (1982). A wolf pack sociogram. In *Wolves of the World: Perspectives of Behaviour, Ecology and Conservation*, ed. F. H. Harrington and P. C. Pacquet, pp. 282–322. Park Ridge, NJ: Noyes Publications.

9 The ethology and epidemiology of canine aggression

RANDALL LOCKWOOD

In *Man Meets Dog* (1953), Konrad Lorenz praised the wonders of domestication that, in a few thousand years, had transformed the wolf into the docile Alsatian dog which his children could playfully and fearlessly torment. He added (p. 75):

> I have a prejudice against people, even very small children, who are afraid of dogs. This prejudice is quite unjustified for it is a completely normal reaction for a small person, at the first sight of such a large beast of prey, at first to be anxious and careful. But the contrary standpoint, that I love children that show no fear even of big, strange dogs and know how to handle them properly, has its justification, for this can only be done by someone who possesses a certain understanding of nature and our fellow beings.

Lorenz admitted in his later years that much of what he had written about dogs was simply wrong. His assumption that domestication had largely purged the wolf of the behavior that made it potentially dangerous to man was one of his more serious errors.

For many years the phrase 'dog bites man' was a cliche for an event that is the antithesis of news, largely because it is such a common occurrence. Recently, however, media around the world have given enormous attention to dog attacks. This has created the popular impression that such attacks have become more numerous or severe.

Dog bites can affect anyone, from commoners to Queen. Recent articles in the *Washington Post* seriously raised the question of whether the Royal corgis should be allowed into the United States, given their well-publicized penchant for biting. From an epidemiological perspective, dog bite *is* a problem of epidemic proportions, affecting more than 1% of the US population annually and accounting for widespread exposure to many zoonotic diseases (Greene, Lockwood & Goldstein, 1990) and more than 20 fatalities each year. Yet it is a problem that for years has been described by public health officials as an 'unrecognized' epidemic (Harris, Imperato & Oken, 1974).

Several factors have led to increased recognition of the problem. First, a growing body of epidemiological reports have clearly described the extent of the issue (Beck, Loring & Lockwood, 1975; Lockwood & Beck, 1975; Berzon, 1978; Beck, 1981; Pinckney & Kennedy, 1982; Sacks, Sattin & Bonzo, 1989). Second, there has been widespread reporting of some of the more shocking fatal dog attacks in the media. Third, a growing number of bite cases have been brought before the courts. In the US, settlements in excess of $1 million and imprisonment of dog owners on charges of manslaughter have not been uncommon. Finally, a significant proportion of fatal and severe bites have been attributed to a relatively small number of breeds including pit bulls and Rottweilers. This has resulted in highly publicized efforts to restrict such breeds, with resulting conflicts between dog owners and authorities.

This chapter will first review the natural history of canid aggression, and some of the biological factors involved in bite incidents. It will then consider the general epidemiological findings for non-fatal attacks and recent dog-bite fatalities. Finally, some possible solutions to this problem will be proposed.

Why canids bite

Biting is obviously a key component of predatory behavior in canids. However, most social canids show surprisingly low levels of intra-specific aggression. Despite the strong restraint on the use of aggression, biting can occur in many contexts including expressions of dominance, territorial defense, food-competition, protection of young or other pack members, pain-elicited aggression and fear-elicited aggression. Dog attack can occur in any of these contexts, and may also involve components of inter-specific predatory behavior.

It is important to recognize that artificial selection, which has resulted in the production of various breeds of dogs, frequently produces exaggerated physical or behavioral characteristics that would be maladaptive in free-living wild canids. For example, racing breeds such as greyhounds and whippets can outrun most wolves, yet the changes mankind has produced in these animals would render them virtually helpless in the world of the wild wolf.

A major human objective in the production of dog breeds has been the creation of animals more aggressive than their wild ancestors. This has been done to provide protection through inter-specific aggression (e.g. most guarding breeds) or for 'entertainment', in the form of the heightened intra-specific aggression of fighting breeds, including 'pit bull' type dogs.

Although for practical reasons there have been no comprehensive studies of the biology or ethology of fighting breeds, several biological trends have been suggested by veterinarians called upon to treat fighting animals, as well as the experiences of myself and Humane Society field investigators in working with several hundred such animals seized in actions against illegal dog fighting.

Scott & Fuller (1965) reported a genetically based decrease in the latency to show intra-specific aggression in terriers. This simply confirmed a characteristic long-associated with such breeds. Within fighting breeds this characteristic can be even more exaggerated. Among dog fighters, an animal's tendency to attack other animals, despite fatigue or injury, is termed 'gameness'. It is a quality that is strongly selected for by breeders within the 'sport', but which has not been subjected to any formal genetic analysis.

Fighting breeds also appear to have a much higher tolerance of pain, which may be mediated by peculiarities in neurotransmitters or opiate receptor sites. A single anecdotal report of unusual responsiveness to morphine and naloxone in a pit bull (Brown *et al.*, 1987) suggests that there may be physiological differences in the breed, although no definitive studies have been reported in the literature.

In addition to a lowered threshold for attack and higher pain thresholds in many fighting animals, selection for fighting has apparently resulted in the disruption of normal communication in individuals from recent fighting lineages. Under natural conditions, the aggression of wild canids is held in check by a detailed set of postural and facial signals that clearly indicate mood and intent (Fox, 1971*a*; Schenkel, 1967). In addition, aggressive encounters are normally ended rapidly when one individual emits the appropriate 'cut-off' behavior, such as infantile vocalizations (whining, yelping) and submissive displays (Fox, 1971*b*). Dogs from fighting lineages have been under selective pressures that suppress or eliminate accurate communication of aggressive motivation or intent. It is to a fighting dog's advantage for its attack to be unexpected. Many accounts of such attacks on people note that the incident occurred 'without warning'. Similarly, once initiated, such attacks are often not ended by the withdrawal of the opponent or the display of species-typical submissive behavior. Combat involving fighting dogs can continue for several hours and separation of the animals may require the use of a 'parting stick' to physically pry the animals apart.

The extent to which such characteristics are genetically determined within the fighting breeds has been the subject of considerable controversy (Lockwood & Rindy, 1987; Clifford, Green & Watterson, 1990). Although complex behaviors such as pointing, retrieving, herding and livestock guarding are generally accepted to have a strong genetic component, many fanciers of the fighting breeds attribute the comparatively simple lowering of the threshold for aggression to purely environmental influences of irresponsible owners.

It is also important to distinguish between selective influences on inter-specific vs. intra-specific aggression. Dog fighters and advocates of fighting breeds note that, historically, fighting animals that showed aggression to people were generally removed from the gene pool, either by being destroyed or being deemed unsuitable for breeding. It is true that contemporary dogs still employed in fighting are often easily handled by others (such as Humane Society investigators). However, there is no indication that the same selective pressures are in operation since there is currently a market for even the most intractable animals in the guard dog trade.

Clearly, genetic history can influence aggressiveness of breeds and individual dogs, either increasing or decreasing these tendencies. Throughout the history of dogs, many breeds such as the Irish wolfhound and Great Dane have earned a reputation for ferocity, only to become far more docile as trends in breeding shift. Indeed part of the problem with the 'pit bull' controversy is that the lineages of fighting and non-fighting animals within the fighting breeds have been separated for many generations, but have shown relatively little physical divergence. As a result, an American pit bull terrier from recent fighting stock may be physically indistinguishable from an American or English Staffordshire (bull) terrier 50 generations removed from the fighting pits, yet the two animals could be behaviorally very different.

Selective breeding can increase or decrease the tendency for dogs to bite in different contexts. Since the level of aggressiveness can be affected by several factors with likely genetic influence, including basic temperament, timidity and the presence of painful genetic disorders, it is possible for the *lack* of any directional selection in breeding to produce an increased tendency toward aggressiveness. For example, genetic factors underlying fearfulness may increase the likelihood of fear-biting. Other genetic factors contributing to painful congenital physical defects could increase pain-elicited aggression. In the United States at least 50 000 dogs are produced each year in 'puppy mills' for the mass pet trade. Usually the most popular breeds are represented in these intensive breeding operations and any animals of the desired breeds capable of producing young are likely to be bred and sold, regardless of temperament. The result has been the proliferation of physically and behaviorally unsound animals from among the most popular breeds, including those not traditionally associated with aggression to people, such as cocker spaniels, golden retrievers, malamutes and Siberian huskies. This problem has been widely documented in the American media (see Anon., 1990).

Any or all of the influences outlined above can help to account for biological predisposition of a dog toward aggression. Additional biological factors that can influence the tendency toward aggression include the animal's age, sex, reproductive status (intact vs. spayed or neutered) and overall health. However, the likelihood that a particular individual will bite is also strongly influenced by many environmental variables including the training of the animal, the extent of its socialization to people (especially children), the quality of the animal's supervision and restraint, and the behavior of the victim (Lockwood, 1986). This multiplicity of interacting factors in dog bite makes it difficult and often meaningless to base predictions of a particular animal's aggressive behavior on a single characteristic, such as breed.

The epidemiology of dog bite

Having reviewed the factors that can contribute to a dog-bite incident, let us briefly examine some epidemiological findings surrounding this problem. In the United States there is no centralized record-keeping of dog-bite incidents. Communities vary widely in the extent to which these cases are investigated and bites are generally vastly under-reported (Jones & Beck, 1984). However, a general picture of bite epidemiology has emerged from a number of comprehensive surveys including Beck *et al.* (1975), as well as reports from local animal control agencies (Miller, 1986; Moore, 1987 *in lit.*; Oswald, 1991). Additional insights can be obtained from press accounts of dog bite incidents (Lockwood & Rindy, 1987) and the study of the 'worst-case' scenarios, those attacks which involve human fatalities. An overview of such attacks in the last decade is provided by Sacks *et al.* (1989), and in-depth analysis of a smaller number of incidents is provided by Borchelt *et al.* (1983). I will also review the most recent evidence from the Humane Society of the United States (HSUS) investigations of 37 fatal dog attacks occurring during 1989 and 1990.

The victim

Age of victim

Dog bite is a health problem that disproportionately affects children. Beck *et al.* (1975) found that 38% of reported bites in St Louis involved children under nine, who constituted only 15% of the population. Adults over 50 comprised 30% of the city's population, but only 11% of the bites. All other studies show a similar overrepresentation of young children among bite victims.

Fatal attacks show a bimodal age distribution, affecting the very young and the very old. Of the 157 victims of fatal dog attack reported by Sacks *et al.* (1989), 70% were under ten years of age and 22% were less than a year, while 21% were over 50. In the 1989 and 1990 cases, 60% were under five and 25% were over 72. Most of the victims falling outside of these age ranges were in some way debilitated, including one acute alcoholic and another victim attacked while having a seizure. It is interesting to note that this pattern of attacking the very young, the very old, and the infirm is consistent with the usual selection of 'prey' by wild canids, although predation was not considered a primary motivation in many of these incidents.

Sex of victim

Non-fatal dog attack is disproportionately directed against males. In the Beck *et al.* (1975) survey, 65% of the victims were male. Moore (1987 *in lit.*) reported 59% of bite victims in Palm Beach County, Florida, were male. There is no consistent pattern in the case of fatal attacks. Pinckney & Kennedy (1982) reported only 33% of the victims of fatal dog attack to be males in their review of cases from 1975–90. In Sacks *et al.* (1989) 60% of the victims during 1977–88 were male, while HSUS 1989–90 data indicated 48% male victims. This variability may be due to the fact that the majority of fatal dog attack victims are young infants whose behavior played a less important role in the attack than in the far more numerous non-fatal attacks on older children.

Activity of victim

Under principles of Common Law there is the assumption that dogs are harmless unless they have previously demonstrated a vicious propensity. This often leads to the related assumption that victims of dog attack have provoked or otherwise precipitated the attack. However, those studies that have attempted to document the context in which an attack has occurred generally show that bite victims are rarely engaging in activity that could legally be considered provocation (i.e. teasing, tormenting or causing physical injury to the animal, or attempting to commit a crime). In the non-fatal bites surveyed by Beck *et al.* (1975) the victims had no interaction with the dog, or were walking or sitting in 75% of the cases. In 9.6% of the cases, the victim was playing with the dog and in only 6.5% of the cases could the victim's behavior be classified as provocation.

Lockwood & Rindy (1987) compared contexts reported in press accounts of non-fatal attacks by pit bulls ($N = 101$) and all other breeds ($N = 62$). In the pit bull incidents, 58% of victims were walking or had no interaction with the dog prior to attack, 19.8% were bitten coming to the aid of a person or animal that had been attacked, 7.9% were playing with the animal and 5% were provoking the animal. In the cases involving all other breeds, 48.4% involved no direct interaction, 27.4% play and 1.6% provocation.

In their report on fatal attacks, Sacks *et al.* (1989) did not provide details of victim behavior prior to the bite, but they noted that 6.9% of these incidents involved attacks on sleeping infants. The HSUS analysis of 1989–90 fatalities found 20% of the incidents involved attacks on sleeping infants, 43% occurred while the victim was walking near the dog, 30% involved play and 6.7% provocation (victims attacked during commission of a crime).

The dog

Number of animals

Most epidemiological reports do not mention the number of animals involved in non-fatal attacks. It is likely that most of these involve a single dog. Earlier investigations of dog-bite fatalities suggested that these severe incidents were more likely to involve packs of animals (Borchelt *et al.*, 1983). Recently the majority of fatal attacks have involved a single, usually large, animal. Sacks *et al.* (1989) reported that 70% of the fatal attacks from 1979–88 were by individual dogs, 20% were by two and 10% involved groups ranging from 3 to 22. The 1989–90 incidents follow an identical pattern.

Ownership of animals

The popular perception of dog bite is that it is largely a problem caused by stray dogs. Beck *et al.* (1975) pointed out the important distinction between problems caused by true strays (i.e. ownerless or feral animals) vs. *straying*, unrestrained owned dogs. Of the biting animals in that survey, 14.5% were considered stray, 5.9% were owned by the victim or victim's family and the rest were otherwise owned. Sacks *et al.* (1989) identified 70% of the dogs involved in 1979–88 fatalities as owned pets and 27% as strays. In its investigations of 1989–90 incidents, the HSUS made a greater effort to locate owners of the dogs in question for the purposes of filing criminal charges where appropriate. Of the 37 dogs in these cases, 51% were owned by the victim's family and 37% by a friend or neighbor. Only one animal (3%) was a stray with no known owner.

Restraint

Although many bites are attributed to dogs running loose, animal control officers frequently comment on the role of chaining or other restraint in producing an animal that is actually more likely to bite. Such an animal might already have a predisposition to bite (and is therefore chained), but this may only exacerbate the situation by removing opportunities for socialization and by aggravating frustration, defensive aggression and other undesirable behavior.

None of the major epidemiological surveys comment on the nature of the restraint of dogs in non-fatal attacks. In the Lockwood & Rindy (1987) survey, 42.7% of the cases of pit bull attacks involved animals that were fenced, chained or inside prior to the incident. Another 14% involved the dogs jumping fences or breaking chains. For bites involving other breeds, 26.7% of the animals were similarly restrained but only 1% involved breaking restraint.

Sacks *et al.* (1989) reported that 28% of the animals in the fatal attacks they studied were chained at the time. Of the dogs involved in fatal attacks during 1989–90, 26% were chained, 32% in the house and 32% running loose.

Sex and spay/neuter status

Since much canid aggression is under hormonal influence, and since animal control agencies make spaying or neutering of pets a significant priority, it is important to attempt to get evidence on the reproductive status of animals involved in attacks. Serious dog bite seems to be a phenomenon primarily associated with male dogs. In the Beck *et al.* (1975) survey, 70% of the biting animals were male. Moore (1987 *in lit.*) was able to collect more detailed information on biting animals, recording information on breed, sex and reproductive status. Overall, 87% of all biting animals in that survey were males and 60% were unneutered males. Of the remaining 13% of bites attributed to females, half were by unspayed females. These statistics varied somewhat with breed. The breeds most frequently associated with bites also had the highest proportions of bites attributed to males (German shepherds, 86%; pit bulls, 90%; chow chows 92%; and Rottweilers, 98%).

Breed

From an epidemiological perspective, it is difficult to draw scientifically sound conclusions about the relative dangers posed by different breeds. Accurate breed-specific bite rates are hard to obtain. Such statistics require good information for both the numerator (number of bites attributed to a particular breed) and the denominator (number of animals of that breed in the population). This requires comprehensive reports of all bites, reliable breed identification, and detailed information about the demographics of the entire dog population of the area in question. Such numbers are often unreliable since compliance with local dog licensing or registration requirements is usually below 20% in most US communities.

Several epidemiological studies attempted to draw some attention to breeds apparently associated with higher risks. Pinckney & Kennedy (1982) attempted to compute breed-specific bite rates using relative numbers of animals of different breeds registered with the American Kennel Club to compute the denominator, a procedure that is unlikely to reflect the overall United States dog population (Lockwood & Rindy, 1987).

Others have attempted to compute rates based on local registration, licensing or impound figures that are incomplete, but which should more accurately reflect breed representation in local populations. For example, Berzon (1978) reported that German shepherds made up 45% of the dogs listed in Baltimore bite reports, yet comprised only 23% of the animals registered in the city. From Miller (1986) it is possible to compute an index of the extent to which the representation of various breeds in the population of biting dogs in that area (Pinellas County, FL) deviates from their representation among the animals registered in that region. The breeds showing the greatest over-representation in the bite population were pit bulls (17.8% bite population and 3.7% of overall population = 4.81×), chow chows (2.43×), German shepherds (2.02×) and Dobermans (1.37×).

A similar analysis is provided by Moore (1987 *in lit.*), who used registration data to compute the percentage of the registered population of various breeds that are involved in bites. The highest rankings in that survey were pit bulls (12.3%), chow chows (11.4%), German shepherds (6.5%), Dobermans (4.3%) and Rottweilers (4.1%).

The relatively small numbers of animals involved in fatal attacks does not lend itself to this kind of bite-rate analysis in the absence of any national census on dog population. However, the patterns that emerge are consistent with the above findings. Sacks *et al.* (1989) reported that, of the 101 animals in their survey for which breed could be determined, pit bulls and pit bull mixes comprised 43%, German shepherds and shepherd mixes 15%, Siberian huskies, malamutes and mixes 18%, Dobermans 5%, Rottweilers 5% and wolf–dog hybrids 5%. The HSUS analysis of the 39 animals involved in fatal attacks during 1989–90 showed pit bulls and mixes comprised 25.6%, German shepherds and mixes 17.9%, Siberian huskies, malamutes and mixes 15.4%, wolf–dog hybrids 10.2% and chow chows 7.7%. All of these figures are likely to be significantly greater than their representation in the overall dog population of the United States.

Conclusions

Although dog bite is a serious public health problem, it is important to remember that such encounters represent a very small fraction of the hundreds of millions of human–dog contacts that occur each day, most of which are deeply enjoyed. Likewise, the HSUS's focus on the small fraction of dogs implicated in human fatalities should not obscure the fact that these 20 or so animals involved in such attacks each year represent an infinitesimal portion of the American dog population, less than .00004%! The proportion of American humans who kill other human beings is more than 200 times this fraction.

Humankind has made the dog in its image and, increasingly, that image has become a violent one. The breeds of dogs that have been chosen to reflect our aggressive impulses have changed over the millennia. In the last 20 years the choice has moved from German shepherds, to Dobermans, to pit bulls, to Rottweilers to a current surge in problem wolf–dog hybrids.

Problems of irresponsible ownership are not unique to pit bulls or any other breed, nor will they be in the future. Effective animal control legislation must emphasize responsible and humane ownership of genetically sound animals, as well as the responsible supervision of children and animals when they interact (Lockwood, 1988). I believe this can be encouraged in several ways:

1. By strengthening and enforcing laws against dog fighting and the irresponsible use of guard and attack dogs.
2. By eliminating the mass-production of poorly-bred and unsocialized animals in large-scale 'puppy mills'.
3. By introducing and enforcing strong animal control laws that place the burden of responsibility for the animal's actions on its owner.
4. By encouraging programs that educate the public about responsible dog ownership and the problems of dog bite.

It is possible to protect the health and safety of the public and at the same time preserve the rights of responsible dog owners. By placing greater emphasis on responsible and humane animal care, we can go a long way toward solving these problems and preserving the special human–dog relationship that has developed over thousands of years.

References

Anon. (1990). Puppy mill media blitz. *The Humane Society News*, **35**(3), 7–8.

Beck, A. M. (1981). The epidemiology of animal bite. *The Compendium on Continuing Education for the Practising Veterinarian*, **3**, 254–8.

Beck, A. M., Loring, H. & Lockwood, R. (1975). The ecology of dog bite. *Public Health Reports*, **90**, 262–7.

Berzon, D. R. (1978). The animal bite epidemic in Baltimore, Maryland; review and update. *American Journal of Public Health*, **68**, 593–5.

Borchelt, P. L., Lockwood, R., Beck, A. M. & Voith, V. L. (1983). Attacks by packs of dogs involving predation on human beings. *Public Health Reports*, **98**, 57–65.

Brown, S. A., Crowell-Davis, S., Malcom, T. & Edwards, P. (1987). Naloxone-responsive compulsive tail chasing in a dog. *Journal of the American Veterinary Medical Association*, **183**, 654–7.

Clifford, S. A., Green, K. A. & Watterson, R. M. (1990). *The Pit Bull Dilemma: The Gathering Storm.* Philadelphia, PA: The Charles Press.

Fox, M. W. (1971*a*). *The Behaviour of Wolves, Dogs and Related Canids.* New York: Harper and Row.

Fox, M. W. (1971*b*). Socio-infantile and socio-sexual signals in canids: a comparative and ontogenetic study. *Zeitschrift für Tierpsychologie*, **28**, 185–210.

Greene, C. E., Lockwood, R. & Goldstein, E. J. C. (1990). Bite and scratch infections. In *Infectious Diseases of the Dog and Cat*, ed. C. E. Greene *et al.*, pp. 614–20. Philadelphia, PA: W. B. Saunders.

Harris, D., Imperato, P. J. & Oken, B. (1974). Dog bites – an unrecognized epidemic. *Bulletin of the New York Academy of Medicine*, **50**, 981–1000.

Jones, B. A. & Beck, A. M. (1984). Unreported dog bite and attitudes towards dogs. In *The Pet Connection: Its Influence on Our Health and Quality of Life*, ed. R. K. Anderson, B. L. Hart & L. A. Hart, pp. 355–63. Minneapolis: Center to Study Human-Animal Relationships and Environments, University of Minnesota.

Lockwood, R. (1986). Vicious dogs. *The Humane Society News*, **31**(1), 1–4.

Lockwood, R. (1988). Humane concerns about dangerous dog laws. *University of Dayton Law Review*, **13**, 267–77.

Lockwood, R. & Beck, A. M. (1975). Dog-bite injury to St. Louis letter carriers. *Public Health Reports*, **90**, 267–9.

Lockwood, R. & Rindy, K. (1987). Are 'pit bulls' different? An analysis of the pit bull terrier controversy. *Anthrozoös*, **1**, 2–8.

Lorenz, K. Z. (1953). *Man Meets Dog.* Harmondsworth, Middx.: Penguin Books.

Miller, K. (1986). Letter to the Editor. *Community Animal Control*, **5**(4), 7–8.

Moore, D. L. (1987 *in lit*). A study of animal-to-human bites by breed in Palm Beach County, Florida.

Oswald, M. (1991). Report on the potentially dangerous dog problem: Multnomah County, Oregon. *Anthrozoös*, **4**, 247–54.

Pinckney, L. E. & Kennedy, L. A. (1982). Traumatic

deaths from dog attacks in the United States. *Pediatrics*, **39**, 193–6.

Sacks, J. J., Sattin, R. W. & Bonzo, S. E. (1989). Dog bite-related fatalities from 1979 through 1988. *Journal of the American Medical Association*, **262**, 1489–92.

Schenkel, R. (1967). Submission: its features and function in the wolf and dog. *American Zoologist*, **7**, 319–29.

Scott, J. P. & Fuller, J. L. (1965). *Genetics and Social Behavior of the Dog*. Chicago: University of Chicago Press.

10 Canine behavioural therapy

ROGER A. MUGFORD

Introduction

Not so long ago, the notion of dogs needing the attentions of a behaviour therapist would have seemed ludicrous, and even in 1993 the idea still seems strange to some. The field is a mere two decades old, if one takes as its starting point Tuber, Hothersall & Voith's 'modest proposal' (1974), or William Campbell's still excellent *Behavior Problems in Dogs* (1975). Nevertheless, during the past 20 years there has been a remarkable growth in numbers of practitioners and in the sophistication of the methods employed to modify behaviour problems in dogs. If one wonders how dogs managed without animal behaviour therapists in days gone by, the answer is that they suffered or were killed. Unwanted or inappropriate behaviour was, and probably still is, one of the major reasons for early euthanasia in dogs (Stead, 1982), despite the potential for saving these young dogs' lives through the use of behaviour therapy (see Burghardt, 1991).

Social attitudes towards the dog are changing, from one that was functional and relatively emotion-free to the contemporary attitude where owners may believe that their dogs are truly members of the family. Just as society would not sanction the killing of a demanding grandmother, so, to many, the rejection or destruction of a difficult dog is not acceptable *if* there is some practical and affordable alternative. I am certain that most behavioural problems can be resolved to the benefit and safety of animals, their owners and society. But if that is the promise of animal behaviour therapy, what does it consist of in practice?

A definition of animal behaviour therapy (ABT)

Animal behaviour therapy is *the application of scientific principles to modify an animal's behaviour for the ultimate benefit of both the animal and the owner*. The scientific basis of ABT is illustrated throughout this volume, and is 'borrowed' from a wide diversity of disciplines. During the 1960s and 1970s, the application of learning theory and research in operant behaviour achieved wide acceptance amongst psychiatrists and psychologists dealing with disturbed or mentally subnormal people (e.g. Thompson & Grabowski, 1977). These same ideas can, with modification, be applied also to the treatment of behavioural problems in animals. The field of comparative psychology, especially where it deals with the origins of psychopathologies in humans and animals, can also be applied to the treatment of dogs and other companion animals. The pioneering experimental work of the Bar Harbor (Maine) research group (Scott & Fuller, 1965) shed important light on the genetic basis of canine behaviour, although lessons have also been learned from experimentation upon a wide variety of other species (for a review see Keehn, 1979). In addition, knowledge derived from medical psychopharmacology can be applied to the task of treating disturbed and unwanted behaviour in dogs, although a cautious approach should be taken and, generally speaking, drugs used only to complement changes in husbandry and training.

There is often a medical basis to disturbed behaviour in the dog, and all branches of clinical veterinary medicine are relevant to making a differential diagnosis, and in the selection of appropriate treatments. A wide variety of diseases and conditions, such as disturbances of the endocrine and nervous systems, local sources of pain, metabolic disturbances and nutritional imbalances find their primary expression through changes in behaviour (Reisner, 1991). Indeed, this link between physical health, mental states and behaviour makes the veterinary surgeon the logical practitioner of ABT (see later).

Despite a certain amount of overlap, Animal Behaviour Therapy is distinct from dog training, which I would describe as: *a method of modifying the frequency or intensity of specific behaviour, usually in an applied setting, by application of the laws of effect.* This writer is constantly reminded of how easily trainable dogs are, though there does not seem to be a consensus amongst trainers on either theory or practice (Mugford, 1992; Myles, 1991). Sometimes, it seems to the author that dogs are trained despite rather than because of the actions of trainers and owners. Regrettably, there has been relatively little transfer of knowledge between the biological and behavioural sciences and the dog training sector. The increasing use of dogs for applied tasks such as guiding the blind, assisting the deaf and the physically disabled is, however, helping to create a more scientific approach to dog training. (e.g. Johnston, 1990).

Two personal qualities are also needed for the successful practice of ABT. The first of these is common sense. The diversity and unusualness of many of the situations that arise in the relationship between people and dogs means that the therapist is constantly having to innovate and fall back upon *ad hoc* application of commonsensical principles. Small matters of detail, such as changing the latches on cupboard doors so that dogs cannot gain access to food stored in kitchens, may make all the difference between success and failure. Many cases would flounder were it not for the inventive, alert practitioner spotting the obvious.

The second vital quality needed in a prospective ABT practitioner is compassion. Owners are often distressed, confused and harbour an undue sense of guilt, which must obviously be relieved by the therapist. The interpersonal skills needed to communicate that sense of compassion and professionalism can undoubtedly be acquired through instruction, but often one is drawn to the conclusion that whereas some people 'have it' others find compassion a strange bedfellow.

Ethological studies of the dog's wild relatives, such as the wolf *Canis lupus*, and the coyote *Canis latrans*, have undoubtedly given us a better insight into the behaviour of dogs. In particular, it has helped us appreciate what constitutes normal or abnormal behaviour. For example, epimeletic vomiting (food regurgitation) is a normal aspect of parental behaviour seen in most wild canids (Beuler, 1974). It has also been observed in feral dogs as a means of transporting food. Nevertheless, this normal activity still arouses comment amongst some veterinarians and pet owners when lactating bitches are observed to vomit for their puppies.

Ethological theory can also have a profound practical impact upon how dogs are viewed and treated. The concept of dominance hierarchies, for example, has become a controversial issue amongst ethologists, particularly regarding its importance as a factor governing social interactions and relationships in species such as wolves (see Lockwood, 1979). Attachment and bonding provide the emotional basis for forming alliances or coalitions between dogs, and between dogs and people, and these processes have been highlighted by the present writer as being a more important factor in the social behaviour of domestic dogs than dominance hierarchies (Mugford, 1992).

There is a widespread belief that many of the behavioural problems presented by dogs arise from mistakes by their owners. This is epitomized by Barbara Woodhouse's (1978) choice of title '*No Bad Dogs*' for a popular training book. The implication that dogs' problems are the fault of their owners is still held widely amongst dog trainers and some veterinarians.

In the author's own practice the possibility of owners precipitating behavioural problem in pets has been routinely monitored during consultations with several thousands of clients whose dogs were referred for treatment. In 1984, an *ad hoc* evaluation was made of the social-psychological health of 100 dog owners, and these assessments were then related, where possible, to their pets' behaviour. Even in 1984, when animal behaviour therapy was in its infancy and the author's practice only recently established, 84% of the sample were subjectively classified as normal with appropriate and healthy attitudes towards their pets. They were not, in my opinion, the principal cause of their dog's disturbed behaviour. Of the remaining 16%, 10% could not be classified because of insufficient information, leaving only 4% of owners whose actions were the cause of their pets' problems (see also O'Farrell, 1989 and Chapter 11).

This issue is important in the practice of ABT because to blame the owner precipitates an unproductive sense of guilt and failure, both of which are unnecessary obstacles to making progress in behaviour modification. It is the author's policy not to share such suspicions of blame with clients unless that will assist in the overall therapeutic process. In the author's practice, 41% of clients own a second or additional dogs which do not present behavioural problems, and 85% of clients have previously owned dogs which were problem-free. Owners are thus being seen as no more important a factor in ABT than in other spheres of veterinary practice.

Dogs of any breed or mixture of breeds can present their owners with a behavioural problem. Nevertheless, judging from the numbers registered with the Kennel Club, it appears that certain breeds are consistently over-represented while others are under-represented. The breeds referred to the author's practice most frequently are, in order: German shepherd dogs, Labrador retrievers, golden retrievers, cocker spaniels and Border collies, although more cross breeds are seen than any single pure breed of dog. Consistently under-represented in our sample of canine patients are flat-coated retrievers, Dobermanns, Yorkshire terriers, poodles and a few others. Particular breeds also show a marked tendency to present a particular type of unwanted behaviour. German shepherds, for instance, are consistently over-represented with problems of territorial aggression, as they are also over-represented in most surveys of dog bites (see Wright, 1991).

Within the author's practice, golden retrievers and cocker spaniels are the most commonly referred breeds showing aggression towards their owners. In many cases, these attacks are unprovoked and unexpected. Labradors and cross breeds are significantly less likely to present any aggression-related problem than either golden retrievers, cocker spaniels or German shepherds, but they are significantly more likely to present separation-related problems such as destroying the house when left alone (see later). The origins of these substantial breed differences are often difficult to determine, but once quantified they may be used as a basis for selecting appropriate breeds for particular owners (see Hart & Hart, 1988; Hart, Chapter 5).

In addition to the genetic factors influencing the expression of behaviour and behavioural problems, early upbringing can be critical. The author has recently analysed a sample of 1864 dogs presenting some kind of behavioural problem and, of these, 220 (approximately 12% of the total) displayed separation-related problems. Only 10% of purebred dogs obtained directly from breeders presented separation-related problems, whereas 55% of purebred dogs originating from so-called 'puppy farms' or 'puppy mills' present such problems. One can only speculate as to what may be the critical factor(s) in the early lives of puppies from puppy farms, although transport stress, and/or low levels of sensory and social stimulation during the puppies' first 6–8 weeks of life are likely to be important (see Serpell & Jagoe, Chapter 6).

Trauma and misfortune can be a significant factor in the ontogeny of behavioural problems. Sudden onset sound phobias, for example, often arise after a single exposure to a loud noise (Tuber *et al.* 1974), and sometimes it is just bad luck that the affected dog was at a particular place at the particular time of a thunder storm or a fire cracker.

The type and prevalence of behavioural problems may also depend on a dog's gender. For most aggression-related problems, higher incidences are consistently reported in male dogs compared to bitches of the same breed (Borchelt, 1983; Lockwood, Chapter 9). Overall, 70% of the author's practice concerns male dogs, and only 30% bitches. There is no association, however, between sex or breed of dog and the incidence of noise phobias, obsessive compulsive disorders or chasing livestock.

Finally, the content and timing of meals can have a significant effect upon the behaviour of dogs, both in everyday and in problematical situations (Mugford, 1987). A few individuals have an extreme and pathological response to high protein diets, though other nutrients may also be implicated (Ballarini, 1990).

Methods and equipment of animal behaviour therapy

The first priority for treatment of behavioural problems in animals is to take an accurate clinical and behavioural history. Clinical examination should include thorough external examination for signs of trauma, ectoparasites, general health and that of the major organ systems, especially the heart and lungs. Ideally, a comprehensive metabolic profile would also be obtained. At this stage, a differential diagnosis may rule out an acquired, environmental or 'behavioural' basis to the problem, and a clinical approach is then justified. Assuming that the problem has a behavioural basis, it is vital that an accurate history be taken. A simple questionnaire requesting details of sex of dog, breed, source, feeding habits, exercise, training history, type of house, access to garden and so on can be administered. When investigating the particular problem about which the owner is complaining, one needs to determine the following issues: When did it start? What is its frequency? Where does it occur? When does it occur? Who is present? What happens just before? What happens afterwards? Who is concerned? What remedies have been attempted? What are the consequences of failure to correct the behavioural problem?

Sometimes, the mere discipline of creating a comprehensive clinical and behavioural history naturally identifies the solution to a problem, and no further direct testing or experimentation is required. For example, history alone is usually sufficient to define appropriate treatment for most house-soiling problems, but direct testing and handling benefits treatment of more complex social and interactive problems, such as those in which dogs are aggressive towards other dogs. In such cases, there is no substitute for testing the dog in a variety of environments, and exposing it to other animals when accompanied by different family members or the behaviour therapist. Minutes of practical 'hands-on' experience can substitute for hours of discussion.

There is no single method that can be used for all cases receiving behavioural therapy, and a considerable variety of approaches will be described in later sections of this chapter. Nevertheless, there are some general principles that can be applied to many dog cases.

Errorless learning

The idea behind the errorless learning paradigm is to create a situation in which it is difficult for the animal to make mistakes. Tasks are learned in a structured fashion, with maximum emphasis upon reward for successful performance, rather than upon punishment for failure. For instance, the objective may be to teach a recall to the dog's name and a 'come' command. The environment in which such a simple task is taught should ideally prevent escape (e.g. in a fenced tennis court) and the dog could wear a lengthy, trailing lead. Thus, the opportunities for ignoring the owner or trainer's commands has been reduced and the task should be error-free.

Response substitution training

As an alternative to direct punishment for performance of an undesired behaviour, one can entrain a variety of postures and actions for dogs to perform at times that they would otherwise perform the undesired behaviour. For instance, supposing the dog attacks his mistress's ankles on each occasion she hurries through the home to answer the telephone. In such situations, one can train the response of 'sit' on a mat, positioned within sight but out of range of the telephone. The dog is trained to rush to his mat in anticipation of, say, a titbit given *after* the telephone call has been completed. Similar 'response substitutes' can be applied to the problems of territorial aggression, attacks upon dogs and a variety of other situations (see later).

Instrumental learning

Dogs tend to perform a range of both desired and undesired behaviour spontaneously. Positive behaviour is often not noticed, or is taken for granted by owners during their everyday relationship with their pet. Undesirable behaviour however, such as barking or biting may be reinforced inadvertently by the owner's responses. A more systematic approach to observation and reinforcement (shaping) of desirable behaviour is a productive strategy for owners to adopt in training their pets. This approach to dog training has only recently been explored comprehensively (Johnston, 1990; Mugford, 1992) and the approach is in complete contrast to traditional dog training methods that are based upon Konrad Most's theory of compulsion (Most, 1910).

Equipment

Most dogs come for therapy sessions to the author's premises equipped with a collar or choke chain and lead, and not much else. They usually leave festooned with equipment to improve safety margins, assist in control of boisterous dogs, interrupt undesired behaviour, rewards for promoting desirable behaviour, toys, massage devices, audio tapes, diets, behavioural protocols and so on! It is the author's consistent experience that the use of physical aids can dramatically hasten and simplify achievement of the desired behavioural goals. In our survey of 1864 referred cases, 81% were resolved or markedly improved within three months and 40% improved immediately after having been seen by a therapist. There are few other areas of veterinary practice where such good success and rapid response to treatment can be achieved.

A selection of equipment commonly employed in canine behaviour therapy is listed below:

1 *Choke chains*. Chokers are almost synonymous with the procedures of dog training, yet their use and abuse has recently attracted criticism (Mugford, 1981; Myles, 1991), to the extent that there is now a broad consensus that the choker is best discarded. The difficulties that most owners encounter with choke chains arise from mistiming the 'check' in relation to their commands. Chokers deliver pain, which is wrong

from the animal welfare standpoint; it delays behaviour modification and can injure the dog.

2 *Half chokes*. Half chokes prevent the 'slipping' of the collar over the dog's head but if properly adjusted do not strangulate it.
3 *Headcollars*. Halters for dogs have only been available since 1984. They are a practical expression of a concept from the world of horses, with its emphasis upon schooling and control. By gaining control of the head, one also controls the body and it must follow. In the author's behavioural practice, the advent of canine halters has dramatically improved the success rate for managing aggressive, powerful and boisterous dogs. They relieve owners of the need to acquire well-timed handling skills and powerful muscles, allowing owners to concentrate upon directing their pet's behaviour.
4 *Extending leads*. These create a compromise between safety and control at a distance, yet allow relative freedom for the dog to perform normal behaviour. Restriction to a traditional one-metre fixed lead often creates abnormal levels of aggression and protectiveness in dogs, which often stops when they are off the lead. There are several designs available, but the author recommends those that have the most resonant sound box when the ratchet is 'snapped' and are of the strongest construction. Alternatively, a trailing rope lead of 5–10 metres long can be employed for the same distant 'schooling' of dogs, but more skill is required to prevent tangling of the rope.
5 *Sound alarms*. Gas-propelled and electronic alarms are available, some ultrasonic, other very loud and within the audible range. There are a variety of applications, but they are usually used as 'orienting stimuli' to distract dogs and especially to interrupt unwanted behaviour such as fighting.
6 *The tin can*. More useful than the above, in the author's opinion, is the ubiquitous soft drinks can with a few pebbles to rattle inside. It can be used as a novel stimulus to interrupt undesired behaviour, and thereafter as a conditioned stimulus to signal any number of response substitutes. For instance, an initial superstitious fear of the rattle can be induced by throwing it near dogs when they pull ahead on an extending lead and ignore the 'heel' command. Thereafter, it can be the powerful non-vocal and owner-independent conditioned stimulus that interrupts undesired behaviour such as barking, jumping up, digging, threat behaviours and so on.
7 *Food*. Food has been widely employed in experimental studies of animal learning, yet strangely in dog training it is often frowned upon. In the author's experience, the use of titbits is a logical and attractive element of the behaviour therapist's armorarium: to put dogs at ease or to reinforce desirable behaviours. After the successful acquisition of new behaviour, however, the frequency and predictability of food rewards can be reduced using a variety of reinforcement schedules (Johnston, 1990).
8 *Muzzles*. There are many situations in which muzzles provide instant relief from unwanted behaviour while a longer term behaviour-modification strategy is devised. Examples would include dogs that habitually eat faeces, self-mutilate or bite. It is vital that muzzles be sized and fitted appropriate to the individual dog and that the muzzle does not restrict free panting (for thermoregulation), and ideally should permit drinking and the taking of titbits during training. In terms of animal welfare, muzzles were always a compromise between the safety function and tolerance by the dog, and this issue has recently been under review in the United Kingdom where a British Standards Institute specification for muzzles has now been approved.

Specific therapies

The most frequent and worrying behavioural problem presented to the author and other referring behaviour specialists is that of canine aggression. Aggression is not a unitary concept, having a variety of underlying physiological and motivational causes and social consequences. In the sections that follow, the types of canine aggression will be described in terms of the victim and the context in which they occur, rather than making presumptions about the underlying causation or function.

Aggression to owner

Turk was a three-year-old, entire, male German shepherd dog, who had been owned by Mr and Mrs M for two years. He had been adopted from his first owners knowing that he had bitten a teenage boy, but Mr and Mrs M felt that as they had previously owned German shepherds, were childless and in their opinion knowledgeable about dogs, they would cope.

Mr M had been bitten on three occasions in the months prior to consultation, each of the bites being serious and the last requiring reparative surgery to his hand. The attacks always occurred in the presence of Mrs M, always in the evening, always prior to meals, always when the dog was between husband and wife and, on two out of three occasions, as he entered the lounge. Interestingly, Mr M had a strong and effective relationship with Turk. He did all the training, exercise and in other contexts had no problems at all with him. By contrast, Mrs M was semi-invalided with arthritis and Turk was her near-constant companion when Mr M went to work as a mechanic.

The case was referred to the author because Turk's owners had enquired whether or not it would be possible to have his teeth removed by their veterinarian. Euthanasia was never considered as a possible option. The following specific advice was given:

1. Mrs M was to ignore the dog on all occasions prior to and during Mr M's return from work.
2. Turk was not to be acknowledged when he initiated friendship with Mr or Mrs M, rather be rejected, but at a later stage invited back after a 'sit' command, to receive physical and verbal interactions at moments of the owner's choosing.
3. Turk was kept off chairs and excluded from the bedroom, and for the first two weeks was also excluded from the lounge (where two attacks had occurred).
4. The dog was fed twice per day (as usual by Mr M) on a high fibre complete dry diet of 18% protein content.
5. Turk was trained to a choke chain but still pulled. A headcollar was recommended, which remained fitted with lead attached while in the house, to provide an easy point of contact for control by Mr M should another emergency arise.
6. Boisterous tug-of-war games, which Mr M and Turk enjoyed, were to cease. They were replaced by more structured throw-fetch sequences to be played both indoors and out with Mr M always remaining in control.
7. A 'response substitute' of Turk running to and lying in his bed was trained to interrupt any high risk moment when the dog interposed himself between Mr and Mrs M. The bed was repositioned in the hall.
8. Handling, grooming and dental care of Turk was to be performed on a routine basis by Mr M to emphasize his physical superiority over Turk.
9. The usual obedience commands were to be administered indoors, for instance making the dog 'wait' and to walk behind Mr M through doorways.
10. Depending on the outcome of the above, progestagen therapy (medroxy progesterone acetate, MPA) was suggested, though in practice it proved to be unnecessary.
11. Last resorts were considered, and included euthanasia. Since this was discounted by Mr and Mrs M, the lowering of canine teeth followed by root canal filling (Shipp, 1991) was recommended, if necessary.

Mr and Mrs M reported a marked change in Turk's body language during the ensuing week, 'returning to his old self and fondness for Mr M'. Most crucial of all advice was the diminished affection given by Mrs M, which was more important than any of the activities performed by Mr M. It is characteristic in such cases that only one family member is affected, who is usually of the same sex as the dog.

The situations in which people may be bitten by their own pet are often highly specific, for instance in defence of food, a particular object, or at a particular place or time. The animal is thus only a danger for a small proportion of the time it is with the victim, and there are usually many opportunities for devising non-threatening behavioural contingencies. Rarely is direct physical punishment recommended, primarily because it places people unnecessarily at risk of being bitten.

Aggression towards visitors

Most dogs bark but do not bite when people visit or 'trespass' upon their territory. This is a useful and appreciated function of the modern house dog, but the balance between the two is always a fine one. Fortunately, territorial behaviour is easily modified to favour greater tolerance of strangers.

Maurice was an eight-year-old giant schnauzer who had always barked incessantly at the ringing of the doorbell on arrival of visitors. Until two weeks prior to consultation, however, he had never bitten a visitor. The owners had taken precautions against such an occurrence for most of the preceding eight years, believing that he would bite if given the opportunity.

Because barking was such an intense feature of Maurice's territorial behaviour, the following strategy was adopted:

1 He was fitted with an Aboistop (Dynavet, France), a device which automatically releases citronella scent at the moment a dog barks. After two trials, it suppressed Maurice's barking reliably.
2 Maurice was trained to run up the stairs, to the fourth step, to sit and watch but not approach visitors as they entered. This 'response substitution' blocked approach and attack at the door.
3 Visitors were equipped with food, initially to toss at Maurice, later to offer on an open hand.
4 Family members were to reject Maurice before and during the time of visitors being in the house, and some guests were asked to walk Maurice outdoors.

Maurice's problem was instantly resolved. However, additional measures that can assist in undermining territorial behaviour are to increase off-territory exercise, in the short term to have the dog muzzled, and to change the point of entry of visitors to the house (e.g. back door rather than front). The most important feature in any strategy to modify territorial behaviour in dogs is that owners devise a strong and effective controlling relationship, but not one where they become the resource or the subject whom the dog defends.

Fighting dogs

Jason was a recently neutered, four-year-old, male cross breed, who attacked entire, male dogs but not castrates or bitches. The problem was gradual in onset, beginning shortly after puberty when he was himself attacked by another male. Jason was a skilful fighter, who had badly injured a small Yorkshire terrier. After that he was castrated, but the hoped-for benefits did not materialize so he was referred to the author for treatment.

The following straightforward recommendations were given:

1 Confident control of Jason was recommended by the owners employing a halter and extending lead for as long as necessary.
2 Trained opponent dogs were set before Jason, and any threats or attempts to attack were punished by his being led into a sideways-on posture.
3 Jason was sensitized to the rattle tin can, thereafter the owner was told to remain quiet and calm in all contacts with other dogs. In the event of crisis, a personal protection aerosol was to be used.
4 The habits of walking were to be changed, from avoiding dogs to now walking towards them.
5 In the second stage of therapy, Jason could be released off-lead, but wearing a muzzle and with a trailing lead attached to his regular collar.

In the third and final stage of treatment, Jason was allowed to be unmuzzled but kept on the trailing lead in the event of an emergency.

Patients like Jason are treated on a frequent basis by the author, and the technique outlined above has a high probability of resolving intermale aggression when combined with castration and/or complementary progestagen therapy. Preliminary evidence from operating puppy playgroups (Mugford, 1992) suggests that early social contact with other dogs is a vital ingredient to maintaining social tolerance of dogs during adulthood.

Sibling rivalry

Amanda and Mimi were five-year-old, cross-breed littermate sisters whom the lady owner had had since puppies. On the night of her returning home after receiving news of a catastrophic business failure and while emotionally distressed, the dogs began fighting in the hall. She was badly injured in the process of separating them. Two further fights occurred after which the dogs were separated; one remained at home, the other went to boarding kennels for five months. The case was handled during a home visit by the author as follows:

1 Both dogs were muzzled.
2 Short term, no social contact, display of affection nor customary sleeping on the owner's bed was permitted.
3 The owner was equipped with devices to interrupt a fight which did not place her at personal risk. These were a personal attack alarm and a compressed CO_2 fire extinguisher.
4 One dog, the slightly larger Mimi, was, after 2–3 weeks without incident, to be given slightly greater rights and privileges than the smaller Amanda – i.e. first to be greeted, first to be fed, first to be released off the lead, etc.
5 Joint exercise was to be resumed off territory, but still muzzled if off their leads.
6 Strict enforcement of obedience commands was recommended to the owner after the initial crisis period of 2–3 weeks had elapsed.
7 In the event of a recurrence of fighting, the 'underdog' Amanda was to be given progestagen therapy (MPA by injection).

There were four attempted fights during the first 24 hours; three more in the ensuing seven days. The dogs were kept muzzled for a total of four weeks before switching to the use of headcollars and trailing leads. At the time of writing (nine months since commencement of treatment) the dogs have not fought again.

A key factor in this strategy is to change the equal and democratic approach of caring owners towards their pets. A strong predisposing factor to serious aggression within pairs of dogs is that they are related and of the same sex. The sensible avoidance strategy is to select opposite sex, different breed pairs. The association between pairs of dogs fighting and their owner being emotionally upset is a frequent experience of the author's practice, though the precise cause may often not be identifiable.

Attachment problems

Just as attachment or love is the basis of our appreciation for the company of dogs, so it can also be a major cause of behavioural problems. Examples of dogs with attachment problems include those that howl when left alone, destroy objects by scratching and chewing, or display a breakdown in customary toileting habits. In addition, there may be psychosomatic and physical responses to separation, such as self-mutilation and air swallowing. Attachment or dependence is also a significant factor in many expressions of aggressive behaviour in dogs, when they may try to prevent owners leaving the house or 'protect' him or her from approaches by people and other animals.

Several factors may contribute to the development of attachment problems, including genetics and where the dog was obtained (see earlier), age at acquisition, rehoming experiences, or use of punishment in early toilet training (McCrave, 1991). In the author's experience, separation-related problems respond well to training; indeed we have a 75% success rate when monitored over a three-month period. A general strategy for treatment is as follows:

1 A cooler relationship should be devised by the owner, with a specific target of, say 20–30% of the time in the house separated by a closed door.
2 To achieve the above, a systematic programme of brief separations, of gradually increasing duration, might be necessary.
3 At the moment of departure, either during the training phase or when an extended absence from the house occurs, the attitude of the owner should be especially rejecting and offhand, thereby reducing the contrast between the owners being home or away.
4 The routine of departure should so far as possible be disrupted, for instance owners could leave wearing different clothes, from different points of exit and at unusual times.
5 On returning home, owners should not react to the apparently 'guilty' expression of the dog but be pleasant, no matter what the damage done by the dog when it was alone.
6 It is often helpful that the dog be given the free run of the house, rather than be confined to a small area. However, for expensively destructive dogs initial confinement to a training crate may be justified. Alternatively, the pet might be put into a secure out-house.
7 Links should be devised that remind the dog of his owner's presence when he is away; for instance, worn clothes on which to lie, tape recordings of voices, radio and TV etc.
8 Exercise prior to separation can be a helpful factor, as may feeding the dog before leaving it

alone. Such dogs rarely eat when left but, for some, a bone or chew can be a worthwhile distraction.

9 Drug therapies can be very useful in separation-related problems, if animal husbandry and behavioural approaches prove insufficient. Some alternatives are suggested by Mader (1991) and McCrave (1991). In the author's experience, diazepam, amitriptyline and, in rare cases where there is cardiac and respiratory involvement, beta blockers such as propanolol may help.

Barking

The vocal repertoire of dogs is often limited compared with that of the wolf and, in some individuals, it may acquire a stereotyped quality. The usual contexts in which barking occurs include: response to alarm or distress, defence of territory, play solicitation and hunting. These different contexts justify slight variations in approach to the control of barking, but broadly speaking the following strategies apply:

1 Barking is often reinforced inadvertently by some action of the owner, for instance touching or talking to the dog. Clearly this should stop.
2 Punishment for barking can be a reliable and effective means of suppressing this behaviour, *whilst the owner is present*. Suitable approaches to punishment will vary with the particular individual, but might include a spray-jet of water, the rattle tin can (see earlier) or an ultrasonic beam. Physically hitting the dog, or shouting, is often not effective because such contacts may inadvertently reinforce the undesired behaviour.
3 The Aboistop (Dynavet, France) is a remarkably effective device that acts primarily through distraction as the dog attempts to locate the source of odour. It is also the author's impression that the aroma of citronella has a calming effect upon some individuals. This device, however, may not be appropriate if the dog is barking while distressed (i.e. separation anxiety). Application of management rules given earlier should precede use of the Aboistop, together with attempts at environmental enrichment by offering toys, or more human and canine company.

Automated electric shock collars are marketed widely outside the United Kingdom. With such devices, the dog's barking activates a microphone linked to the shock circuit. It is the author's experience that these devices are inhumane, often ineffective, and may profoundly damage the dog's sense of safety.

Finally, the procedure of surgical devocalization is sometimes advocated, but it is not acceptable amongst the veterinary professions of most European countries.

Obsessive compulsive disorders

The performance of behaviour such as pacing, head flicking, tail chasing, excessive licking, and light and shadow chasing are described as obsessive compulsive disorders (OCDs), primarily because they are repeated and seemingly do not have a productive outcome. Such behaviour is performed by many wild and domesticated species when confined and is especially common amongst animals kept in zoos or under intensive farming systems. There is general agreement that conflict and severe social and environmental deprivation are triggering factors for the performance of such behaviour in the dog (Luescher, McKeown & Halip, 1991), as it is in elephants, polar bears or pigs. The first priority for treatment is to improve the quality of the animal's life and remove obvious sources of stress and conflict. In addition, the following procedures can be helpful:

1 Changing the environment so far as is possible, for instance taking the dog out of its home setting and to the outdoors, to work or out in the car.
2 Interesting and exhausting toys should be provided for the dog, such as manipulative rubber toys with which many dogs tend to develop a competing but harmless compulsion to chew and chase.
3 Modification of the diet to frequent small meals of a high-fibre composition can be helpful, since most observers describe an increased frequency of such compulsive behaviour patterns when animals are hungry.
4 The use of narcotic antagonists to modify

stereotypic behaviours has been described on an experimental basis by Dodman *et al.* (1988). Pharmacological options were reviewed by Luescher *et al.* (1991), but they are still at an early stage of development and presently give only short-term relief. In the author's experience, the injectable morphine antagonist naltrexone can provide a useful diagnostic aid to confirm that performance of a particular stereotypic behaviour is influenced by the endogenous opiate system.

Coprophagia

Eating their own or other animals' faeces is a normal activity in young, growing dogs and it probably provides a useful supplementary source of energy when food is in short supply. For many companion dogs, set point for body weight is considerably higher than their owner's aesthetic notion of an appropriate weight. Such dogs tend to be chronically underfed and may therefore attempt to supplement their diet by eating faeces. Nevertheless, the habit often disgusts human onlookers and can be ameliorated as follows:

1. Frequency of meals should be increased to whatever can be consistently maintained, say three to four times per day.
2. The fibre content of the diet should be increased by either selecting a known bulky, complete, dry diet or by adding supplementary fibre from, for instance, bran, boiled green vegetables, or even shredded paper.
3. If the dog is eating its own faeces, it should be trained to defaecate on command by the owner, for which it then receives a food reward. Faeces should be collected and removed.
4. Institutionalized or kennelled dogs are more likely to exhibit coprophagic behaviour. A more interesting and active life, with toys and company should be offered.
5. There may be a nutritional or clinical basis to coprophagia in a dog that suddenly commences eating faeces. In rare cases, there may be a pancreatic insufficiency (Houpt, 1991), although there would usually be other clinical signs to indicate such a condition. Certain compounds, particularly those rich in sulphur amino acids and ferrous sulphate decrease the palatability of the dog's faeces and can be useful where the animal is eating its own stools.

Finally, coprophagia can be reliably suppressed using a conditioned aversion paradigm, in which lithium chloride is administered to the animal shortly after consuming faeces (Mugford, 1977). The procedure, however, should be regarded as a last resort since coprophagia is not life-threatening.

Fearfulness

The expression of fear in novel or threatening situations is a desirable, adaptive trait in wild animals. Much of the focus of domestication of wild animals has been to reduce this fearfulness (Mugford, 1989). In the case of domestic dogs, owners often expect their pets to be unafraid and confident in a remarkable range of circumstances and levels of stimulation, which for some individuals may prove excessive. Fearfulness in the dog may result from a combination of genetic, neurobiological (Shull-Selcer & Stag, 1991) and ontogenetic factors such as early experience (Scott & Fuller, 1965). Dogs presenting either generalized fears or specific phobias can be a difficult challenge for animal behaviour therapy and refractory to treatment. Such behaviour, however, provides an interesting opportunity for applying techniques to animals that were originally devised for the treatment of disabling fears and phobias in humans. For example, systematic desensitization, flooding and counter-conditioning. In addition, combined psychopharmacological therapies can greatly assist in the management of the fearful patient. A typical treatment scheme for a fearful dog would be as follows:

1. Owners must be schooled not to reinforce expressions of fearfulness. Rather, they should try to be cheery and attempt to distract the dog from attending to the fear-evoking situation.
2. The behaviour of the dog should be closely observed and every expression of extrovert or tolerant behaviour should be praised or otherwise rewarded.
3. Levels of stimulation should be chosen which, so far as possible, are tolerated by the dog, so that it can learn to relax. The most common specific fear or phobia is directed at loud sound stimuli, which can be presented in a systematic fashion from tape recordings played upon good

audio equipment (see Shull-Selcer & Stag, 1991).

4 Escape by running away naturally reduces fearfulness, but becomes counter-productive by deepening the fear and exposing the animal to danger. During training sessions, the dog should be controlled on an extending lead linked to either a harness or headcollar, so that the dog cannot traumatise itself by yanking on the neck.

5 Veterinary assistance should be sought in relation to medium and long-term drug therapies. In the author's practice, we have sometimes obtained good results using background medication of a beta blocker such as propanalol, combined with an initial anti-anxiety compound such as diazepam or amitriptyline.

The obvious avoidance strategies for excessive fearfulness in dogs are firstly to select a breed or strain of dog that is not nervous, and secondly to expose the young puppy to a wide variety of stimulating circumstances that assist development of coping strategies in later life (see Serpell & Jagoe, Chapter 6).

Feeding problems

Regulation of food intake by the dog is a complex matter, easily disrupted by variations in feeding practices and when diets of widely differing palatabilities and caloric densities are given (Mugford & Thorne, 1980; Thorne, Chapter 7). Excessive food intake, even to the point of clinical signs of hyperphagia, is not uncommon in dogs. Indeed, Anderson (1973) and Edney & Smith (1986) have estimated that approximately one-third of the British population of pet dogs is obese. Inappetance, or finicky feeding, tends to be clustered in certain breeds of dogs, notably poodles, Irish setters, German shepherds and a few others. A genetic basis for the regulation of food intake and adiposity is suggested by such breed differences, as it is by the characteristic gluttony of breeds such as the Labrador retriever and beagle.

Management of most feeding problems can be achieved by using good principles of animal husbandry and dietetics, though chronically inappetant dogs may be assisted using an appetite-stimulating drug such as Mysoline, B-vitamins or, in a crisis, certain steroids.

Toileting problems

As we have seen elsewhere in this volume (Bradshaw & Nott, Chapter 8), there is a strong communicatory element to both urination and defaecation by the domestic dog and its wild relatives, with particular significance for marking at the *edge* of territory. If dogs behaved like marmosets that routinely anoint the *core* of their territories with urine, they would not perhaps have become so popular as housepets.

Most problems of undesired urination and defaecation arise from mistiming of meals or exercise and are readily resolved by using commonsense principles of animal husbandry. For instance, if the problem is one of urination or defaecation overnight, the most productive strategy would be to shift the time of feeding and associated prandial drinking to early in the day, with frequent exercise and rewards for urination and defaecation during the late evening. Undesired toileting in the daytime would justify the opposite strategy of feeding in the evening.

The composition and, especially, fibre levels of diets can be an important factor in determining the volume of faeces and thus the frequency of defaecation. Excessive drinking beyond essential physiological requirements may be stimulated by high salt levels or clinical causes of polydipsia such as diabetes, and incontinence may be a secondary consequence of this behaviour. Rarely, the author has encountered excessive drinking that has no clinical or endocrine basis but is an entirely learned or emotional response. Psychogenic polydipsia arising from unresolved conflict was the key factor in a remarkable case of a spaniel that drank up to 15 litres of water per day and urinated a nearly equal volume indoors. A comprehensive medical/behavioural approach led to the conclusion that this was not due to a clinical condition but, rather, an oral compulsion (see earlier section) for which the overall lifestyle of the dog needed modification and enrichment. A tricyclic antidepressant (amitriptyline, 2 mg/kg/day) was administered, together with a systematic, gradual decrease in availability of water during frequent but restricted bouts of drinking.

It should also be emphasized that there is marked

sexual dimorphism in the manner and context of urine marking by the dog, and castration of male dogs reliably depresses the frequency of inappropriate urine marking.

Who should practice animal behaviour therapy?

No accurate survey data are available on the proportion of dog owners who experience serious and disabling behavioural problems with their pets, although this author estimates that the numbers are probably substantial, judging from the growth of his own practice and the number of other referral centres that have been created around the world. Behaviour problems are also a common reason for rehousing or requesting the euthanasia of pets. Who then should the troubled pet owner turn to for help?

In the past and, to some extent, the present, dog trainers have been looked upon as the natural source of advice and knowledge about dog behaviour and training. Unfortunately, the quality and validity of such advice is often poor and sometimes it is extraordinarily mistaken. Nevertheless, many trainers have extended their knowledge about animal behaviour, physiology and the like, and work to a good standard.

During the 1980s, a number of behavioural and psychologically trained individuals, including this author, established referral practices to deal with behavioural problems. The field seemed to be a natural one for employment of graduates from the behavioural sciences who were seeking a career working with animals. Unfortunately, the present writer has found that such graduates need further clinical qualifications if they are to make a comprehensive assessment of cases, or prescribe necessary medical therapies including the use of drugs. In the author's practice, which employs veterinarians for behaviour therapy and which deals only with referred cases from veterinarians in general practice, we estimate that 63% of our canine caseload benefit from the involvement of a veterinary surgeon during diagnosis or subsequent treatment.

The conclusion to be drawn is, I believe, that veterinary surgeons in practice should be the primary source of information and advice about behavioural problems in companion animals. According to Burghardt (1991) behavioural considerations should contribute to virtually all branches of veterinary practice, and should include: 'assisting clients with pet selection, providing preventative behavioural services, diagnosing and treating behavioural disorders in pets, becoming involved with pet-facilitated therapy, and providing effective help to clients experiencing pet loss'.

Behaviour therapy for dogs is an illustration of how science can be taken out of the laboratory and applied to the benefit of a species that, in the past, has been the subject of considerable experimentation and abuse (e.g. Pavlov, 1927; Seligman, Maier & Geer, 1968). The principle that inconvenient or ill-behaving dogs are not just discarded or killed but helped, reformed and tolerated is an important sign that human society may be acquiring civilized attitudes to animals in general. Animal behaviour therapy is then a meeting ground for science, welfare, sentiment and professional practice.

References

Anderson, R. S. (1973). Obesity in the dog and cat. In *The Veterinary Annual*, ed. C. S. G. Grunsell & F. W. G. Hill, pp. 182–6. Bristol: Wright.

Ballarini, G. (1990). Animal psychodietetics. *Journal of small Animal Practice*, **31**, 523–32.

Borchelt, P. L. (1983). Aggressive behavior of dogs kept as companion animals: classification and influence of sex, reproductive status and breed. *Applied Animal Ethology*, **10**, 45–61.

Bueler, L. E. (1974). *Wild Dogs of the World*. London: Constable.

Burghardt, W. F. (1991). Behavioral medicine as a part of a comprehensive small animal medical program. *Veterinary Clinics of North America: Small Animal Practice*, **21**, 343–52.

Campbell, W. E. (1975). *Behavior Problems in Dogs*. Santa Barbara, CA: American Veterinary Publications.

Dodman, N. H., Shuster, S. D., White, D. V. M., Court, M. H., Parker, D. & Ross, D. (1988). Use of narcotic antagonists to modify stereotypic self-licking, self-chewing and scratching behavior in dogs. *Journal of the American Veterinary Medical Association*, **193**, 815–19.

Edney, A. T. B. & Smith, P. M. (1986). The study of obesity in dogs visiting veterinary practices in the United Kingdom. *The Veterinary Record*, **118**, 391–6.

Hart, B. L. & Hart, L. A. (1988). *The Perfect Puppy: How to Choose Your Dog by its Behavior*. New York: W. H. Freeman.

Houpt, K. A. (1991). Feeding and drinking problems. *Veterinary Clinics of North America; Small Animal Practice*, **21**, 281–98.

Johnston, B. (1990). *The Skilful Mind of the Guide Dog: Towards a Cognitive and Holistic Model of Training*. Reading, Berks.: Guide Dogs for the Blind Association.

Keehn, J. D. (1979). *Origins of Madness: Psychopathology in Animal Life*. Oxford, Pergamon Press.

Lockwood, R. (1979). Dominance in wolves: useful construct or bad habit? In *The Behavior and Ecology of Wolves*, ed. E. Klinghammer, pp. 225–44. New York: Garland STPM Press.

Luescher, V. A., McKeown, D. B. & Halip, J. (1991). Stereotypic or obsessive-compulsive disorders in dogs and cats. *Veterinary Clinics of North America: Small Animal Practice*, **21**, 401–13.

Mader, A. R. (1991). Psychotherapic drugs and behavioural therapy. *Veterinary Clinics of North America: Small Animal Practice*, **21**, 329–42.

McCrave, E. A. (1991). Diagnostic criteria for separation anxiety in the dog. *Veterinary Clinics of North America: Small Animal Practice*, **21**, 247–55.

Most, K. (1910). *Abrichtung des Hundes (Training Dogs)*, reprinted (in English) 1984. London: Popular Dogs.

Mugford, R. A. (1977). External influences on the feeding of carnivores. In *The Chemical Senses and Nutrition*, ed. M. Kare & O. Maller, pp. 25–50. New York: Academic Press.

Mugford, R. A. (1981). Where to put your choker. *International Journal for the Study of Animal Problems*, **2**, 249–51.

Mugford, R. A. (1987). The influence of nutrition on canine behaviour. In *The Veterinary Annual*, ed. H. E. Carter, **28**, 1046–54.

Mugford, R. A. (1989). Detection and control of fear in small companion animals. In *The Detection and Control of Fear in Animals*, ed. T. E. Gibson, pp. 61–71. London: BVA Animal Welfare Foundation.

Mugford, R. A. (1992). *Dog Training the Mugford Way*. London: Hutchinson/Stanley Paul.

Mugford, R. A. & Thorne, C. (1980). Comparative studies of meal patterns in pet and laboratory housed dogs and cats. In *Nutrition of the Dog and Cat*, ed. R. S. Anderson, pp. 3–14. Oxford: Pergamon.

Myles, S. (1991). Trainers and chokers; how dog trainers affect behaviour problems in dogs. *Veterinary Clinics of North America: Small Animal Practice*, **21**, 239–46.

O'Farrell, V. (1989). *Problem Dog Behaviour and Misbehaviour*. London: Methuen.

Pavlov, I. P. (1927). *Conditioned Reflexes*. London; Oxford University Press.

Reisner, I. (1991). The pathophysiologic basis of behavior problems. *Veterinary Clinics of North America: Small Animal Practice*, **21**, 207–27.

Scott, J. P. & Fuller, J. L. (1965). *Genetics and Social Behavior of the Dog*. Chicago: University of Chicago Press.

Seligman, M. E., Maier, S. F. & Geer, J. (1968). Alleviation of learned helplessness in the dog. *Journal of Abnormal and Social Psychology*, **73**, 256–62.

Shipp, A. D. (1991). Crown reduction – disarming of biting pets. *Journal of Veterinary Dentistry*, **8**, 4–6.

Shull-Selcer, E. A. & Stagg, W. (1991). Advances in the understanding and treatment of noise phobias. *Veterinary Clinics of North America: Small Animal Practice*, **21**, 353–67.

Stead, A. C. (1982). Euthanasia in the dog and cat. *Journal of small Animal Practice*, **23**, 37–43.

Thompson, T. & Grabowski, J. (1977). *Behavior Modification of the Mentally Retarded*. New York: Oxford University Press.

Tuber, O. S., Hothersall, O. & Voith, V. L. (1974). Animal clinical psychology: A modest proposal. *American Psychologist*, **29**, 762–6.

Woodhouse, B. (1978). *No Bad Dogs*. Aylesbury, Bucks; Hazell Watson & Viney.

Wright, J. C. (1991). Canine aggression toward people; bite scenarios and prevention. *Veterinary Clinics of North America: Small Animal Practice*, **21**, 299–314.

11 Effects of owner personality and attitudes on dog behaviour

VALERIE O'FARRELL

Introduction

Is there a connection between an owner's personality or attitudes and his or her dog's behaviour? As a clinical psychologist attempting to treat dog behaviour problems, this issue is of particular interest to me. In popular dog literature (e.g. Woodhouse, 1978) such a connection is often assumed: a dog's tendency to bite or growl at people may be attributed to its owner's lack of moral fibre; its nervousness in the show ring may be put down to 'fear transmitted down the lead'. Breeders, I have noticed, seem particularly likely to make these attributions, perhaps in an effort to deflect blame from hereditary traits. As a result, the owners of problem dogs are often filled with guilt because they feel they must be responsible for the dog's problem. These feelings are often exacerbated if the owner cannot think of a definite causative event, such as a traumatic separation or fright. In such cases, owner's tend to blame flaws in their own personalities.

Case histories

Anecdotal evidence that seems to demonstrate a connection between owner personality and dog behaviour is not hard to find. Here are two examples from my own case records:

CASE 1. Miss A, a retired physiotherapist, particularly requested that I visit her at home for a consultation because she suffered from agoraphobia. The problem was her nine-month-old corgi puppy, Pearl, who had developed a similar fear of going out of the house. In Pearl's case, this fear extended to the garden: she would go out to urinate and defaecate, but then dash back into the house without lingering to play. She was reluctant to go for walks around the block on a lead, and could not be allowed to run freely in parks because of her tendency to dash for home. Miss A attributed the phobia to two traumatic events. At four months of age, Pearl fell off a low wall in the garden onto her head. Around the same time, she was accidentally shut in a room by herself during a window cleaner's visit. Miss A found her afterwards, trembling behind a sofa. She noticed that thereafter Pearl was particularly afraid of metallic noises that resembled the clanking of a window cleaner's ladder. Miss A told me that some of her friends had made a link between her own agoraphobia and Pearl's problem: they told her that her own fear of going out had made Pearl afraid.

CASE 2 Miss B was a school teacher who lived with her elderly mother. She consulted me about Peter, their ten-year-old corgi, who was, as she put it, 'a bit irritable'. This irritability took the form of growling, barking, and, if given the opportunity, biting. He became aggressive if approached while feeding or while resting on the furniture. The most bizarre manifestation of his irritability happened when the telephone rang. If he managed to reach it first, he knocked the receiver off the hook and barked and growled into it. If Miss B got there first, she had to hold her conversation standing on a table where he could not reach her. The 'irritability' seemed so all-pervasive that it was hard to see anything likeable about Peter. Mrs and Miss B were devoted to him, however. He was 'fussy about his food' so they lovingly prepared special meals of chicken or liver. They seemed to take pleasure in gratifying his every whim: playing with him when he brought them his toys, letting him in and out to the garden constantly on demand. 'I suppose', said Miss B coyly, 'you'll say it is all my fault for spoiling him'.

However persuasive we may find such anecdotes, scientifically they do not constitute reliable evidence of a link between owner personality and dog behaviour: the conjunction of a neurotic or submissive personality in the owner and a phobic or aggressive problem in the dog might have happened by chance. What is needed is evidence that certain kinds of dog/owner pairs occur more frequently than would be predicted by chance. A small study of mine goes a little way towards providing such evidence.

Empirical evidence

Fifty dog owners attending the Small Animal Clinic of the Royal (Dick) School of Veterinary Studies for veterinary treatment were interviewed. They completed a questionnaire that covered the following areas. First, I asked each owner in some detail about his dog's behaviour, concentrating on possible problems. To obtain information that was as objective as possible, I asked specific questions about the frequency of occurrence of observable behaviour. For each kind of problem behaviour, I also asked the owners to rate separately how much of a problem it was for them. The importance of this distinction will become obvious later. Second, I asked the owners to rate their attitudes towards their dogs on various scales: how much they missed the dog when they were away from it, how upset they would be if anything happened to the dog, and so on. Third, I asked the owners about aspects of their own behaviour that seemed likely to reflect their attitudes towards their dogs: for example, whether the dog slept in the bedroom, what they did to the dog if it misbehaved and what they fed the dog. Finally, the owners filled in a short personality inventory: the Neuroticism scale of the Eysenck Personality Inventory. This has been shown to differentiate reliably between patients receiving treatment for neurotic illness and normal subjects (Eysenck & Eysenck, 1964).

The data were analysed by calculating, across all subjects, the correlations between scores on the different items. Various subsets of items were then subjected to a technique called Principal Components Analysis (PCA). This is a statistical procedure that simplifies a correlation matrix by extracting from it factors or dimensions which underlie the relationships between the different variables. PCA of the items concerned with the dog's behaviour revealed two main factors (Table 11.1). The first factor was labeled 'aggression towards people', and the second, 'displacement activities', as these seemed to best describe the actions or activities associated with the two factors. For example, sexual mounting of people or inanimate objects is often seen in response to a conflict situation such as the arrival of visitors, while destructiveness when left alone is usually a response to the frustration of not being able to accompany the owner, hence their classification as displacement behaviour (O'Farrell, 1992).

The items concerned with owner attitudes were also subjected to PCA and the items loading significantly on the first two factors are shown in Table 11.2. The first factor seems to represent degree of attachment to the dog, whereas the items loading on Factor 2 might be said to reflect the degree of anthropomorphic emotional involvement.

For each subject, a score was also calculated for each of the four factors by summing the subject's score for the relevant items, weighted by their factor loadings. The inter-correlations between these scores, as well as with the Neuroticism score, are shown in Table 11.3. Contrary to the popular stereotype, neurosis was not found to be associated with phobic behaviour in dogs ($r = 0.027$). It was, however, correlated with the extent to which the owner found the dog's fear a problem ($r = 0.32$, $P < 0.02$). Although these results can only be regarded as suggestive, they do provide a basis for speculation that there are relationships between owner attitudes and personality, on the one hand, and dog behaviour on the other. Anthropomorphic emotional involvement seems to be related to certain forms of aggression, and, although neurosis in the owner is not associated with fears in the dog, it may be related to what I term 'displacement activities'.

These findings are of interest because they agree with both clinical experience and (with the exception of the relationships between phobias and attitudes) with popular stereotypes. In the light of this evidence, it would be wrong to assume a causal relationship between Miss A's fears and those of her pet. Rather, it seemed to me that Miss A's phobia was helpful to Pearl. By empathizing with Pearl, Miss A did not force the dog into phobic situations and so increase her fear. She was, on the other hand,

Table 11.1. *Principal components analysis of dog behaviour*

Factor 1 aggression towards people (13.0% of variance)	
Growling at anyone	0.87*
Growling at family members	0.85*
Growling at visitors	0.57**
Pestering for attention	0.43*
Biting people	0.42*
Growling or biting when disturbed	0.42*
Growling or biting when patted	0.42*
Not over-excited with visitors	0.39*
Factor 2 displacement activities (10.3% of variance)	
Sexual mounting of people or inanimate objects	0.79**
Destructive when left alone	0.78**
Biting people	0.39*
Pestering for attention	0.35*

** = $P < 0.001$; * = $P < 0.01$.

Table 11.2. *Principal components analysis of owner attitudes*

Factor 1 Attachment (19.5% of variance)	
1. Would feel desperate if dog died	+0.78
2. Dislikes being parted from dog	+0.73
3. Dog sleeps in bedroom	+0.65
Factor 2 Anthropomorphic emotional involvement	
1. Feeds dog specially prepared food (e.g. chicken, scrambled egg)	+0.89
2. Feeds titbits often	+0.88
3. Likes a dog to be loving and dependent	+0.49

extremely concerned about Pearl's fear and sought advice even though the problem seemed to be gradually improving.

With the relationship between the dog's aggression and the owner's anthropomorphic emotional involvement, however, there would appear to be an element of truth to the popular stereotype of the indulgent owner and the small, snappy dog. Certainly, these dyads are encountered in clinical practice, the case of Miss B and Peter being a good example. One explanation for this association arises from the fact that aggression by a dog towards its owners is most often associated with dominance, i.e. aggression towards a perceived subordinate who seems to be mounting a dominance challenge (by patting or disturbing it). A dog who is disposed to assert itself in these contexts, by virtue of genetic or hormonal factors, is more likely to perceive itself as dominant over its owners, if they acquiesce to its demands (O'Farrell, 1992). An owner who wants her dog to love her is more likely to be acquiescent in this way, and the feeding of titbits is one example of this sort of behaviour. The association between feeding specially prepared food and dominance aggression is interesting given the putative links between dietary factors and aggressiveness (see Mugford, Chapter 10). Such food, consisting mainly of chicken, scrambled egg or minced beef, would presumably contain a higher proportion of protein than most commercial dog food and the protein would be of a higher quality. However, in the absence of firm evidence of a causal link between diet and behaviour, it seems equally likely that dominant dogs use food and feeding as a dominance issue (as Peter did), and that this may involve not eating immediately at the owner's behest. The owner may then interpret this behaviour as rejection of the food and, eager to please the dog, may offer a more palatable alternative which the dog cannot resist eating immediately.

The finding that owner neurosis is positively correlated with displacement activities in the dog also fits in with a popular stereotype: this time of the neurotic owner of a yappy, over-excited dog. It also accords with my own clinical experience, although other clinicians may not necessarily agree. I find that most dogs who engage in marked displacement activities, such as destructiveness in the owner's absence, or whose life is taken over by a stereotypy, such as tail-chasing, either have a history of traumatic separations (e.g. rescue dogs; see Serpell & Jagoe, Chapter 6) or have owners whose own anxiety levels are high. The following is an example of such a case:

CASE 3. Mrs C came to consult me about her two-year-old Alsatian, Rambo. She was pale, puffy and dishevelled; she talked non-stop in an agitated and incoherent way. Rambo's problem was his 'restlessness'. When he and Mrs C were at home together, he tried to follow her everywhere around the house, even into the lavatory. When he was with her, he would dash to and fro, twining round her legs. If she shouted at him, he would sit quietly for a moment and then start again. If she left the house, he would become really frantic and she usually found some damage when she returned. The Venetian blinds had been destroyed, presumably in his attempts to look for her out of the window.

I asked if anything had happened around the time of the onset of this behaviour. She said she had left work because of a 'breakdown'. She had panic attacks, in which she sweated, her heart raced, and she thought she was going to die.

A possible explanation of this association is not so obvious as in the case of dog aggression, but I suggest that psychoanalytic theory, in particular Object Relations Theory, can provide one. This theory states that, in neurosis, the patient projects onto external objects characteristics of his/her inner objects which he/she has difficulty in managing (see Fairbairn, 1952). When this happens in a close relationship, such as a marriage, the person is likely to behave

Table 11.3. *Owner attitudes and personality and dog behaviour: intercorrelations between factor scores*

	Aggression	Displacement activities	Attachment	Anthropomorphic involvement	Neurosis
Aggression	–	+0.27	+0.08	+0.31*	+0.01
Displacement activities		–	–0.09	+0.27	+0.34**
Attachment			–	–0.19	+0.12
Anthropomorphic involvement				–	–0.15

$^{**}P < 0.02$, $^{*}P < 0.05$.

towards the other individual not so much in terms of that individual's own needs as in terms of the emotions which the patient has projected on to him/her (e.g. Dicks, 1967). If the object of projection is a dog, then I suggest it is likely to be subjected to inconsistent punishment and/or reward, and that this is likely to produce a state of conflict in the dog, and the sort of 'restless' behaviour displayed by Rambo. A further clinical example may help to illustrate these processes in action.

CASE 4. Minnie, an 18-month-old Alsatian, was brought to me by Mrs D and her 14-year-old son. Mr D was away working on an oil rig. Minnie's problem was that she spent her day chasing her tail. This had started about two weeks previously, when the family had returned from holiday and Minnie from kennels. The D's had felt unhappy about the kennels: Minnie had stayed there before but this time the manager had changed, and he had an odd, brusque manner.

According to Mrs D, Minnie spent all her waking hours in tail-chasing, pausing only to eat, urinate and defaecate: she did it when alone in the garden and Mrs D heard her bumping into the furniture as she did it alone at night, downstairs. The whirling must have been stimulus-dependent to some extent, however: when left alone in a small yard at the clinic she did not do it.

The D's felt that the problem had been caused by a traumatic experience in the kennels. But other aspects of the case led me to think that the D's themselves were contributing to it. The tail-chasing had not started immediately when Minnie returned from the kennels, but four or five days later. Also, Mrs D was absolutely distraught and in extreme distress. She repeatedly voiced the opinion that Minnie was dying, in spite of the fact that Minnie looked in good condition and had not even mutilated her tail. She also said repeatedly that Minnie would have to be put down. Her son sat silent and tense on the floor, cuddling Minnie.

As the consultation went on, it emerged that Mrs D was due to undergo coronary bypass surgery in a few weeks. During her recent holiday, she had become dangerously ill and had been admitted to the hospital as an emergency patient.

It seemed to me that Minnie's problem was in part due to the fact that Mrs D was projecting her own feelings onto Minnie. Mrs D was in an intolerable situation: faced with a life-threatening illness and major surgery, all she could do was wait. In an attempt to cope with her anxiety, she projected her fear of death onto Minnie. She could then deal with the anxiety in two ways: on the one hand by being solicitous and comforting to Minnie whenever she engaged in the abnormal behaviour; on the other hand, by planning to get rid of Minnie and, with her, the problem. She and Minnie were thus caught in a vicious circle. The rewarding of the stereotype, interspersed with bursts of hostility, increased its frequency. The fact that Minnie's problem did not resolve but seemed to be getting worse had the effect of increasing Mrs D's anxiety.

Conclusions

Some evidence has been put forward in this chapter that certain kinds of owner personalities and attitudes are associated with certain kinds of behaviour problems in the dog. It has also been suggested that this association is a causal one. It should be emphasized that this link will not necessarily be found in every instance of displacement activity or dominance aggression in a dog, or that other factors, such as genetic or hormonal influences are of no importance. Further research is clearly needed, preferably involving direct observation of dog/owner dyads. One of the most useful aims of such research would be to tease out the kinds of behaviour patterns on the part of the owner that are most directly linked with problem behaviour in the dog. This chapter has concentrated on explanations couched in terms of psychoanalytic theory, but the explanatory potential of other theories would also be worth exploring.[1]

[1] Cain (1983), for example, provided evidence for the usefulness of Systems Theory – it seems likely that a dog's behaviour is affected not only by its relationship with an individual owner, but also by relationships and interactions involving the whole family. Personal Construct Theory (see Kelly, 1955) also offers the advantage of a measurement technique, the repertory grid, that is accessible to dog owners and might be a useful clinical tool.

References

Cain, A. O. (1983). A study of pets in the family system. In *New Perspectives on Our Lives with Companion Animals*. ed. A. H. Katcher & A. N. Beck, pp. 72–81. Philadelphia, PA: University of Pennsylvania Press.

Dicks, H. V. (1967). *Marital Tensions*. London: Routledge & Kegan Paul.

Eysenck, H. J. & Eysenck, S. B. G. (1964). *Manual of the Eysenck Personality Inventory*. London: University of London Press.

Fairbairn, R. (1952). *Object Relations Theory and Personality*. New York: Basic Books.

Kelly, G. (1955). *The Psychology of Personal Constructs*. New York: Norton.

O'Farrell, V. (1992). *Manual of Canine Behaviour*, 2nd edn. Cheltenham, Glos.: British Small Animal Veterinary Association.

Woodhouse, B. (1978). *No Bad Dogs*. Aylesbury, Bucks.: Hazell, Watson & Viney.

III Human–dog interactions

12 Dogs as human companions: a review of the relationship

LYNETTE A. HART

Introduction

People have been closely associated with dogs – or with their wolf ancestors – for many thousands of years, although the precise origin of the relationship is still the subject of speculation (see Clutton-Brock, 1980 and Chapter 2). Evidence from Epipalaeolithic and early Neolithic sites indicates that humans probably began taming wolves at least 12 000 years ago, and it is clear that by the time of the ancient Egyptians several distinct breeds of dogs already existed (Clutton-Brock, 1976). Currently in the United States, more than 50 million dogs reside in roughly 38% of all households (Market Research Corporation of America, 1987; American Pet Products Manufacturers Association, 1988). Of these, the vast majority are kept as social companions.

Are dogs special?

Comparisons with other species

Among the array of different species that serve as companion animals, dogs are in many ways exceptional. As early as the turn of this century, a large survey of children's school essays about pet animals had already demonstrated the dog's outstanding popularity (Bucke, 1903). The children in this survey emphasized the highly personalized attention provided by their dogs with phrases such as 'he likes me', 'guards me', 'follows me', 'protects me',' 'barks when I come home from school', 'is good to me' and so on. The children also appreciated the dog's ability to express love and affection by jumping up, running around, wagging its tail and soliciting play. In addition, many mentioned how the dog kept them company and played with them when they were feeling lonely or sad. Many of these early observations have now been confirmed by the results of more recent studies.

In a telephone survey of 436 Rhode Island residents, for example, Albert & Bulcroft (1987, 1988) found that dogs were the most popular pet. Sixty per cent of pet owners had at least one dog, and dogs were the most desired pet among non-owners. Owners who selected dogs as their favorite pets reported feeling more attached to their pets than did people whose favorite pets were cats or other animals. Dogs also seemed to be more adept at playing affectionate and emotionally supportive roles than other animals, leading the authors to suggest that perhaps dogs interacted with their owners in ways resulting in higher levels of attachment. The survey found that dog owners spent more time actively interacting with their pets – grooming them, walking them, giving them special treats – than did cat owners (see Fig. 12.1). Dog owners were also more willing to spend any amount on veterinary treatment than cat owners, although more cat than dog owners admitted to sleeping with their pets.

In another study based on observations of people and their pets interacting at home, Miller & Lago (1990) found that interactive behavior, such as whining, begging, making noise, obeying and being near the owner, occurred at far greater frequency with dogs than with cats (see Fig. 12.2). Dogs also interacted actively with unfamiliar persons whereas cats tended to be calm and aloof (Fig. 12.3). Owners

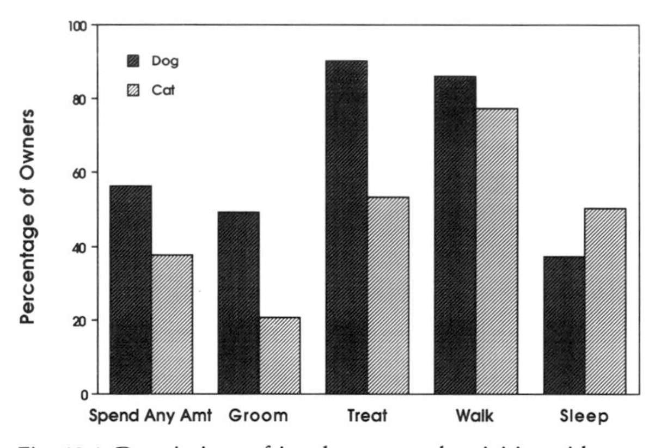

Fig. 12.1. Descriptions of involvement and activities with dogs and cats as reported by owners. After Albert & Bulcroft (1987).

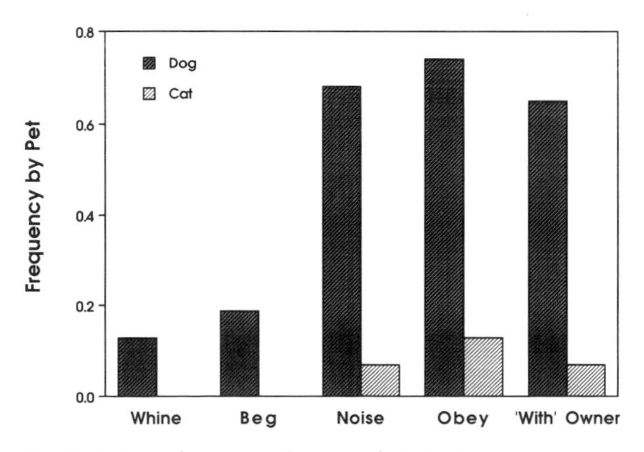

Fig. 12.2. Mean frequency that specific behaviors were performed by dogs and cats observed with their owners and a stranger at home. After Miller & Lago (1990).

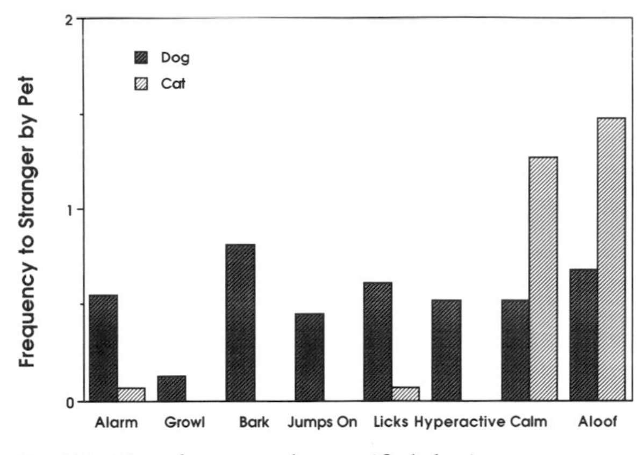

Fig. 12.3. Mean frequency that specific behaviors were directed toward a stranger by dogs and cats observed with their owners at home. After Miller & Lago (1990).

issued a mean of 0.48 orders to their dogs, as compared with 0.07 orders given to cats. However, owners told an average of 1.87 stories about their cats but only 1.32 stories about dogs. In a previous study, Lago, Knight & Connell (1983) had reported higher levels of both behavioral and physical intimacy between dogs and their owners than between people and cats, although some cat owners were also extremely attached to their pets. Seventy-one per cent of dog owners also regarded themselves as dominant compared to only 57% of cat owners. Overall, these studies suggest that dogs are better at adjusting their interactions to the owner's demands than other companion animals. Dogs exhibit highly coordinated behavior; standing, moving and sitting in synchrony with their owners to an extent rarely observed in cats.

When dog owners in Melbourne, Australia, were asked to supply a list of adjectives describing their dogs, and these were subsequently subjected to factor analysis, three major factors emerged: acceptance/trust, love/friendship and intelligence/obedience (Salmon & Salmon, 1983). These owners felt that the main benefits of dog ownership were companionship, protection and happiness or pleasure. Three-quarters of them felt a need to be physically protected by a dog, and the same number believed that their dog helped to protect their home from burglary.

Comparisons with human companions

Two previous studies have attempted to compare the importance of pets with that of human family members. The first, without distinguishing the species of pet, surveyed a convenience sample of 62 respondents of whom 8% reported feeling closer to the pet than to anyone else in the family. However, a much higher number (44%) reported that the pet received the most strokes of anyone in the family, and that it served as the focus of favorable attention. Many of these pet owners seemed to find it easier to offer affection to their animals than to other family members. In many cases, the animal was also highly emotionally involved with the people in the households surveyed. Eighty-one per cent of respondents reported that pets reacted to anxiety and tension within the family by developing diarrhea, gastric upsets or epileptic seizures (Cain, 1983).

The second study sought to assess the relative closeness that dog owners felt towards their dogs by asking them to represent their significant relationships pictorially using a technique known as the Family Life Space Diagram. More than one-third of these owners placed the dog closer to themselves than to any other family member (Barker & Barker, 1988). Taken together, the results of these two studies suggest that, for about a third of owners, the dog's importance ranks on a par with that of human members of the family.

When Davis (1987*a*) asked preadolescents the reasons why their families had acquired a dog, most referred to a sort of pet deficit – simply needing a pet – as the main reason. Entertainment, the parents' need for a pet, companionship and love were also mentioned but by fewer children (see Fig. 12.4). In another study (Bryant, 1985, 1986), children of a similar age mentioned play/companionship, love/

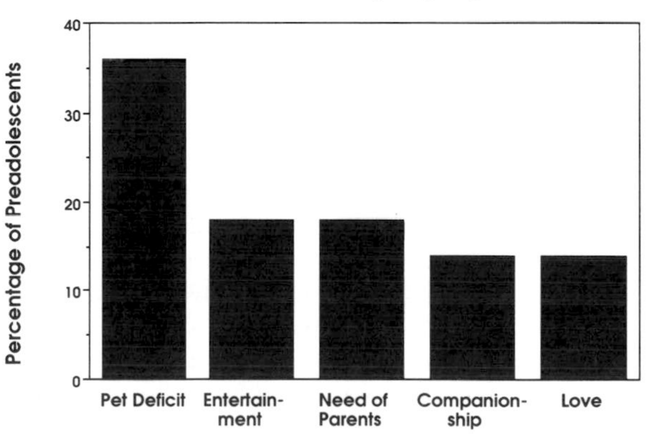

Fig. 12.4. Reasons families acquire a dog as reported by preadolescents. After Davis (1987*a*).

affection, physical qualities, good temperament, entertainment, reliable friend and opportunities for nurturance when asked what their own pets provided. When describing neighborhood pets, however, love and reliable friendship were omitted from the list of special traits. The three most frequent interactions with dogs revealed in a survey of ten-year-olds were: playing with, exercising and talking to dogs (MacDonald, 1981).

Why dogs are special

Displays of affection

Many behavior patterns of dogs seem especially designed to elicit attachment. Dogs are naturally affectionate, a trait that is more characteristic of some breeds than others (Hart & Hart, 1988), and they can even be instructed to provide affection. It is standard practice, for example, to teach service dogs that assist people who use wheelchairs to provide their owners with displays of affection in response to a verbal command (Mader, Hart & Bergin, 1989).

Darwin (1873) described the specific behavior patterns dogs use to express affection. They include: lowering the head and whole body with the tail extended and wagging from side to side, drawing the ears back alongside the head, rubbing up against the owner, and attempting to lick the owner's hands, face or ears. These ritualized greeting signals indicate to owners that the dog is pleased to see them. Dogs seek out their owners for mutual contact, and provide affection that is not contingent upon the owner's success or appearance. In this way, dogs may provide their owners with feelings of unconditional acceptance and, at the same time, enhance the person's attachment to the dog (Catanzaro, 1984; Voith, 1985). The unconditional nature of the dog's affection may also allow owners to direct or redirect anger at the dog without putting the entire relationship at risk.

Loyalty and devotion

Certain traits make dogs ideally suited to be human companions. They develop specific attachments for individuals, and remain near or in physical contact with their owners as if attached by an invisible cord. They also tend to be active during the daytime when people are active and, with appropriate training, they defer to us as dominant social partners. More important, however, are dogs' extraordinary powers of non-verbal expression by which they signal their love and regard for humans.

In order to assess the satisfaction of dog owners with various aspects of their pets' behavior, Serpell (1983) invited 57 urban dog owners to rate both their own pets and a hypothetical 'ideal' dog on 22 different behavioral traits. The traits with the highest ratings and the least variability between owners included expressiveness, enjoyment of walks, loyalty/affection, welcoming behavior and attentiveness. These ratings also corresponded closely with the owners' 'ideal' ratings. Traits with more average ratings and considerable variation between owners included playfulness, attachment to one person, friendliness to other people, territoriality, friendliness to other dogs, attitude to food and sense of humor. Serpell (1983) concluded from this that owners varied in their preferences for these traits and that they were therefore less important for insuring compatibility in the relationship.

A further important asset of dogs, although it is one they share in common with other pets, is that they lack the power of speech and are therefore unable to offer advice, judgement or criticism. Nevertheless, they are affectionate and empathic so their friendship tends to be seen as sincere, reliable and trustworthy, while at the same time lacking many of the threats associated with human friendships (Serpell, 1986*a*).

Play

According to one study (Stallones *et al.*, 1988), 95% of pet owners regard their pets as friends. A similar proportion of dog owners reported playing often with their pets, as compared with only 73% of cat owners. Similarly, when asked to respond to the statement, 'the dog gives me an outlet for playfulness', 80% of 259 Swedish dog-owners agreed (Adell-Bath *et al.*, 1979). In another study involving observations of people walking their dogs, some type of game with the dog was observed on 36% of walks (Messent, 1983). In general, more fetch-type games were played with medium-sized to large dogs than with small ones.

Surprisingly little is known about the amount of time people spend playing with their animal companions. In a survey of Swiss pet owners, Turner

Fig. 12.5. Attraction of young children to animals. Young toddlers respond to both mechanical and live dogs, but a real dog elicits the stronger interest (Kidd & Kidd, 1987). Photograph: Joan Borinstein.

(1985) found that dog owners reported spending an average of 17.5 hours per week interacting with their pets while cat owners reported an average of only 10 hours. However, when 96 California veterinary students were asked to estimate the amount of time they spent interacting with their pets, the dog owners averaged 35.3 hours per week and the cat owners averaged 33.2 hours. For dog owners, 44% of this time was estimated as play, as compared with 36% for cat owners (J. Angus, personal communication).

Touch

A study of three- to four-year-old children's interactions with dogs revealed that 67% of these interactions involved body contact with the dog, such as putting a hand on the dog, patting it or hitting it. In contrast, vocal and verbal behavior comprised only 9% of the interactions (Millot & Filiatre, 1986). In a subsequent study touching was again the most frequent behavior shown in the presence of a dog, accounting for 40% of all child–dog interactions (Filiatre *et al.*, 1988).

In an analysis of 1105 photographs of dogs or cats in a family setting submitted to a national photographic contest, Katcher & Beck (1985) found that 97% of the pictures illustrated people and animals touching each other, generally with the heads of the animal and human close together. Over 92% showed a dyadic relationship, with one person and one animal occupying the center of the photograph. Touching was also a primary mode of interaction with a dog in a study of nursing home residents (Neer, Dorn & Grayson, 1987). Of the nine different types of interaction recorded involving the dog, grooming and touching were the two most commonly employed by residents.

The value of dogs for different types of people

Albert & Bulcroft's (1987, 1988) Rhode Island study found that households with children at home tended to have more pets than either widows or families with an 'empty nest', or with an infant. However, feelings of attachment to the pet were lowest in families where children were at home. Although pet ownership was highest among households containing large families, attachment to pets was highest among people living alone and among couples who did not have children living at home. The authors noted that the single, divorced and widowed individuals and childless couples who were most attached to their pets also expressed more anthropomorphic attitudes to their pets, particularly in relation to dogs. In a longitudinal study of older people (a population that experiences increasing losses), Lago, Connell & Knight (1985) found that persons who stayed at home and spent more time with the animal also became more attached and formed a stronger relationship with it.

An 'invisible cord' often seems to connect a dog to its owner (Serpell, 1986*a*). Almost invariably, dogs are more attentive to their owners than their owners are to them. In a study of ten families' interactions with their dogs, the associations between the dog and the adult family members were found to differ between families with and without children (Smith, 1983). In childless families the people and the dog interacted more readily, more frequently and in a more complex fashion, and the dog spent more time in close proximity to someone. In contrast, dogs in families with children interacted less frequently per person per hour, even when the children were not physically present.

The results of another survey (Salmon & Salmon, 1983) suggested that dogs satisfied more of the needs of widowed, separated and divorced people than those of people at other stages of life. Apparently, the needs of these people were not being met fully by a family network, and hence the dog was playing a more important part in their lives. The dog was more of a close friend, more like a child, made them feel safer and provided them with greater opportunity for exercise than it did for people with intact families. Among older childless couples, 73% believed that walking the dog had encouraged conversations with people, as compared with only 48% of people at other stages of life.

Findings from a survey of 1144 elderly, married women in Maryland pointed to the risks of having a pet and not being very attached to it (Ory & Goldberg, 1983). Thirteen per cent of women who were not attached to their pets reported that they were unhappy, as compared with 6% for attached pet owners and 5.5% for nonowners (see Fig. 12.6). Furthermore, for a greater percentage of these married women in the nonattached group, the spouse failed to serve as a satisfactory confidant (31.4%, as compared with 23.0% of the attached group, and 20.3% of the nonowners). These data support the view that women who have pets but who are not attached to them are also significantly worse off in their relationships with people. In a related finding, Brown, Shaw & Kirkland (1972) found that low affection for dogs was associated with low affection for people. In the case of men, low affection was also associated with a low desire for such affection.

The foregoing studies suggest that the mutual attachment between dogs and humans tends to

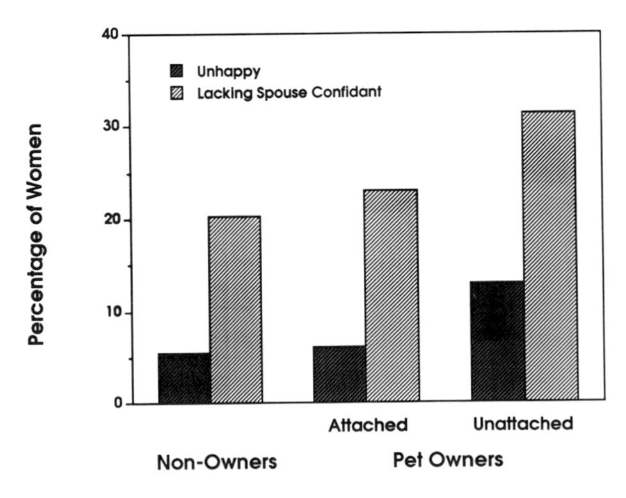

Fig. 12.6. Percentage of married women who described themselves as being unhappy and whose spouse was not a confidant, expressed as a function of pet ownership and attachment to the pet. After Ory & Goldberg (1983).

increase with time spent together. For people who form close attachments to people and/or dogs, a dog may become a central focus of attention and love when the person's other social contacts are diminished.

Socialization effects of dogs for people

The idea that dogs facilitate human social interactions seems almost self-evident. In a study of 259 Swedish dog owners, 83% of those questioned agreed with the statement, 'My dog gives me the opportunity of talking with other people' (Adell-Bath *et al.*, 1979). Seventy-nine per cent also agreed with the statement, 'The dog makes friends for me'.

As scientific exploration of pet ownership has diversified, more evidence has emerged of what Mugford (1980) has termed 'the social significance of pet ownership.' In this farsighted paper, Mugford initially focuses on the significance of companionship by animals in fostering two major motivating needs for humans: affiliation and self-esteem. He then addresses additional psychological benefits of pet ownership, including the fact that they play, give and accept love, provide emotional security and serve as child substitutes. After reviewing the available literature, he concludes that the practical outcome of pet ownership, particularly ownership of dogs, is to increase the owners' extraversion and so promote social interactions within both the home and the community.

Observing people walking in a London park, Messent (1984) found that the company of a dog greatly facilitated the walkers' conversations with strangers. Another study conducted around the same time showed that the presence of an animal in drawings caused the people in the drawings to be perceived as more satisfied, friendly, industrious, wealthy, happy, generous and comfortable (Lockwood, 1983).

The social value of canine companionship and partnership is most obvious with persons who use wheelchairs and have service dogs (Hart, 1990). A series of studies have been conducted to assess whether service dogs can enhance the social attractiveness and acceptance of people with disabilities. In retrospective interviews, disabled individuals with service dogs estimated a median of eight friendly approaches from adults per shopping trip, while only one friendly approach was estimated if a dog was not present (Hart, Hart & Bergin, 1987). In a prospective study, the spontaneous responses of strangers to people in wheelchairs with or without service dogs were compared (Eddy, Hart & Boltz, 1988). As shown in Fig. 12.7, wheelchair-bound adults with service dogs received more social acknowledgments from passersby than those unaccompanied by dogs. Studies of disabled children have documented similar effects (Mader *et al.*, 1989). The increased social acceptance of disabled children with service dogs occurred even on the school playground – where able-bodied children were accustomed to seeing the dog every day – as well as in a shopping mall where the dog was a novelty. A dog can therefore normalize social responses to those individuals who often would be ignored or avoided because of a disability, shyness or physical unattractiveness.

Dogs seem to display an inexhaustible willingness to form and sustain partnerships with humans. This is illustrated most dramatically by the partnership between service dogs and people in wheelchairs. The dog and its owner come to be seen by other people as a team, more predictably together than any mother and child, marital couple or pair of siblings. With such closeness, the dog–person team enjoys certain advantages typical of any party whose members are perceived to be together (Goffman, 1971). This closeness is made evident to both the owner and to onlookers by the dog's alert attention and responsiveness to the owner's commands.

Fig. 12.7. Socializing effect of service dogs. When they had service dogs, people using wheelchairs were frequently approached by passersby.

In order to investigate the conversations of able-bodied people walking their dogs, Rogers, Hart & Boltz (1989) invited a sample of dog walkers to carry small tape recorders. It was found that all walkers talked to their dogs, as well as asking them questions. Dog walkers exchanged greetings and casual conversations with passersby, but tended to engage in fewer long, involved stories than persons walking without dogs. Consequently, dog walkers uttered fewer verbs in the past tense than did other walkers; conversations of dog walkers were in the here-and-now. About 80% of nouns uttered by people walking dogs referred to the dog. Among passersby speaking with a dog owner, about 25% of the nouns referred to the dog, and this was true whether or not the dog was present. Thus, the dog served an important role as both a conversational partner and as a focus of conversations with passersby.

The experience of talking and playing with a pet, especially a dog, may educate a child in some of the subtleties of social relationships. In a study of German adolescents, pet owners were found to be more skilled at decoding human, nonverbal facial expressions (Guttmann, Predovic & Zemanek, 1985). This effect was particularly noticeable among boys. This ability appeared to be associated with greater social acceptance, since pet owners were selected more frequently by others to be confidants, companions and partners. Pet-owning children were also more willing to establish new friends. When children were asked in another study to rate their own social competence, their self-ratings were positively associated with the number of pets in the family and most felt that their pets had helped them to make friends (Serpell, 1986*b*).

Foster children are especially vulnerable socially because they lack the secure position in the family that is characteristic of most children. Replies to a written questionnaire by 60 foster parents suggested that foster children may obtain more from relationships with companion animals than their own children, the parents or the family as a whole (Hutton, 1985). According to parents' ratings, the foster children showed particular benefits in the areas of improved relationships, feeling at ease with people, breaking down social barriers, communicating, having something to talk about and improved mood.

They also reported that the children appreciated having someone who would listen and not give away their secrets.

Among the many studies that have explored the therapeutic role of visiting or residential dogs in nursing home settings, few have investigated the possible social effects on either staff or patients. One early Australian study reported that patients spent less time alone following the introduction of a resident dog than previously (Hogarth-Scott, Salmon & Lavelle, 1983). Ninety-one per cent of patients appreciated that the dog was something to talk about, and 86% of staff felt that the dog was something they could share with patients. Another study monitored the proportion of time that both patients and staff displayed solitary behavior (not engaged in dyadic or group interaction) before and after a resident dog was introduced to a nursing home (Winkler *et al.*, 1989). The frequency with which the staff displayed solitary behavior declined after the dog's arrival, shifting from about 50% to about 41%. Patients engaged in solitary behavior about 77% of the time, both before the dog's arrival and 22 weeks later. The dog became more closely attached to the staff than to the patients, and this coincided with the staff increasing their dyadic interactions with other staff.

Savishinsky (1986) investigated the content of conversations among elderly patients during pet visitation to three geriatric facilities. Animals enhanced the level of conversation among certain residents. Stories of pet loss were commonly recounted and frequent reminiscing was related to the presence of animals. The human and animal visits were said to produce a family atmosphere, momentarily recreating a family. Each of these facilities introduced their own permanent resident animals soon after initiation of the visiting programs.

In recognition of the dog's ability to enhance communication, Lapp (1991) has proposed that dogs be used to bridge the communication gap between young, middle-class student nurses and vulnerable older adults. The strategy of including a dog within the therapeutic milieu might also improve the inferior medical treatment often given to elderly patients (Green, 1981; Greene *et al.*, 1986).

Although animals cannot supplant relationships with other people, they can help to relieve the isolation and partially normalize the social lives of lonely people. A study of elderly women who were living alone or with other persons, some with and some without a pet, found that pets only made a difference for those living alone (Goldmeier, 1986). Among women who lived alone, those with pets were significantly more optimistic, less agitated and less lonely and dissatisfied.

Such studies provide abundant evidence that dogs serve as social companions, as well as easing human social interactions by providing a topic of relaxed and entertaining conversation. Since social contact is the mechanism for nurturing self-esteem in people, the socializing effects of dog companionship are among the most important indirect benefits for people.

Physical and psychological benefits of canine companionship

People often mention that they feel a sense of calmness and security around their pets (Katcher, 1985). Those who are seriously ill may report that animals distract them from worry or pain (McCulloch, 1983). In addition to these somewhat commonplace observations regarding animals' contributions to human health, scientists have over the past decade conducted formal studies into the psychological, physiological and therapeutic impact of relationships with dogs.

Developmental benefits

Dogs can provide significant companionship for children as well as for adults. They respond to demands and offer uncritical sympathy; they may serve as transitional objects and sources of security for a young child venturing away from the mother. Children commonly use their pets for comfort when they are feeling bored, lonely or unhappy (Kidd & Kidd, 1985). Disturbed children seem to rely especially heavily on animals as a source of support. In one study, delinquent adolescents were found to be twice as likely to talk to their pets and three times as likely to seek out their pet's company when lonely or bored (Robin *et al.*, 1983).

In a study of 213 children with pets and 44 lacking pets, a major benefit of pet ownership that emerged was mutuality; a term that refers to 'involvement with another', 'needing social support', and 'needing to care about other people or living things' (Bryant, 1990). Involvement with dogs, cats or rodents was found to generate greater mutuality than involvement with fish. The exclusivity of the relationship with a pet was also the subject of positive comment by children. The disadvantages of pet-keeping mentioned by the children revolved around empathic concerns regarding the pet's death, welfare, needs and care. Overall, dogs were the favorite of 51% of children, while cats were the favorite of 27%.

The fact that children reveal their emphatic concerns for pets supports the notion that children may sometimes learn how to care for others from their experiences with animals (see Melson, 1990). Some evidence suggests that children who lack younger siblings may compensate, in effect, by spending more time with animals. Children who lack younger siblings seem to perceive their pets with more positive affect (Bryant, 1986; Paul & Serpell, 1992) and children with pets appear to be better informed about how adult animals care for their young than do non-pet owners (Melson & Fogel, 1989). When Melson (1988) interviewed mothers of young school children, she found that children without younger siblings played more with their pets. Evidence that children adjust their involvement with animals according to their family structure was also found in a French study. Children without siblings showed more frequent and longer interactions with a dog than children with siblings (Filiatre, Millot & Montagner, 1985, 1986).

In a study of kindergarten and day-care children, Nielsen & Delude (1989) used a variety of animals to capture children's attention including a tarantula, a cockatiel, two breeds of rabbits, two breeds of dog and various toy animals. Only the dogs elicited displays of intimacy, however, with 21% of the children hugging or kissing them, three times as much by boys as by girls.

When dogs confer enthusiastic affection on someone, that person is likely to feel accepted and viewed as a good person by the dog. In a study of twenty-two 10–12-year-olds, 65% of the children believed that the dog thought the child was a wonderful person (Davis, 1987*b*). This positive reflected appraisal provided by the dog would tend to support the development of a stronger and more positive self concept. Especially for adolescents, a clear and consistent message from another that they matter helps them infer that they are significant (Rosenberg & McCullough, 1981).

Psychological benefits

Therapists find that emotionally disturbed children or adults who have been hurt in their relationships with people relate more easily to animals (Levinson, 1969). If the disturbed person opens up to an animal that is associated with a therapist, the animal may provide acceptance for the therapist, making it possible to establish a connection much sooner (Lapp, 1991). In a study of rejected, neglected or abused children in foster care, it was found that children deficient in reality testing, self-control and empathy, could, in the presence of a dog, quickly realize that they were having an effect on the outside world. The mere presence of dogs was often sufficient to 'elicit laughter, lively conversation, and excitement among even the most hostile and withdrawn of the children' (Gonski, 1985).

For both children and elderly adults, the opportunity to care for an animal may have value in giving the person an experience of mattering to another. Melson's (1988) study of preadolescent children found a substantial correlation between caretaking and emotional involvement. Another study reported that elderly pet owners described themselves with a higher number of positive adjectives than did non-owners (Kidd & Feldmann, 1981). Pet owners also scored higher on a nurturance scale, indicative of their helpfulness and general benevolence toward others. Evidence of higher morale among pet owners was also found in a study of 37 people receiving medical care through the US Veterans Administration (Robb, 1983), although a higher proportion of pet owners than non-owners refused to participate in the study.

During interviews with 11 persons suffering from AIDS concerning the benefits they derive from their animals, Carmack (1991*a*) found that the animal made it possible for a person to focus on the here and now and be more distracted from their illness. In a study of nursing home patients with Alzheimer's disease, more than 25% of the patients showed some behavioral improvements during twice-weekly visits from trained dogs and their volunteer handlers (Schultz, 1987). In an eight-week study at a psychiatric facility where puppies and handlers made weekly visits, Francis (1985) reported improvements in measures of patient's social interaction, psychosocial function, life satisfaction, mental function, depression, social competence and psychologic well-being. Dogs may also have particular value in assuring people that they are secure from harm. People frequently obtain a dog to serve as a watchdog, and a recent prospective study of new cat and dog owners found that, among dog owners, a lasting and statistically significant reduction occurred in fear of crime (Serpell, 1990).

Physiological health

Several studies have now documented the immediate physiological effects on humans of canine companionship (see reviews in Baun, Oetting & Bergstrom, 1991; Friedmann, 1990). While the effect of the presence of a dog on cardiovascular function is possibly the most extensively studied aspect of human–animal interactions, the findings are difficult to interpret and there is no evidence that, say, petting a dog induces beneficial health effects over the long-term. What has been demonstrated in several studies is a reduction in blood pressure for normotensive or hypertensive persons when they pet a dog, with a stronger effect for hypertensive individuals (Katcher *et al.*, 1983). Children were found to show a drop in blood pressure when a dog simply entered the room (Friedmann *et al.*, 1983). In another study pet owners who had suffered a heart attack were found to show improved survivorship after one year, as compared with persons who also had suffered an attack but did not have a pet (Friedmann *et al.*, 1980). This finding, however, has not been replicated.

Touch may play a major role in the lowering of blood pressure while petting a dog (see Vormbrock & Grossberg, 1988). Interestingly the feeling of calmness associated with petting or being with a dog may also be experienced by the dog. A dog's heart rate drops when it is being petted by a person (Lynch & McCarthy, 1969), and monkeys have been reported to show a drop in both heart rate and blood pressure when sitting in close proximity to each other (Manuck, 1987).

Physical and general health

When eight walkers were instructed to take identical walks both with and without their dogs, their walks were found to be slightly but significantly longer when the dog was present (Messent, 1983). Dramatic increases in walking were also found among new dog owners in England in a prospective study of dog and cat adoption (Serpell, 1991), providing strong support for the view that dogs can benefit owners by increasing the amount of exercise they take. Serpell's study also found that dog owners developed a heightened sense of security and self-esteem, and improvements in general health that continued through ten months following adoption.

Further evidence for effects on general health was reported in a United States study of 938 Medicare enrollees, where pet owners reported fewer doctor contacts during a one year period than non-owners (Siegel, 1990). Stressful life events were not related to doctor visits among respondents with pets, whereas they were related among non-owners. Significantly this finding only held true for owners when the different types of pets were considered separately. Dog owners spent more time outdoors with their pets, more time talking to their pets and more time overall with their pets than other people they knew. They felt more attached to their pets, and the positive aspects of pet ownership more strongly outweighed the negative ones than for owners of other pets. Dog owners were more likely to mention that their pets made them feel secure and provided entertainment.

Dogs as therapists

In animal-assisted therapy, dogs and other animals are made available to people as a form of therapeutic intervention. The practice of taking visiting animals into nursing homes is perhaps the best known example, although this type of therapy has been offered to various groups of people with special needs. Methodological problems are inherent in most studies of the benefits of animal-assisted therapy and frequently the orientation of the earlier studies was biased toward obtaining positive results (Beck & Katcher, 1984). Situations such as these, where both animals and their handlers visit an institutional environment together and where patients may have severe medical or mental problems, present a particular challenge to the scientific investigator.

Generally, studies have not investigated systematically whether the observed effects of animal-assisted therapy are due to the animal, its handler or simply to novelty. To explore this question, Hendy (1987) compared the behavior of nursing home residents under four different visiting conditions; no visit, human visitors, dog visitors, and human and dog visitors. Each of the three visiting programs were equally effective at increasing alertness and smiling for many nursing home residents, whereas the group with no visit did not show an improvement. Visiting with a dog may, however, be more rewarding for the visitor, given the social lubricant effect of the animal.

Recent studies of special populations have sought to identify and assess possible therapeutic effects of animal-assisted therapy. In a study of autistic children, where a therapist was accompanied by a dog during 18 therapy sessions, isolation behavior by the children declined sharply during the sessions (compared with presession measures), and assumed an intermediate level during follow-up observations a month after treatment concluded (Redefer & Goodman, 1989). The occurrence of social interactions increased several fold during therapy, but declined again to an intermediate level in the follow-up.

Dogs as buffers against grief and stress

Systematic studies of the health effects of animals, especially dogs, have often focused on vulnerable individuals. Among the most vulnerable adults are those who have recently lost a spouse. In one study of bereaved elderly persons with extremely few confidants, pet ownership and strong attachment to the pet were associated with significantly less depression (Garrity *et al.*, 1989).

In another study of recently widowed women, Bolin (1987) found that nonpet owners reported a deterioration in health after loss of a spouse, whereas dog owners reported no such deterioration, as long as their health was good to begin with. Non-owners rated their health as good before the death and poor afterwards, and expressed despair, social isolation and death anxiety. Five of the 34 dog owners reported that the dog was a greater source of comfort than relatives and friends, and 15 described the dog as somewhat important. If the dog were to die, 19 of the women said they would be extremely upset, 15 would be somewhat upset and no one anticipated being not very upset. In another study of widows it was found that non-owners of pets reported more symptoms, especially those with psychogenic components, and higher drug use, compared with pet owners (Akiyama, Holtzman & Britz, 1986–87).

People who enjoy close relationships with their dogs appear to be buffered against many of the vicissitudes of life. Increasing documentation shows that stresses have less impact and that general health is more stable for dog owners than non-owners. Additionally, the dog facilitates its owner exercising regularly and thus supports physical conditioning.

Problems resulting from dog ownership

While dogs capture human hearts with their unfailing affection and attention, they also present society with some challenging problems. Many owners apparently are unprepared for the investment of time and money that dog ownership requires (Case, 1987). Zoonoses, behavior problems, street sanitation and a seeming oversupply, at least of certain dogs, are among the problems that can be minimized through education and other efforts by local communities.

People thinking of acquiring a dog can improve the likelihood of a successful partnership with their animal if they carefully consider beforehand how well the dog will match their lifestyle. Making an effort to identify a suitable breed, and then exploring the background of the prospective dog, can increase the probability of a compatible relationship (see e.g. Hart, Chapter 5). Not surprisingly, owners who receive a dog as a gift tend to mention more problems than other owners, and they are less likely to seek a replacement dog if their present one were to die (Salmon & Salmon, 1983). People who receive a dog as a gift may develop less of an attachment to it than people who have made a conscious personal decision to acquire a dog.

Attachment of adults to animals, and the likelihood that people may keep pets as adults seems to be influenced by childhood experience. In a survey of 120 adults, contact with pet animals during childhood was shown to be significantly associated with the tendency to keep pets, usually of the same species, as an adult (Serpell, 1981). People who had kept pets in childhood were more likely to consider getting a pet, or to already have one, and current pet owners tended to keep the same type of animal as they had during childhood. Studies in Germany also found that pet ownership during childhood influences choices regarding pet ownership in adulthood (Bergler, 1988). Previous experience of pet keeping should therefore be an important consideration during the process of pet selection.

Canine behavior problems

Behavior problems with dogs, such as disobedience, aggression and separation anxiety, can inject sufficient conflict into the relationship that it becomes intolerable for the human and ultimately outweighs any benefits that are being sought. Aggression accounts for the majority of canine behavior problem cases presented at clinics in the United States and Britain (see Mugford, 1981, 1985; Houpt, 1983; Hart & Hart, 1985). In one United States study, the incidence of aggression was considerably higher among male than female dogs (Houpt, 1983), and an English study reported that intact males were significantly over-represented for all types of behavior problem (Mugford, 1981). Aggression directed toward both owners and strangers was sometimes severe and unpredictable enough to warrant recommending euthanasia. Establishing a social convention of routinely castrating male dogs might be expected to reduce the number of dangerous behavior problems in dogs (but see also Lockwood, Chapter 9).

Little is known regarding how environmental factors may influence the prevalence of behavior problems in dogs, although owner attitudes and behavior are likely to be important (see e.g. O'Farrell, 1986 and Chapter 11). In a survey of 308 dog-owning households in Melbourne, Australia, it was found that in 35% of household no one disciplined the dog (Salmon & Salmon, 1983), and in an open-ended question about the responsibilities of owning a dog, only 2% specifically mentioned training the dog as a responsibility. Dog owners were also poorly informed about the laws concerning pets. Forty-six per cent did not know that their municipality had a law against dogs chasing people, and 32% were unaware of the law against dogs roaming unattended.

Canine behavior problems can also affect an owner's ability to travel and socialize with friends. When the owner loses dominance over the dog, or places the dog in the center of the social network, this may influence the owner's other activities within the community (Miller & Lago, 1990). Reducing travel and severing relationships with friends who did not like the dog are compromises that are exacerbated when a dog develops behavior problems (Catanzaro, 1984).

Zoonoses, allergies and bites

The most serious dog-related hazard is bites. Most animal bites go unreported (Beck & Jones, 1985; see also Lockwood, Chapter 9), yet approximately 1% of all United States emergency room visits are caused by animal bites (Weber *et al.*, 1984). Dogs are responsible for a large majority of the bites that require medical attention, and zoonotic infection may occur in 5–15% of bites (Chretien & Garagusi, 1990; Schantz, 1990). The victim's own dog or a neighbor's dog is responsible for 80% of the dog bites (Moss & Wright, 1987). Letter carriers who own dogs are more likely to get bitten than those without dogs, suggesting that dog owners become less cautious of dogs in general and are more likely to approach strange ones (Lockwood & Beck, 1975). Surprisingly, previous experience of being bitten by a dog was found in one study to be no more frequent among fearful than non-fearful adults (DiNardo *et al.*, 1988). Presumably, most dog bites are avoidable and, in many cases of serious attacks, owners fail to take appropriate steps to prevent the dog becoming a problem (Lockwood & Rindy, 1987).

People generally are poorly informed about the health hazards associated with keeping animals, although those who visit veterinarians are better informed than others (Fontaine & Schantz, 1989). The unwitting new pet owner may have no idea that a puppy from a pet store, for instance, is likely to be carrying intestinal parasites that can also affect humans (Stehr-Green *et al.*, 1987). For all zoonoses, children suffer a higher risk because of their closer physical contact with animals. Developing comprehensive infection control for pets, pet policies and surveillance plans are ways to reduce zoonoses, as well as being more careful in pet selection and taking precautions for individuals who are allergic. Some of these methods are being refined by programs that advise and assist immuno-compromised persons with their pets (Gorczyka, 1990).

Potential risks of zoonoses and allergies from exposure to animals pose a particular concern when animals are brought into institutional facilities for therapeutic purposes. One study of staff attitudes at a nursing home before and after pet visits found that staff were favorably disposed toward the program and that concerns about health risks declined after their experience with the program (Kranz & Schaaf, 1989). Guidelines for screening animals and managing visitation have been prepared that set forth procedures for conducting such programs responsibly (Lee *et al.*, 1985). Apparently, when these precautions are implemented, visitation programs encounter only occasional problems (Stryler-Gordon, Beall & Anderson, 1985).

Problems for dogs

Between five and ten million dogs are euthanized each year in the United States, leading to a current focus on what is termed the problem of pet overpopulation (Nassar & Fluke, 1988). Many newly adopted dogs are quickly discarded. One study concluded that nearly 64% of all dogs obtained as puppies in the United States are disposed of by their owners within a year of acquisition (Arkow & Dow, 1984). Typical stated reasons for getting rid of dogs were lifestyle changes and behavioral problems. Another study reported that 20% of surrendered dogs had behavior problems (described by Rowan & Williams, 1987), while another study reported a figure of 26% (Arkow, 1985). Of these, 59% of the owners said they would keep the dog if the problem could be resolved.

Perhaps even more disturbing are cases where dogs are neglected or actively abused (see also Hubrecht, Chapter 13). Frequently a pattern of abuse occurs within families where both the children and animals are abused (DeViney, Dickert & Lockwood, 1983). The possibility of veterinarians and pediatricians alerting each other to situations where abuse appears likely could perhaps prevent some suffering of both children and animals, although effective and confidential methods for administering such procedures are yet to be developed.

Pet loss

As children grow up and leave home and their parents age, the family home is often scaled down to an apartment. Privately-owned apartments in the United States commonly prohibit pets so people making a move to an apartment often face the prospect of parting with their animals, even though one study of elderly pet owners in apartments found that pets did not cause problems (Hart & Mader, 1986). Many people also fail to acquire pets because of housing limitations (Catanzaro, 1984). Declining health, advancing age and less spacious living arrangements all have been identified in a longitudinal study of 316 elderly as factors that made it more difficult to own a pet (Lago *et al.*, 1985). The convention of not allowing pets in apartments strikes harshly at individuals who live alone and who might be expected to welcome and benefit particularly from animal companionship. Support of pet owners by the community can also be improved by developing more accessible exercise areas for dogs. Some communities also have volunteer programs that provide foster care for pets when their owners are ill.

When discussing the death of a pet dog, people frequently express a profound degree of mutual interdependence with their canine companion. The following example is typical:

> . . . Kojak and I lived alone together, and in our younger days he went to work with me at the Pillar Point Air Force Tracking Station, where I was a Security Guard, until I became disabled with Parkinson's disease. Then we both stayed home to grow old together. Kojak was 15 years, 7 months and 14 days old and had been well until his last year, 1989. Then he went downhill, despite weekly check-ups with Dr Smith, and with my own disability it was becoming increasingly difficult for me to take proper care of him.
> . . . I was in the room with him for the euthanasia. I said goodbye to my good and loyal friend, told him to be a good dog where he was going and to wait for me, that I would be along in time to join him.
> . . . The Pedro Point staff technician baked me a banana loaf bread and the entire staff chipped in and bought me a huge basket of goodies from the Hickory Farms store. The sympathy cards and letters poured in. It is true that Kojak had a good life, and a long one for a Samoyed, but a glance at our picture will convey the closeness that has made separation so difficult, particularly since my age and disability and financially meager means make me hesitate to replace my dog.
> J. Corson, personal communication, 1991.

It is often not appreciated that a companion dog may participate in family life for many years. Table 12.1 presents the median ages of dogs at time of death for pet owners who contacted the Pet Loss Support Hotline at the University of California, Davis (see Mader & Hart, 1992), and as indicated in the records of four veterinary hospitals and one pet cemetery. The overall median age of death was 13 years for all dogs and there were no significant differences among the different facilities. In contrast, dogs delivered to animal shelters by their owners average less than two years old (see Arkow & Dow, 1984).

When a dog dies after providing more than ten years of consistent companionship and sharing in daily and landmark life events, the human companion inevitably experiences a profound loss that

Fig. 12.8. J. Corson and Kojak while they were in good health. Photograph: David Hester.

Table 12.1. *Ages of dogs at time of death. Callers to the hotline were similar to clients of veterinary clinics and a pet cemetery in having relationships of many years with their dogs. No significant difference was found among these different facilities. In contrast, dogs delivered to animal shelters averaged less than two years old*

	Dogs' ages in years	
Institution	*N*	Median
Pet loss support hotline	294	13
Pet cemetery	162	13
Veterinary clinics		
No. 1	117	13
No. 2	38	12
No. 3	22	14
No. 4	273	13
Animal shelter	917	1.7

is unwelcome and perhaps surprising in its intensity. As in any relationship, the person becoming attached to another assumes a vulnerability to the possible loss. The person's role in deciding in favor of euthanasia may also exacerbate feelings of guilt and grief. Given the shorter lifespan of dogs such losses are likely to occur several times within a dog owner's lifetime. The pain of such losses cannot be circumvented, but support is available to offer choices to grieving people and to ease their pain (Hart, Hart & Mader, 1990). Drawing on resources such as pet loss support groups or hotlines, and seeking support from friends and relatives, can lessen the impact of painful events involving the loss of a beloved companion animal.

An animal's role often assumes more importance in a smaller family. Quackenbush (1981) reported a correlation between dependency upon an animal and living alone. He also observed an apparent exacerbation of grief when an animal had been in the family for many years (cited in Beck, 1984). The percentage of owners seeking assistance from a veterinary hospital social worker increased with the dog's age. Among owners of dogs under six years of age, only 3.4% visited the social worker, whereas with older dogs the rate was at least 26.9% of owners. As in human relationships, attachment to an animal appears to enrich and deepen over time. Persons whose pets have served as a major source of affection, intimacy, companionship, and nurturance are especially vulnerable to grief when the animal dies (Carmack, 1991*b*). Among elderly people who have come to be overly dependent on a special relationship with an animal, often as a substitute for a human relationship, the grief may be further intensified.

Acknowledgements

Outstanding bibliographic assistance was provided by Sara Christensen and Jena Meyerstein. James Serpell, an anonymous reviewer and Kathy Berchin offered helpful editorial suggestions. A contribution from Kal Kan Pet Foods provided support toward the preparation of this paper. Dr Bruce Cammack, Dr James Harris, Dr Tom Kendall, Dr Paul Palmatier, the Adobe Animal Hospital and the Sacramento Pet Cemetery generously shared from their facilities information regarding the longevity of dogs.

References

Adell-Bath, M., Krook, A., Sandqvist, G. & Skantze, K. (1979). *Do We Need Dogs? A Study of Dogs' Social Significance to Man*. Gothenburg; University of Gothenburg Press.

Akiyama, H., Holtzman, J. M. & Britz, W. E. (1986–87). Pet ownership and health status during bereavement. *Omega*, **17**, 187–93.

Albert, A. & Bulcroft, K. (1987). Pets and urban life. *Anthrozoös*, **1**, 9–23.

Albert, A. & Bulcroft, K. (1988). Pets, families, and the life course. *Journal of Marriage and the Family*, **50**, 543–52.

American Pet Products Manufacturers Association. (1988). *A Nationwide Survey of Pet Owners*. American Pet Products Manufacturers Association, Inc., 60 East 42nd Street, New York, NY 10165.

Arkow, P. S. (1985). The humane society and the human-companion bond: reflections on the broken bond. *Veterinary Clinics of North America: Small Animal Practice*, **15**, 455–66.

Arkow, P. S. & Dow, S. (1984). The ties that do not bind: a study of human–animal bonds that fail. In *The Pet Connection: Its Influence on Our Health and Quality of Life*, ed. R. K. Anderson, B. L. Hart & L. A. Hart, pp. 348–54. Minneapolis: Center to Study Human–Animal Relationships and Environments, University of Minnesota.

Barker, S. B. & Barker, R. T. (1988). The human–canine bond: closer than family ties? *Journal of Mental Health Counseling*, **10**, 46–56.

Baun, M. M., Oetting, K. & Bergstrom, N. (1991). Health benefits of companion animals in relation to the physiologic indices of relaxation. *Holistic Nursing Practice*, **5**, 16–23.

Beck, A. M. (1984). Population aspects of animal mortality. In *Pet Loss and Human Bereavement*. Ames; Iowa State University Press.

Beck, A. M. & Jones, B. A. (1985). Unreported dog bites in children. *Public Health Reports*, **100**, 315–21.

Beck, A. M. & Katcher, A. H. (1984). A new look at pet-facilitated psychotherapy. *Journal of the American Veterinary Medical Association*, **184**, 414–21.

Bergler, R. (1988). *Man and Dog: The Psychology of a Relationship*. Oxford: Blackwell Scientific Publications.

Bolin, S. E. (1987). The effects of companion animals during conjugal bereavement. *Anthrozoös*, **1**, 26–35.

Brown, L. T., Shaw, T. G. & Kirkland, K. D. (1972). Affection for people as a function of affection for dogs. *Psychological Reports*, **31**, 957–8.

Bryant, B. (1985). The neighborhood walk: sources of support in middle childhood. *Monographs of the Society for Research in Child Development*, **50**(3), Serial No. 210.

Bryant, B. (1986). The relevance of family and neighborhood animals to social–emotional development in middle childhood (Abstract). *Living Together: People, Animals, and the Environment*, p. 68. Renton, WA: Delta Society.

Bryant, B. K. (1990). The richness of the child-pet relationship: a consideration of both benefits and costs of pets to children. *Anthrozoös*, **3**, 253–61.

Bucke, W. F. (1903). Cyno-psychoses. Children's thoughts, reactions, and feelings toward pet dogs. *Pedagogical Seminary*, **10**, 459–513.

Cain, A. O. (1983). A study of pets in the family system. In *New Perspectives on Our Lives with Companion Animals*, ed. A. H. Katcher & A. M. Beck, pp. 72–81. Philadelphia: University of Pennsylvania Press.

Carmack, B. J. (1991*a*). The role of companion animals for persons with AIDS/HIV. *Holistic Nursing Practice*, **5**, 24–31.

Carmack, B. J. (1991*b*). Pet loss and the elderly. *Holistic Nursing Practice*, **5**, 80–7.

Case, D. B. (1987). Dog ownership; a complex web. *Psychological Reports*, **60**, 247–57.

Catanzaro, T. E. (1984). The human–animal bond in military communities. In *The Pet Connection: Its Influence on Our Health and Quality of Life*, ed R. K. Anderson, B. L. Hart & L. A. Hart, pp. 341–7. Minneapolis: Center to Study Human–Animal Relationships and Environments, University of Minnesota.

Chretien, J. H. & Garagusi, V. F. (1990). Infections associated with pets. *American Family Physician*, **41**, 831–45.

Clutton-Brock, J. (1976). The historical background to the domestication of animals. *International Zoo Yearbook*, **16**, 240–4.

Clutton-Brock, J. (1980). The domestication of the dog with special reference to social attitudes to the wolf. *Carnivore*, **3**, 27–33.

Darwin, C. (1873). *Expression of Emotions*. New York: Greenwood Press (1969).

Davis, J. H. (1987*a*). Pet care during preadolescence: developmental considerations. *Child: Care, Health and Development*, **13**, 269–76.

Davis, J. H. (1987*b*). Preadolescent self-concept development and pet ownership. *Anthrozoös*, **1**, 90–4.

DeViney, E., Dickert, J. & Lockwood, R. (1983). The care of pets within child abusing families. *International Journal for the Study of Animal Problems*, **4**, 321–9.

DiNardo, P. A., Guzy, L. T., Jenkins, J. A., Bak, R. M., Tomasi, S. F. & Copland, M. (1988). Etiology and maintenance of dog fears. *Behaviour Research and Therapy*, **26**, 241–4.

Eddy, J., Hart, L. A. & Boltz, R. P. (1988). The effects of service dogs on social acknowledgements of people in wheelchairs. *Journal of Psychology*, **122**, 39–45.

Filiatre, J. C., Millot, J. L. & Montagner, H. (1985). New findings on communication behavior between the

young child and his pet dog. In *The Human–Pet Relationship*, pp. 51–7. Vienna: IEMT.
Filiatre, J. C., Millot, J. L. & Montagner, H. (1986). New data on communication behaviour between the young child and his pet dog. *Behavioral Processes*, **12**, 33–44.
Filiatre, J. C., Millot, J. L., Montagner, H., Eckerlin, A. & Gagnon, A. C. (1988). Advances in the study of the relationship between children and their pet dogs. *Anthrozoös*, **2**, 22–32.
Fontaine, R. E. & Schantz, P. M. (1989). Pet ownership and knowledge of zoonotic diseases in DeKalb County, Georgia. *Anthrozoös*, **3**, 45–9.
Francis, G. M. (1985). Domestic animal visitation as therapy with adult home residents. *International Journal of Nursing Studies*, **22**, 201–6.
Friedmann, E. (1990). The value of pets for health and recovery. In *Pets, Benefits and Practice*, Waltham Symposium 20, ed. I. H. Burger, pp. 8–17. London: BVA Publications.
Friedmann, E., Katcher, A. H., Lynch, J. J. & Thomas, S. A. (1980). Animal companions and one year survival of patients after discharge from a coronary care unit. *Public Health Reports*, **95**, 307–12.
Friedmann, E., Katcher, A. H., Thomas, S. A., Lynch, J. J. & Messent, P. R. (1983). Social interaction and blood pressure. Influence of animal companions. *Journal of Nervous and Mental Disease*, **171**, 461–5.
Garrity, T. F., Stallones, L., Marx, M. B. & Johnson, T. P. (1989). Pet ownership and attachment as supportive factors in the health of the elderly. *Anthrozoös*, **3**, 35–44.
Goffman, E. (1971). *Relations in Public*. New York: Basic Books.
Goldmeier, J. (1986). Pets or people: another research note. *The Gerontologist*, **26**, 203–6.
Gonski, Y. A. (1985). The therapeutic utilization of canines in a child welfare setting. *Child and Adolescent Social Work Journal*, **2**, 93–105.
Gorczyca, K. C. (1990). AIDS, zoonoses, and the human–animal bond: a survey of providers' knowledge and attitudes (Abstract). *Living Together: People, Animals, and the Environment*, p. 35. Renton, WA: Delta Society.
Green, C. (1981). Fostering positive attitudes toward the elderly: a teaching strategy for attitude change. *Journal of Gerontological Nursing*, **7**, 169–74.
Greene, M. G., Adelman, R., Charon, R., Hoffmann, S. (1986). Ageism in the medical encounter: an exploratory study of the doctor–elderly patient relationship. *Language & Communication*, **6**, 113–24.
Guttmann, G., Predovic, M. & Zemanek, M. (1985). The influence of pet ownership on non-verbal communication and social competence in children. In *The Human–Pet Relationship*, pp. 58–63. Vienna: IEMT.
Hart, B. L. & Hart, L. A. (1985). *Canine and Feline Behavioral Therapy*. Philadelphia: Lea & Febiger.
Hart, B. L. & Hart, L. A. (1988). *The Perfect Puppy: How to Choose your Dog by Its Behavior*. New York: W. H. Freeman.
Hart, L. A. (1990). Pets, veterinarians and clients: communicating the benefits. In *Pets, Benefits and Practice*, Waltham Symposium 20, ed. I. H. Burger, pp. 36–43. London: BVA Publications.
Hart, L. A., Hart, B. L. & Bergin, B. (1987). Socializing effects of service dogs for people with disabilities. *Anthrozoös*, **1**, 41–4.
Hart, L. A., Hart, B. L. & Mader, M. (1990). Humane euthanasia and companion animal death: caring for the animal, the client, and the veterinarian. *Journal of the American Veterinary Medical Association*, **197**, 1292–9.
Hart, L. A. & Mader, B. (1986). The successful introduction of pets into California public housing for the elderly. *California Veterinarian*, **40**, 17–21.
Hendy, H. M. (1987). Effects of pet and/or people visits on nursing home residents. *International Journal of Aging and Human Development*, **25**, 279–91.
Hogarth-Scott, S., Salmon, I. & Lavelle, R. (1983). A dog in residence. *People–Animals–Environment*, **1**, 4–6.
Houpt, K. A. (1983). Disruption of the human–companion animal bond: aggressive behavior in dogs. In *New Perspectives on Our Lives with Companion Animals*, ed. A. H. Katcher & A. M. Beck, pp. 197–204. Philadelphia; University of Pennsylvania Press.
Hutton, J. S. (1985). A study of companion animals in foster families: Perceptions of therapeutic values. In *The Human–Pet Relationship*, pp. 64–70. Vienna: IEMT.
Katcher, A. H. (1985). Physiologic and behavioral responses to companion animals. *Veterinary Clinics of North America: Small Animal Practice*, **15**, 403–10.
Katcher, A. H. & Beck, A. M. (1985). Safety and intimacy: physiological and behavioral responses to interaction with companion animals. In *The Human–Pet Relationship*, pp. 122–8. Vienna; IEMT.
Katcher, A. H., Friedmann, E., Beck, A. M., & Lynch, J. (1983). Looking, talking, and blood pressure: the physiological consequences of interaction with the living environment. In *New Perspectives on Our Lives with Companion Animals*, ed. A. H. Katcher & A. M. Beck, pp. 351–62. Philadelphia: University of Pennsylvania Press.
Kidd, A. H. & Feldman, B. M. (1981). Pet ownership and self-perceptions of older people. *Psychological Reports*, **48**, 867–75.
Kidd, A. H. & Kidd, R. M. (1985). Children's attitudes towards their pets. *Psychological Reports*, **57**, 15–31.
Kidd, A. H. & Kidd, R. M. (1987). Reactions of infants and toddlers to live and toy animals. *Psychological Reports*, **61**, 455–64.
Kranz, J. M. & Schaaf, S. (1989). Nursing-home staff attitudes toward a pet visitation program. *Journal of the American Animal Hospital Association*, **25**, 409–17.
Lago, D. J., Connell, C. M. & Knight, B. (1985). The effects of animal companionship on older persons

living at home. In *The Human–Pet Relationship*, pp. 34–46. Vienna; IEMT.

Lago, D. J., Knight, B. & Connell, C. (1983). Relationships with companion animals among the rural elderly. In *New Perspectives on Our Lives with Companion Animals*, ed. A. H. Katcher, & A. M. Beck, pp. 329–40. Philadelphia: University of Pennsylvania Press.

Lapp, C. A. (1991). Nursing students and the elderly: enhancing intergenerational communication through human–animal interaction. *Holistic Nursing Practice*, **5**, 72–9.

Lee, R. L., Zeglan, M. E., Ryan, T., Gowing, C. B. & Hines, L. M. (1985). *Guidelines: Animals in Nursing Homes*. Renton, WA: Delta Society.

Levinson, B. M. (1969). *Pet-Oriented Child Psychotherapy*. Springfield, IL: Charles C. Thomas.

Lockwood, R. (1983). The influence of animals on social perception. *New Perspectives on Our Lives with Companion Animals*, ed. A. H. Katcher & A. M. Beck, pp. 64–71. Philadelphia: University of Pennsylvania Press.

Lockwood, R. & Beck, A. M. (1975). Dog bites among letter carriers in St. Louis. *Public Health Reports*, **90**, 267–9.

Lockwood, R. & Rindy, K. (1987). Are 'pit bulls' different? An analysis of the pit bull terrier controversy. *Anthrozoös*, **1**, 2–8.

Lynch, J. J. & McCarthy, J. F. (1969). Social responding in dogs: heart rate changes to a person. *Psychophysiology*, **5**, 389–93.

MacDonald, A. J. (1981). The pet dog in the home: a study of interactions. In *Interrelations between People and Pets*, ed. B. Fogle. Springfield, IL: Charles C. Thomas.

Mader, B. & Hart, L. A. (1992). Establishing a model pet loss support hotline. *Journal of the American Veterinary Medical Association*, **200**, 270–4.

Mader, B., Hart, L. A. & Bergin, B. (1989). Social acknowledgements for children with disabilities: effects of service dogs. *Child Development*, **60**, 1529–34.

Manuck, S. B. (1987). Coronary disease: discovering the behavioral connection (Abstract). National Institutes of Health Technology Assessment Workshop: Health Benefits of Pets, September 10–11, 1987.

Market Research Corporation of America. (1987). *Anthrozoös*, **1**, 123.

McCullochy, M. J. (1983). Companion animals, human health and the veterinarian. In *Textbook of Veterinary Internal Medicine; Diseases of the Dog and Cat*, 2nd edn, Vol. 1, ed. S. J. Ettinger, pp. 228–35. Philadelphia, PA: Saunders.

Melson, G. F. (1988). Availability of an involvement with pets by children: determinants and correlates. *Anthrozoös*, **2**, 45–52.

Melson, G. F. (1990). Fostering inter-connectedness with animals and nature: the developmental benefits for children. *People, Animals, Environment*, **8**, 15–17.

Melson, G. F. & Fogel, A. (1989). Children's ideas about animal young and their care: a reassessment of gender differences in the development of nurturance. *Anthrozoös*, **2**, 265–73.

Messent, P. R. (1983). Social facilitation of contact with other people by pet dogs. In *New Perspectives on Our Lives with Companion Animals*, ed. A. H. Katcher & A. M. Beck, pp. 37–46. Philadelphia; University of Pennsylvania Press.

Messent, P. R. (1984). Correlates and effects of pet ownership. In *The Pet Connection*, ed. R. K. Anderson, B. L. Hart, & L. A. Hart, pp. 331–40. Minneapolis: CENSHARE, University of Minnesota.

Miller, M. & Lago, D. (1990). Observed pet-owner in-home interactions: species differences and association with the pet relationship scale. *Anthrozoös*, **4**, 49–54.

Millot, J. L. & Filiatre, J. C. (1986). The behavioural sequences in the communication system between the child and his pet dog. *Applied Animal Behaviour Science*, **16**, 383–90.

Moss, S. P. & Wright, J. C. (1987). The effects of dog ownership on judgments of dog bite likelihood. *Anthrozoös*, **1**, 95–9.

Mugford, R. A. (1980). The social significance of pet ownership. In *Ethology and Non-Verbal Communication in Mental Health*, ed. S. A. Corson & E. O'L. Corson, pp. 111–22. Oxford: Pergamon.

Mugford, R. A. (1981). Problem dogs and problem owners: the behavior specialist as an adjunct to veterinary practice. In *Interrelations between People and Pets*, ed. B. Fogle, pp. 295–315. Springfield, IL: Charles C. Thomas.

Mugford, R. A. (1985). Attachment vs. dominance: alternative views of the man–dog relationship. In *The Human–Pet Relationship*, pp. 157–65. Vienna: IEMT.

Nassar, R. & Fluke, J. (1988). *Animal Shelter Reporting Study*. Denver, CO: American Humane Association.

Neer, C. A., Dorn, C. R. & Grayson, I. (1987). Dog interaction with persons receiving institutional geriatric care. *Journal of the American Veterinary Medical Association*, **191**, 300–4.

Nielsen, J. A. & Delude, L. A. (1989). Behavior of young children in the presence of different kinds of animals. *Anthrozoös*, **3**, 119–29.

O'Farrell, V. (1986). *Manual of Canine Behaviour*. Cheltenham, Glos.: British Small Animal Veterinary Association.

Ory, M. G. & Goldberg, E. L. (1983). Pet possession and life satisfaction in elderly women. In *New Perspectives on Our Lives with Companion Animals*, ed. A. H. Katcher & A. M. Beck, pp. 302–17. Philadelphia: University of Pennsylvania Press.

Paul, E. S. & Serpell, J. A. (1992). Why children keep pets: the influence of child and family characteristics. *Anthrozoös*, **5**, 231–44.

Quackenbush, J. E. (1981). Social work in a veterinary

hospital: a response to owner grief reactions. *Archives of the Foundation of Thanatology*, **9**, Abstract 56.

Redefer, L. A. & Goodman, J. F. (1989). Brief report: pet-facilitated therapy with autistic children. *Journal of Autism and Development Disorders*, **19**, 461–7.

Robb, S. S. (1983). Health status correlates of pet–human association in a health-impaired population. In *New Perspectives on Our Lives with Companion Animals*, ed. A. H. Katcher & A. M. Beck, pp. 318–27. Philadelphia: University of Pennsylvania Press.

Robin, M., Ten Bensel, R., Quigley, J. S. & Anderson, R. K. (1983). Childhood pets and the psychosocial development of adolescents. In *New Perspectives on Our Lives with Companion Animals*, ed. A. H. Katcher & A. M. Beck, pp. 436–43. Philadelphia: University of Pennsylvania Press.

Rogers, J. W., Hart, L. A. & Boltz, R. P. (1993). The role of pet dogs in casual conversations of elderly adults. *Journal of Social Psychology*, **133**, 265–77.

Rosenberg, M. & McCullough, B. C. (1981). Mattering: inferred significance and mental health among adolescents. *Research in Community and Mental Health*, Vol. 2, ed. R. G. Simmons, pp. 163–82. Greenwich, CT: JAI Press.

Rowan, A. N. & Williams, J. (1987). The success of companion animal management programs: a review. *Anthrozoös*, **1**, 110–22.

Salmon, P. W., & Salmon, I. M. (1983). Who owns who? Psychological research into the human–pet bond in Australia. In *New Perspectives on Our Lives with Companion Animals*, ed. A. H. Katcher & A. M. Beck, pp. 244–65. Philadelphia: University of Pennsylvania Press.

Savishinsky, J. S. (1986). The human impact of a pet therapy program in three geriatric facilities. *Central Issues in Anthropology*, **6**, 31–41.

Schantz, P. M. (1990). Preventing potential health hazards incidental to the use of pets in therapy. *Anthrozoös*, **4**, 14–23.

Schultz, D. J. (1987). Special design considerations for Alzheimer's facilities. *Contemporary LTC*, November, 48–56, 112.

Serpell, J. A. (1981). Childhood pets and their influence on adults' attitudes. *Psychological Reports*, **49**, 651–4.

Serpell, J. A. (1983). The personality of the dog and its influence on the pet-owner bond. In *New Perspectives on Our Lives with Companion Animals*, ed. A. H. Katcher & A. M. Beck, pp. 57–63. Philadelphia: University of Pennsylvania Press.

Serpell, J. A. (1986*a*). *In the Company of Animals*. New York: Blackwell.

Serpell, J. A. (1986*b*). Social and attitudinal correlates of pet-ownership in middle childhood (Abstract). *Living Together: People, Animals, and the Environment*, p. 127. Renton, WA: Delta Society.

Serpell, J. A. (1990). Evidence for long term effects of pet ownership on human health. In *Pets, Benefits and Practice*, Waltham Symposium 20, ed. I. H. Burger, pp. 1–7. London: BVA Publications.

Serpell, J. A. (1991). Beneficial effects of pet ownership on some aspects of human health and behaviour. *Journal of the Royal Society of Medicine*, **84**, 717–20.

Siegel, J. M. (1990). Stressful life events and use of physician services among the elderly: the modifying role of pet ownership. *The Journal of Personality and Social Psychology*, **58**, 1081–6.

Smith, S. L. (1983). Interactions between pet dog and family members: an ethological study. In *New Perspectives on Our Lives with Companion Animals*, ed. A. H. Katcher & A. M. Beck, pp. 29–36. Philadelphia: University of Pennsylvania Press.

Stallones, L., Marx, M., Garrity, T. F. & Johnson, T. P. (1988). Attachment to companion animals among older pet owners. *Anthrozoös*, **2**, 118–24.

Stehr-Green, J. K., Murray, G., Schantz, P. M. & Wahlquist, S. P. (1987). Intestinal parasites in pet store puppies in Atlanta. *American Journal of Public Health*, **77**, 345–6.

Stryler-Gordon, R., Beall, N. & Anderson, R. K. (1985). Facts and fiction: health risks associated with pets in nursing homes (Abstract). *Journal of the Delta Society*, **2**(1), 73–4.

Turner, D. C. (1985). The human–cat relationship: methods analysis. In *The Human–Pet Relationship*, pp. 147–52. Vienna: IEMT.

Voith, V. L. (1985). Attachment of people to companion animals. *Veterinary Clinics of North America: Small Animal Practice*, **15**, 289–96.

Vormbrock, J. K. & Grossberg, J. M. (1988). Cardiovascular effects of human–pet dog interactions. *Journal of Behavioral Medicine*, **11**, 509–17.

Weber, D. J., Wolfson, J. S., Swartz, M. N. & Hooper, D. C. (1984). *Pasteurella multocida* infections: Report of 34 cases. *Medicine*, **63**, 133–54.

Winkler, A., Fairnie, H., Gericevich, F. & Long, M. (1989). The impact of a resident dog on an institution for the elderly: effects on perceptions and social interactions. *The Gerontologist*, **29**, 216–23.

13 The welfare of dogs in human care

ROBERT HUBRECHT

Introduction

The most common role for the domestic dog in Western countries is that of companion animal or pet. Dogs, however, are also used as working and laboratory animals, and may live alongside human society in feral or semi-feral populations. In each of these different types of relationship, situations arise in which dog welfare may be compromised. Most people would agree, on basic moral or humanitarian grounds, that we have a responsibility to care for dogs, no matter why and for whatever purpose we keep them. But in order to provide such care, we need to make sensible decisions concerning what constitutes acceptable systems of housing, methods of transport, medical care, euthanasia, and so on. Ownerless, feral dogs do not impose the same moral responsibilities as owned animals, but welfare problems still exist and need to be addressed.

Serpell (Chapter 16) discusses the ambivalent attitudes to dogs that exist in many societies, including those of Europe and North America. This chapter explores one manifestation of this ambivalence. On the one hand, dogs in the West are often treated as one of the family and there is a concomitant public concern for their welfare. On the other hand, despite a growing body of research into farm or zoo animal welfare, the welfare of the domestic dog has been largely ignored as a topic of scientific inquiry.

Pet ownership and canine welfare

In 1992 it was estimated that there were 7.3 million pet dogs in the United Kingdom (PFMA Profile, 1993). So far, no one has attempted to assess the welfare of dogs living in human households, primarily because such a task would be exceedingly difficult to carry out and interpret in view of the many differences between households (Althaus, 1989). It is probably true to say, however, that most pet dogs lead relatively contented lives in which they are housed in comfort, adequately fed and exercised, cared for when sick and generally well looked after. Nevertheless, many dogs suffer from excessive or misguided care. Over-feeding, for example, can lead to obesity and associated health problems, particularly in some breeds (Anderson, 1973). Owner-exacerbated behavioural problems may also put the welfare of many pet dogs at risk (see Fox, 1968; O'Farrell, 1989 and Chapter 11; Mugford, Chapter 10).

Neglect and cruelty

Historically, dogs have often been poorly treated. Aristotle thought that animals were inferior to humans because they lacked the power of rational thought (Grant, 1989). In the seventeenth century, the followers of Descartes argued that animals were essentially automata and so could not feel pain. Although there were notable campaigners for compassion to animals, the overall effect of this anthropocentric worldview was that cruelty to animals was relatively commonplace (Hume, 1962; Serpell, 1986; Ryder, 1989). Not all of this cruelty can be attributed to speciesism, however, since views on human pain and suffering were also less compassionate than they are today. By the eighteenth and nineteenth centuries, attitudes towards animals in Britain were changing and the first animal protection legislation was enacted. In 1822 the first Act to prevent cruelty to animals was passed, and in 1835 bull-baiting and dog-fighting were both banned (Ritchie, 1981; Ryder, 1989). Since then, there has been a considerable change in public opinion and pet dogs are now protected by a wide range of legislation (Crofts, 1984; Sandys-Winsch, 1984). Nonetheless, it would be naive to imagine that legislation has eliminated cruelty. Britain is supposed to be a nation of dog lovers. Yet, as Table 13.1 demonstrates, the neglect and abandonment of dogs accounts for the vast majority of convictions obtained by the Royal Society for the Prevention of Cruelty to Animals (RSPCA). It is true that convictions for the more horrifying types of cruelty such as dog-fighting and deliberate cruelty are relatively few, but then the participants are likely to take steps to avoid detection.

Breed defects

As many as 800 true breeding types of dog are estimated to be in existence around the world (Fox & Bekoff, 1975). Until the nineteenth century, most breeds were developed for particular working characteristics and therefore the worst excesses of modern breeding were not seen. However, the modern desire to obtain 'perfect' breed types has resulted in selection for exaggerated characteristics

Table 13.1. *RSPCA Convictions 1990–2*

Convictions	1990	1991	1992	Total
Neglect	1172	1065	1231	3468
Abandoned	103	87	73	263
Dangerous dog offences	5	3	98	106
Dog fighting or baiting	52	37	4	93
Abandon in car during hot weather	34	17	22	73
Ill-treating	19	20	20	59
Improper killing	16	12	16	44
Keep while disqualified	14	11	13	38
Transport in a suffering manner	35		1	36
Beating	12	7	6	25
Illegal operation	15	6	2	23
Apply restrictive ligature (tail docking)	1	10	5	16
Failure to control	5	2	8	15
Kicking	6	2	7	15
Terrify	1	5	3	9
Permit to attack dog/cat	2	4	1	7
Shooting and wound	2	1	2	5
Draw carriage on a highway	4			4
Stabbing	1		3	4
Breeding offences	3			3
Injuring	1	1	1	3
Scalding			3	3
Baiting	2			2
Conveyance offence			2	2
Improper killing (attempted)			2	2
Selling unlicensed			2	2
Strangled on tether		2		2
Confine in small kennel			1	1
Confine in unsuitable container			1	1
Criminal damage			1	1
Exposure to risk of injury			1	1
Guard dog offence			1	1
Improper tethering			1	1
Transit offence			1	1
Unlicensed pet shop				
Total	1505	1292	1532	4329

that may adversely affect the welfare of certain breeds (Wolfensohn, 1981). The turned in eyelids (entropion) of breeds, such as the chow chow, selected for diamond-shaped eyes, cause considerable discomfort and may need surgical correction. The opposite condition (ectropion), which can lead to conjunctivitis, occurs most frequently in breeds selected for their mournful look, such as the bloodhound. Respiratory defects tend to be common in breeds with excessively short muzzles such as the bulldog, while selection for long backs, as in the dachshund and basset hound, may lead to spinal problems (Wolfensohn, 1981; Macdonald, 1985; Council for Science and Society, 1988; Willis, 1989).

In some breeds, genetic defects have arisen as accidental by-products of selection for other, apparently unrelated, traits. Careful breeding has produced good results in reducing the prevalence of some of these conditions. Progressive retinal atrophy (PRA), for example, affects a number of breeds including the Border collie, but adoption of eye tests for collies entered for sheepdog trials has resulted in a drop in prevalence from around 12 to 2% (Wolfensohn, 1981). As Willis (1989) pointed out, however, the genetics of these traits may be complex and expert advice from geneticists is often required. Undesirable behaviour characteristics may also arise in some breeds from selection for the 'look' rather than the personality and behaviour of dogs. (Borchelt, 1983; Mugford, 1984 and Chapter 10; Willis, 1989). One example of this type of problem is the behaviour known as 'flank-sucking' in Dobermann pinschers (Hart, 1977). The British Kennel Club recognizes such problems (Council for Science and Society, 1988), and it is to be hoped that future breeding will produce improvements.

Non-essential 'cosmetic' surgery

The desire to make dogs conform to some arbitrary physical 'ideal' has also encouraged the development of a number of surgical procedures which are carried out on dogs for non-therapeutic reasons. Such operations undoubtedly affect their welfare. Young (1976) regarded all such procedures, including castration, spaying, tattooing, declawing, devoicing, ear-cropping, tail-nicking and docking, teeth-cutting and ear implants, as 'mutilations' and therefore morally unacceptable. This seems a rather extreme position, however, since obviously some procedures are easier to justify than others. Sterilization, for example, is carried out under anaesthetic, and the small amount of suffering inflicted on the animals is probably outweighed by the reduction in the problem of unwanted puppies. Sterilized animals also tend to live longer and less restricted lives than intact animals. Ear-tattooing may also be necessary as a means of individual identification in large kennels, although alternative and less painful techniques, such as the subcutaneous injection of microchip transponders, are available.

Surgery carried out purely for 'cosmetic' purposes is clearly of no possible welfare benefit to the animals involved, and is now considered unacceptable by the Royal College of Veterinary Surgeons, the RSPCA, the British Veterinary Association and British Small Animal Veterinary Association. In July 1993 it became illegal in the UK for a lay person to dock a puppy's tail, and an EEC provision has been drafted to prohibit cosmetic operations. Nonetheless, the British Kennel Club still supports the practice of tail-docking (Council for Science and Society, 1988; Kennel Gazette, 1992, 1993).

Stray and feral dogs

A Working Party of the Council for Science and Society (1988) estimated that 300 000 to 400 000 dogs were lost, abandoned or strayed each year in the United Kingdom, and data from Battersea Dogs Home show that, in the years between 1980 and 1987, there was a 27% increase in the number of dogs received. Apart from their nuisance value to people, stray dogs often suffer from malnutrition and disease (Rubin & Beck, 1982) and are likely to be involved in traffic accidents. In Britain, the police have a statutory responsibility to round up stray dogs. The police then normally pass dogs on to various animal rescue organizations where they must be kept for at least seven days to allow time for owners to reclaim their pets. In some cases, the conditions under which dogs are kept in animal shelters are far from ideal (see below). Once the statutory seven-day period has elapsed, it is up to the staff at each shelter to decide how long a dog will be housed before being humanely destroyed. The sometimes arbitrary ways in which such decisions are often made have recently been described and discussed by Arluke (1994). Feral dogs – defined here as dogs that do not have owners and which have gone wild – generally derive from the stray dog population (e.g. Gentry, 1983; Boitani *et al.*, Chapter 15). They survive largely by scavenging and may become disease ridden and emaciated as a result (Fox *et al.*, 1975; Macdonald & Carr, Chapter 14). Culling often represents the most humane solution to the welfare problems posed by such animals.

The welfare of working dogs

Apart from their use as pets, dogs have been selected and trained to fulfil a wide variety of different working roles. In the past, some of these uses for dogs involved the infliction of considerable cruelty. The use of dog trains for transport in the eighteenth and nineteenth centuries provides a case in point. According to Ritchie (1981) these dog teams could be tracked for 20 miles or more by the bloody paw marks they left on the road. Campaigning by the RSPCA eventually resulted in the use of dogs for transport being prohibited in 1854. Moral reservations about some of the modern uses of working dogs are also justified, although nowadays most working dogs are probably no worse off from a welfare standpoint than pet dogs, and many enjoy a more varied and active life.

The welfare of laboratory dogs

The dog is a medium-sized mammal that is relatively easy to keep, and which has comparable physiology and similar dietary needs to those of humans. As a consequence, it is widely used in scientific and medical research as a model for human disease, the development of surgical techniques and in toxicology (MacArthur, 1987). In 1989, 12 625 scientific procedures were carried out on dogs in the United Kingdom. By 1991 the number had fallen by 16% to 10 583, and to 9085 procedures in 1992 (HMSO, 1990, 1992, 1993). The beagle is the most frequently used breed because of its temperament and because it has relatively few abnormalities (Ottewill, 1968). The use of the beagle in the laboratory is described in Andersen & Good (1970) while a more up-to-date review of general care of the dog is provided by MacArthur (1987). In the United Kingdom, research on dogs is covered by the Animals (Scientific Procedures) Act 1986, which regulates 'any experimental or other scientific procedure which may have the effect of causing that animal pain, suffering, distress or lasting harm'. As a response to this Act, the Royal Society and the Universities Federation for Welfare (UFAW) jointly published guidelines for the housing and care of animals used in scientific procedures (Royal Society/UFAW, 1987), and the Government has used these guidelines as a basis for its own Code of Practice (HMSO, 1989). The Code provides minimum cage or pen dimensions for dogs of different sizes and housed in varying numbers (Table 13.2). The Code also recommends: that dogs should normally be housed on solid rather than grid floors, that compatible dogs may be housed together in pairs, that dogs need to have regular human contact, and that dogs should be able to exercise with other dogs. While most of these appear to be sound recommendations and are based on the practical experience of experts, there is little scientific evidence to support them.

In the United States, dogs are covered by the Animal Welfare Act. The regulations require that the minimum dimensions of the kennel should be the square of the dog's length plus six inches, and cage height must be six inches taller than the height of the dog (Hetts, 1991). These dimensions clearly represent the minimum necessary to allow a dog to turn around. Further guidelines on the care of dogs are provided by the National Institutes for Health and Institute of Laboratory Animal Resources Guide for the Care and Use of Laboratory Animals (1985). Amendments to the Animal Welfare Act in 1985 have resulted in a directive that requires that standards should be developed for the exercise and socialization of dogs. Again, there was at the time little a priori scientific evidence to justify these amendments, but they have been of benefit in provoking research and general discussion (Campbell *et al.*, 1988; Mench & Krulisch 1990; Clark, Calpi & Armstrong, 1991; Hetts, 1991; Hetts *et al.*, 1992).

Table 13.2. *UK Home Office Code of Practice minimum floor areas (m^2) for the housing of dogs used in scientific procedures*

Weight Kg	In Groups	Singly
<5	1	4.5
5–10	1.9	4.5
10–25	2.25	4.5
25–35	3.25	6.5
>35	4.0	8.0

Determining welfare

There are difficulties with the definition of welfare and the concept has been much debated. Welfare clearly relates to an animal's ability to adjust to, or cope successfully with, the prevailing conditions in which it finds itself. In many cases, animals are capable of adapting to changes in their environment by making appropriate adjustments to their behaviour and physiology. Welfare problems arise, however, when environmental conditions are so extreme that the animal is no longer able to cope successfully.

To satisfy basic welfare criteria, animals need to be kept in conditions providing adequate food, water and ventilation, and protected from the risk of injury or ill health. Beyond these basic prerequisites, the assessment of an animal's social or behavioural needs is more problematic. One approach is to ask whether the environment in question allows the animal to behave in a 'natural' fashion. Another is to ask whether the animal is suffering, i.e. undergoing any of a wide range of unpleasant emotional states (Dawkins, 1980; Duncan & Dawkins, 1983). Carpenter (1980) related welfare to 'the degree to which animals can adapt without suffering to the environments designated by man'. Most people would agree that if an animal is suffering, then its welfare is poor, but determining whether or not an animal is suffering can be difficult (see review in Hetts, 1991).

The problems of measuring dog welfare are essentially the same as those for any other mammal, and therefore the same general techniques can be used. Broom (1988) has defined welfare as being: 'The state of the animal as regards its attempts to cope with its environment'. By studying an animal's physiological and/or behavioural responses to its environment, it is possible to obtain some measure of its success at coping.

Physiological measures

Animals have evolved ways of coping with various environmental pressures that are reflected by measurable changes in their physiology. Two basic physiological systems are involved, although they should be seen as part of an integrated whole (Dantzer, Mormède & Henry, 1983).

1. The sympathetic adrenomedullary axis is activated when an animal is challenged by some sort of environmental stressor and it attempts to rectify the situation as, for example, during an agonistic encounter or on exposure to an unusual or alarming stimulus. Activation of the sympathetic nervous system leads, amongst other things, to increases in heart rate, blood pressure, respiration rate, coagulability of the blood and in the secretion of adrenaline from the adrenal medulla. Deactivation of the sympathetic nervous system is associated with a state of relaxation (Dantzer *et al.*, 1983). The adrenomedullary system has been investigated in the dog by Liang *et al.* (1979), who showed that exposure to a stressful environment resulted in an increase in circulating catecholamines and greater ventricular vulnerability to fibrillation.
2. Alternatively, if coping by active means is not possible and the animal is exposed to severe or prolonged stressors, then it may resort to submissive, vigilant behaviour sometimes involving extreme passivity. Such behaviour is associated with the activation of the hypothalamico–pituitary–adrenocortical axis. This response is characterized by the secretion of adrenocorticotropic hormone (ACTH) releasing factor from the hypothalamus that stimulates the secretion of ACTH from the anterior pituitary. This in turn leads to increased secretion of cortisol from the adrenal cortex. Activation of the sympathetic nervous system may also occur (Schneiderman & McCabe, 1985). Activation of the hypothalamico–pituitary–adrenocortical axis is associated with situations in which an animal is unable to control its environment (Dantzer *et al.*, 1983).

Duncan & Dawkins (1983) have pointed out that physiological responses tend to be graduated, so it may be difficult to decide when a particular degree of physiological change reflects an unacceptable drop in welfare (see e.g. Mendl, 1991). It is easier to determine poor welfare when a long-term failure to cope produces measurable reductions in health. Stress induced gastric ulcers or heart conditions may result and the effectiveness of the immune system may be impaired (Kelley, 1980; Schneiderman & McCabe, 1985). Stress may also affect animals' reproduction and growth adversely, although good reproductive rates, health and growth are not necessarily reliable indicators of good welfare (Dawkins, 1980).

Few studies have attempted to relate dog physiology to welfare (Neamand *et al.*, 1975; Hite *et al.*, 1977; Campbell *et al.*, 1988). These have tended to concentrate on the effects of pen size, and none has found any effect of this variable on a range of blood and urine chemistry values, including plasma cortisol. Knol (1989), however, described a series of experiments on male dogs which showed that, of two potential stressors, immobilization, but not social conflict, produced rises in circulating levels of cortisol. Hence, the type of stressor may be crucial in terms of hormonal response and welfare implications. Recent United States legislation has placed emphasis on the provision of exercise programmes for laboratory dogs, but Clark *et al.* (1991) were unable to demonstrate a substantial effect of either cage size or exercise levels on the physical fitness of dogs housed in standard USA cages.

Behavioural measures

Comparisons with field studies

A common technique used to provide clues to a species' needs in captivity, is to study its behaviour in the wild (Dawkins, 1980). The domestic dog, however, has been subjected to thousands of years of artificial selection that has produced a number of changes to its ancestral behaviour including prolongation of the juvenile period (Frank & Frank, 1982), early sexual maturation and breed-specific truncation of prey-catching behaviour (reviewed in Bradshaw & Brown, 1990). Furthermore, breeds differ in their behaviour (e.g. Scott & Fuller, 1965) and may therefore have different welfare needs.

The closest living ancestor to the domestic dog is believed to be the southern or Indian wolf, *Canis lupus pallipes* (Hemmer, 1990; Clutton-Brock, Chapter 2). However, there have been no comparative studies of behaviour with this subspecies. Its close relative the grey wolf has been studied extensively and is one of the most social members of the family Canidae (reviewed in Nowak & Paradiso, 1983), but it is believed that the southern wolf may be less social (Hemmer, 1990). The behaviour of feral dog populations, or of primitive wild dogs such as the dingo, may also provide insights into the domestic dog's natural requirements. Unfortunately, as Nowak & Paradiso (1983) pointed out, dingoes and feral domestic dogs are the least studied canids with respect to their behaviour and ecology in the wild.

Feral dogs usually roam in packs, although solitary animals are sometimes seen (Scott & Causey, 1973; Daniels & Beckoff, 1989; Macdonald & Carr, Chapter 14). Free-ranging urban dogs also maintain social relationships with conspecifics, although environmental factors affect the details of these relationships. In behavioural terms, a clear distinction also exists between feral dogs and those that have owners, but which roam freely for part of the day (Rubin & Beck, 1982). Several studies have purported to show that free-ranging urban dogs are solitary and lack social structure (Berman & Dunbar, 1983; Daniels, 1983*a*), but Carr (1985), Font (1987) and Macdonald & Carr (Chapter 14) have argued that while they may forage solitarily, they still interact as social groups. The dingo, *Canis familiaris dingo*, has been reported as ranging over areas of up to 5840 hectares in a period of 45 days (Harden, 1985). While not as large as this, the area over which feral domestic dogs travel may also be considerable. Nesbitt (1975), for example, recorded a range of 2850 hectares in the natural conditions of a wildlife refuge. Daily activity patterns also vary. In Nesbitt's study the dogs travelled at all times of day and night, while Scott & Causey (1973) found that packs were most active during the night. Human activities tend to govern free-ranging dog behaviour in urban areas, and peak activity times occur during crepuscular periods (Beck, 1975; Fox, Beck & Blackman, 1975; Rubin & Beck, 1982; Font, 1987). Dingoes also seem to be most active at these times of the day (Harden, 1985). Domestic dogs differ from most large wild canids in that they have switched from the usual monogamous mating system with paternal care (Malcolm, 1983) to a promiscuous mating system (Kleiman & Brady, 1978). Nonetheless, observations of free-ranging urban dogs show that females exercise mate selection and appear to mate only with familiar males (Daniels, 1983*b*).

While these data are useful for comparison, it should be remembered that in the wild most animals' activity patterns are more variable than in captivity (Tester, 1987). Moreover, as Hetts (1991) has pointed out, without appropriate experimental tests, one cannot be certain that an animal in captivity will necessarily suffer if it is denied the opportunity to perform a species-typical behaviour.

Laboratory studies

In view of the difficulties associated with making comparisons between domesticated and wild animal behaviour, an alternative approach may be to compare animals' behavioural responses to living in either poor or good artificial environments. While there are problems in determining what constitutes a good environment, it is perhaps easier to reach a consensus on what constitutes a poor one. Work with many other species has shown that housing in impoverished or restricted physical or social environments can lead to the development of behavioural abnormalities, and this also has been reported to be the case for dogs (Thompson, Malzack & Scott, 1956; Fox, 1965, 1986; Anderson & Good, 1970; Luescher, McKeown & Halip, 1991). Such abnormal behaviour may take the form of a reduced behavioural repertoire, the development of repetitive and apparently functionless behaviour or 'stereotypies' (*sensu* Broom, 1983), such as circling, pacing, whirling, repetitive grooming or self-biting, polydipsia or polyphagia, compulsive staring, and the excessive social facilitation of normal behaviour, such as barking. A difficulty then arises in determining how much abnormal behaviour an animal needs to show to indicate that it is suffering (Duncan & Dawkins, 1983; Mendl, 1991). Broom (1983) has suggested, for example, that stereotypies occupying more than 10% of the animal's waking time are indicative of poor welfare, although one needs to be alert to the possibility that the animal's current behaviour may be a legacy of previous housing (see Mason, 1991*a*, *b*). Ideally, any such behavioural changes require experimental examination to demonstrate stress or poor welfare (Duncan, 1983).

A technique that has been used with other species to help assess the suitability of housing is to give animals some choice over their environment and then manipulate the situation in various ways to determine how much the animal is prepared to 'pay' to have access to a particular environmental resource (Dawkins, 1980, 1988). Such studies have not yet been attempted for any canid. Another approach is to subject animals to known stressful situations and compare the elicited behaviour to that shown in the test housing (e.g.

Duncan, 1983). This is, perhaps, a less humane approach, and has not been attempted with the dog.

It is reasonable to assume that mammals, in common with people, feel pain (see Rollin, 1986) and most people would consider that the welfare of an animal in pain is poor. Those handling dogs should be able to tell whether they are in pain, and Morton & Griffiths (1985) and Taylor (1985) have provided useful check-lists of clinical signs of pain in the dog. These include distinctive vocalizations and body postures, the seeking of cold surfaces, penile protrusion and frequent micturation.

Welfare aspects of dog housing

The current situation

In most laboratories and animal shelters, dog housing designs tend to give higher priority to hygiene, health and ease of use by kennel staff or technicians than to providing for other aspects of canine welfare. Ottewill (1968) provides an example of this attitude in his detailed description of the basic environmental and accommodation requirements of the experimental dog. The paper discusses physical welfare in terms of heating, ventilation, etc., but ways in which housing or pen furniture could be altered or designed to allow dogs to behave normally are virtually ignored apart from a reference to the problem of boredom. Ottewill's paper should be seen as a product of its times, and could no longer be said to represent the standard housing policy. Nevertheless, although modern dog housing varies enormously in detail between different institutions, the basic features remain remarkably conservative. Of necessity, pens are small and bear no relation to, say, the typical ranging behaviour of feral dogs or indeed the sort of daily exercise enjoyed by many pets. In addition, dogs are opportunistic feeders (Kleiman, 1967; Thorne, Chapter 7) and Morris (1970) has argued that such a 'diet generalist' is more likely to crave novelty in its environment. Yet, in order to allow staff easy access and vision, dog pens tend to be simple, with the result that their occupants can view all areas of the pen without moving. Sleeping areas are often merely raised boards or an insulated raised area of the floor, and the dog is not usually able to retreat to a hidden area. Inside and outside runs are usually bare of any type of environmental enrichment. Toys such as chews or marrow bones are sometimes provided, but often these are denied to group-housed dogs on the grounds that they may cause fights. The situation is made even worse when, for quarantine or scientific reasons, dogs have to be housed singly. In such impoverished and abnormal conditions it is not necessarily anthropomorphic to suggest that an animal may become bored (*sensu* Wemelsfelder, 1985) or frustrated (Appleby, 1991). Nor is it surprising that behavioural abnormalities sometimes develop (Fox, 1986; Luescher *et al.*, 1991; Hetts *et al.*, 1992).

Research into dog welfare and housing

Design of housing

Although dogs are well known as pets, little research, apart from studies on physical health, exists on the welfare requirements of dogs in animal housing. Many aspects of dog housing, including height of partitions, number of cages per room, stocking density, provision of sleeping areas, etc., may all be relevant to canine welfare, but previous studies have tended to concentrate primarily on the effects of cage or pen size. Pettijohn, Davis & Scott (1980) working with the telomian, a breed of dog particularly prone to agonistic interactions, found that a 75% reduction in cage size had no apparent effect on the frequency of aggressive interactions, although a substantial increase in levels of illumination tended to reduce aggression (Pettijohn, 1978). Work by Neamand *et al.* (1975), Hite *et al.* (1977) and Campbell *et al.* (1988) indicated that increasing cage size (within the range of 0.58–3.00 m^2) has little or no effect on dogs' activity levels. Indeed, Hughes, Campbell & Kenney (1989) found that dogs housed in larger cages were actually less active. More recently, Bebak & Beck (1993) compared groups of four dogs housed in cages of 2.22 and 7.32 m^2 (exceeding USDA requirements by 1.3 m^2 per dog) and found no effect on aggression or play. These findings may give the impression that cage size is relatively unimportant to dogs. However, the initial cage sizes used in these studies were typically small, as were the experimental increases in area available. Such small increases in available space may be of no practical or biological relevance to dogs. Furthermore, the automated activity recording system used by Hughes *et al.* (1989) cannot distinguish between movements carried out for different purposes (e.g. play, locomotion, etc.), and would therefore ignore potentially beneficial qualitative changes in behaviour.

Using an ethogram of over 50 different canine behaviour patterns (see Table 13.3(*a*)), Hubrecht,

Table 13.3(*a*). *Behaviour categories used in the study*

Category	Definition
Rest	Lying down with eyes open or closed
Sit	Sit on hind legs
Stand	Stand on four legs
Walk	Ambulatory gait
Trot	Trotting gait
Run	Running gait
Hind legs	Standing on hind legs using forelegs against a wall to support the body
Circle	Repetitive circling around pen
Tail chase	Repetitive chasing of tail
Pace	Repetitive pacing usually along a fence
Social pace	Repetitive pacing along fence in parallel with a dog on the other side
Jump	Repetitive jumping so that hind legs leave the ground
Wall bounce	Repetitive jumping at wall, rebounding off it
Flank suck	Repetitive and prolonged auto-grooming of flank
Contact dog[a]	Lying in contact with dog
Amicable dog	Lick, paw or allogroom dog often with tail wag
Threat dog	Snarl, raise hackles to dog
Attack dog	Bite, snap, or chase dog often with aggressive vocalizations
Defensive dog	Evade dog, cower, roll over, lick face
Competitive dog	Defend object or food from dog
Sniff dog	Nose to any area of another dog
Solicit play dog	Bow, short charges with bouncing gait, often barking
Play with dog	Bouncing gait, play face, wrestle, play chase
T-dog	Muzzle placed across neck of another dog
Mount dog	Hetero/homosexual mounting of another dog
Mounted	Focal animal mounted by other dog
Amicable human	Lick, paw, allogroom human, often with tail wag
Threat human	Snarl, raise hackles to human
Attack human	Bite, snap or chase human
Defensive human	Evade human, cower, roll over
Competitive human	Defend object or food from human
Sniff human	Nose to any area of human
Solicit play human	Bow, metaplay with human
Play human	Bouncing gait, play face, wrestle, play chase
Pat dog[a]	Human pat dog
Bark[a]	Staccato vocalizations
Howl[a]	Long drawn out vocalizations
Bark at passers	Recorded where object of barking could be seen
Autogroom	Lick, pull at body/pelage
Dig	Dig at ground with fore paws
Urinate squat	Urinate in squatting position
Urinate raised leg	Urinate with one leg cocked
Kick ground	Scratching ground usually following urination or defecation
Sniff ground	Nose to ground
Eat	Eating food
Drink	Drinking
Coprophagy	Eat own or other dog's faeces
Chew	Chew nonnutritive material, e.g. pebble
Eat grass	Eating grass (only possible at Woodgreen)
Mouth toy	Chew toy
Defecate	
Kennel	In kennel (LGH) only
Staff available[a]	Staff available for interaction with dogs in pen
Public available[a]	Public available for interaction with dogs in pen

Locomotory categories were recorded in the absence of other mutually exclusive behaviours.
[a]Nonexclusive category

Serpell & Poole (1992) conducted comparative observations of dogs housed either singly or in groups at four different laboratory and animal shelter sites. Due to differences in breed, age and history between the sites, behavioural differences may not have been solely due to differences in housing regimen. Nonetheless, consistent trends allow some conclusions to be drawn. The sites differed in terms of cage or pen size, the

Table 13.3(*b*). *Combined behaviour categories used in the study*

Active	Walk, trot, run, hind legs
Active (repetitive)	Circle, pace, social pace, jump, tail chase, wall bounce, flank-suck
Inactive	Sit, rest, stand
Socialize with human	Amicable human, defensive human, play with human, sniff human
Socialize with dogs	Amicable dog, threat dog, attack dog, defensive dog, competitive dog, play dog, solicit play dog, T-dog, sniff dog, mount dog, mounted by other dog
Alimentary	Eat, drink, eat grass, urinate raised, urinate squat, defecate, coprophagy
Others	Dig, sniff-ground, kick grass, chew, mouth toy, autogroom

degree of social interaction possible, and in the physical complexity of the environment provided (see Table 13.4). Although all of the dogs spent much of their time inactive, as shown by their activity budgets (see Fig. 13.1), biologically meaningful differences in pen size had the expected consequences on locomotory behaviour. In the smallest pens at Battersea Dogs Home (2.17 m^2), the dogs spent significantly more time resting and less time walking than at any of the other sites (see Fig. 13.2). Conversely, in the very large pens provided at Wood Green Animal Shelter (744 m^2), the dogs showed significantly more trotting and running behaviour than at the other sites (see Fig. 13.3). Running, however, was a rare activity and was normally seen as a response to some disturbance outside the enclosure. At Wood Green, the dogs also had access to large areas of grass, but this did not appear to make much difference to their behaviour. Activities, such as digging and eating grass, which were only possible in these sorts of enclosures, were very rare.

Items of cage furniture also had significant effects on the dogs' behaviour. In the Laboratory Group-housed (LGH) pens, food was provided in pellets from a hopper lowered into the pen for up to 2 hours per day. This feeding method differed to that at all other sites, where food was provided in bowls. The dogs in LGH spent much longer feeding and drinking than at any of the other sites (see Fig. 13.4), and it was apparent that they took small quantities from the hopper, often carrying the food away for chewing. This result suggests that, as with many other species (see e.g. Anderson & Chamove, 1984), choice of appropriate feeding methods might help to alleviate boredom in captivity. There was some limited interference between the dogs, but the feeding period was adequate for all the dogs to eat their fill. LGH dogs were also provided with a kennel within their pen, and this also proved to be an important resource. The dogs spent an average of 35% of their time inside the kennel, and it was used most in the afternoon (51% of the time observed). The dogs may have preferred the kennel because it had a hard floor rather than the slatted one used in the rest of the pens, but the kennel was also used during social play and as a refuge from the attentions of other dogs.

Table 13.4. *Study sites, ages of subjects and duration of stay in housing at time of observation*

Site	Type of accommodation	Type of pen	Floor area of pens (m^2)	No. of dogs per pen	Mean age (months)	Mean duration (days)
Wood Green	Shelter	Outside run	744.00	5–11	24[a]	20
Battersea	Shelter	Indoor	2.17	1	26[a]	81
Laboratory Group Housing (LGH)	Laboratory	Indoor-slatted floor	6.69	5	7	298
Laboratory Single Housing (LSH)	Laboratory	Indoor	4.17–6.83	1	29	548

[a]Ages at Woodgreen and Battersea were estimated by veterinary staff.

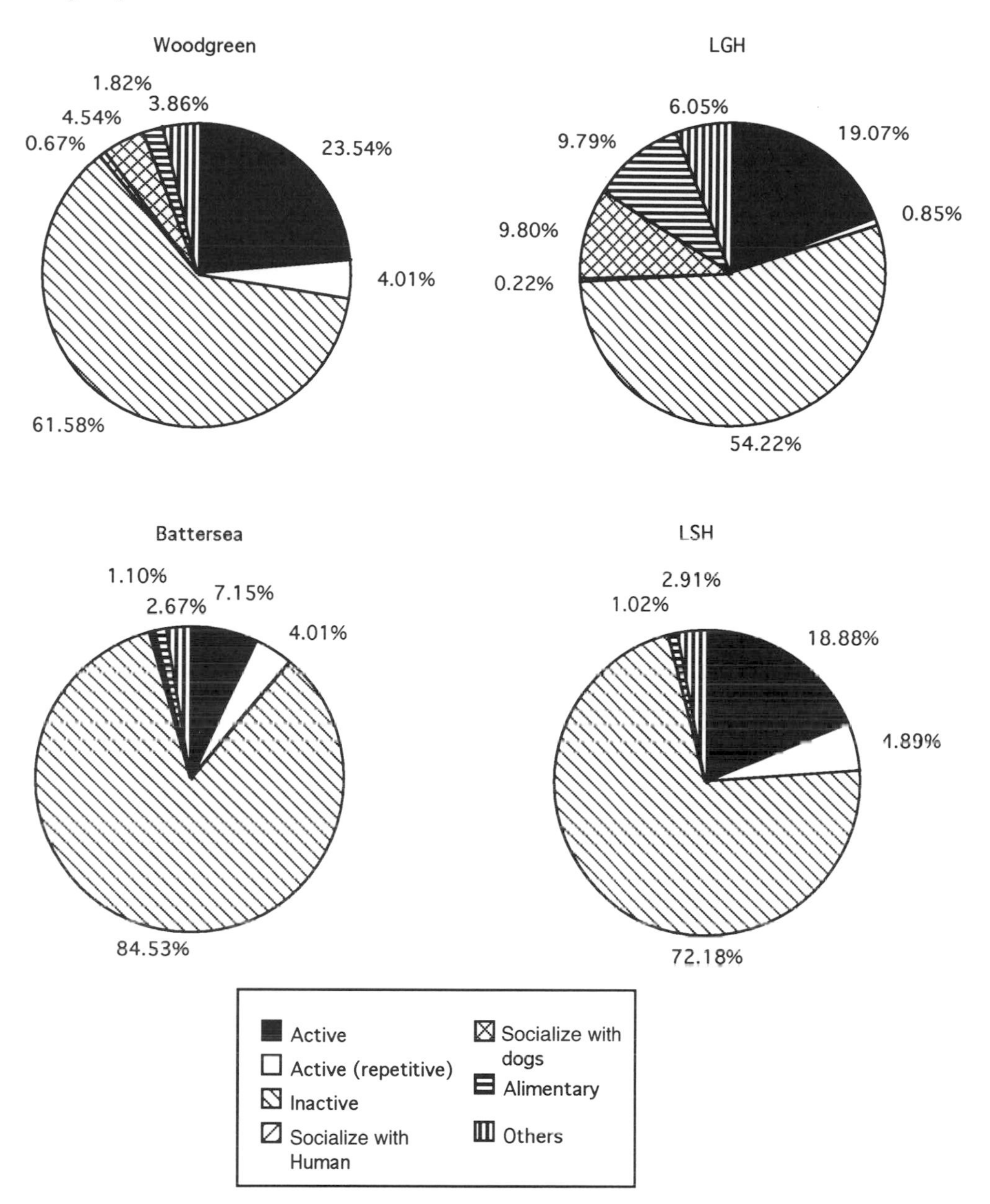

Fig. 13.1. Dog activity budgets in animal shelter and laboratory housing (for composition of behaviour categories see Table 13.3(*b*)).

Kennels may be valuable because of the extra degree of control they give the dogs over their social and physical environment, and it is relevant that recent research by Jeppesen & Pedersen (1991) has shown that the addition of a nest-box to the cages of farmed silver foxes reduces their fearfulness and increases their willingness to explore a novel environment. Moreover, foxes with access to nest-boxes have lower base levels of serum cortisol and higher lymphocyte counts.

In an attempt to increase the complexity of laboratory housing for dogs, Hubrecht (1993) recently experimented with the installation of high platforms accessible by steps (see Fig. 13.5). These provided the dogs with a vantage point from which they could see outside their pens, and also effectively gave them

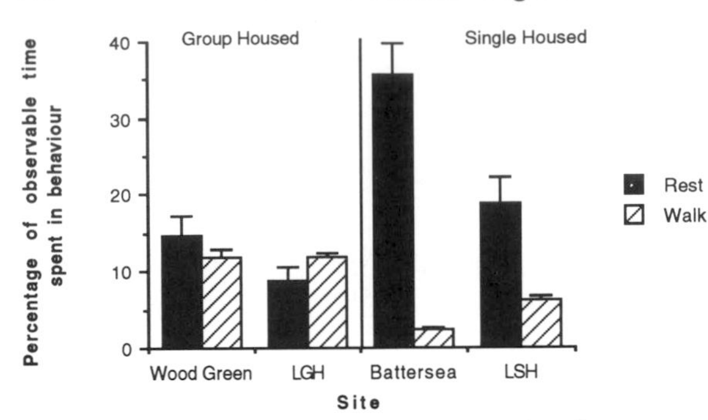

Fig. 13.2. Percentage of time spent resting and walking in four housing conditions.

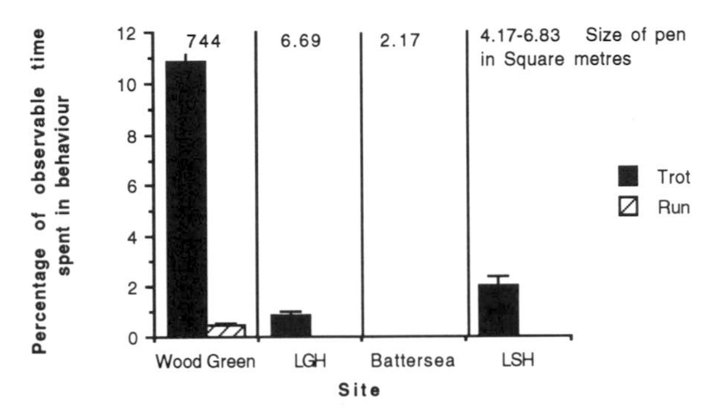

Fig. 13.3. Percentage of time spent trotting & running in four housing conditions.

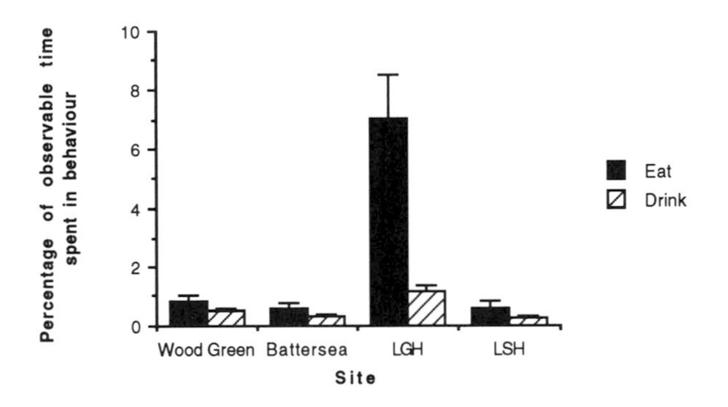

Fig. 13.4. Percentage of time spent eating & drinking.

more floor area and a more interesting three dimensional environment. On average, dogs spend over 50% of their time on these platforms.

Intra-specific social contacts and welfare

It is well known that lack of control over events in one's environment can result in a failure to cope (e.g. Seligman, 1975). The provision of social contacts may help to increase an animal's sense of control, and will

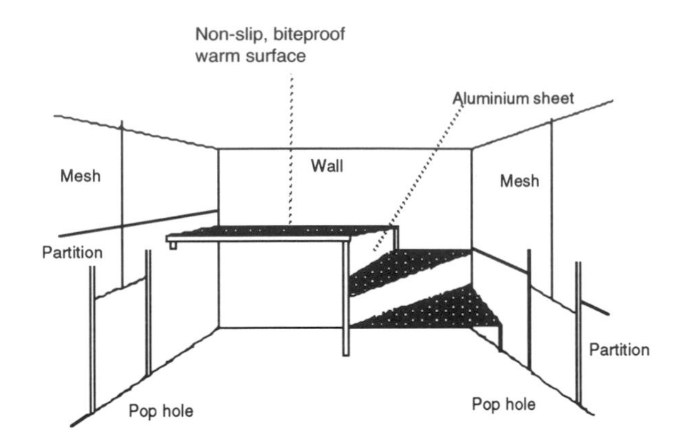

Fig. 13.5. Environmental enrichment design for laboratory dog housing incorporating a high platform accessible by steps.

also result in a more complex environment overall. As in many other mammals, early social isolation may cause severe behavioural deficits in the domestic dog (Fox, 1965, 1986). Isolation from conspecifics may also lead to deficits even when human social contact is provided (Fox & Stelzner, 1967). Little previous research has examined the influence of conspecific interactions on the welfare of dogs, although Hughes *et al* (1989) stated that housing in pairs reduced activity and Solarz (1970) asserted that good visibility between cages tended to stimulate allelomimetic behaviour such as socially-facilitated barking. More recent work by Hetts *et al.* (1992) described behavioural changes under various housing regimens; the most bizarre behaviour being seen under conditions of greatest social isolation.

Housing dogs in groups has two important consequences: not only are the dogs provided with a more complex social environment, but also, generally, a larger pen size. These two variables tend to confound each other, although the data of Hubrecht *et al.* (1992) suggested that housing in groups is associated with more walking and less time resting (see Fig. 13.2). Housing in groups obviously enables dogs to behave socially (see Fig. 13.6), and also has the side effect of making the olfactory environment more interesting. This is demonstrated by the greater amount of time group-housed dogs spend sniffing the ground (see Fig. 13.7). Although these activities account for relatively small proportions of the dogs' total activity budgets (4.5–9.7%), they are neverthe-

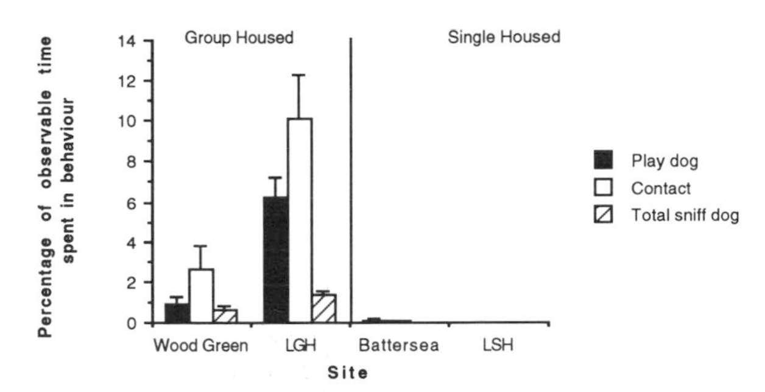

Fig. 13.6. Percentage of time spent in social behaviour.

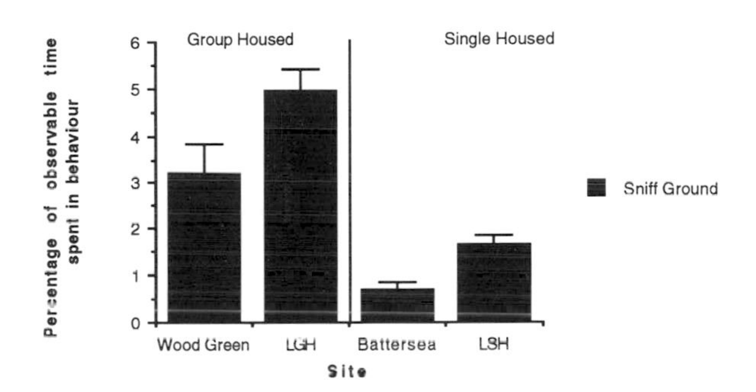

Fig. 13.7. Percentage of time spent sniffing ground.

less likely to help to alleviate boredom. The same study also suggested that group housing is associated with lower cumulative totals of repetitive or stereotyped behaviour patterns, such as circling and pacing (see Fig. 13.8). Circling was the predominant stereotyped behaviour in the smaller pens, while in the larger pens at Wood Green pacing was more common. This effect of cage size has also been noted in other species (see e.g. Meyer-Holzapfel, 1968).

Where dogs are housed in adjacent pens with wire

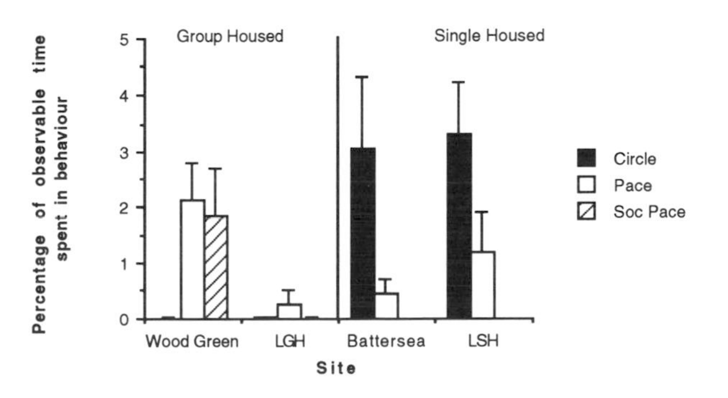

Fig. 13.8. Percentage of time spent in repetitive or 'stereotypic' behaviour.

fencing between them, social pacing may result (see Fig. 13.8). This activity involves one dog running up and down the length of the fence in parallel with another dog on the other side, and often barking at it. Social pacing was reminiscent of other stereotyped behaviour, but the apparent social element places it in a distinct category. The least amount of stereotypic or repetitive behaviour was seen at the LGH site. This may have been because the dogs were younger at this site, or because the relatively high stocking density may have prevented the development of repetitive locomotory behaviour through physical interference, i.e. individual attempts to circle or pace would probably have been interrupted by some form of social interaction. The interpretation of relative durations of stereotypies shown by caged animals is not straight forward, as they are not always indicative of the animal's current welfare status (see Mason, 1991*a*, *b*). Nonetheless, the fact that more than 10% of the dogs at three of the four sites spent over 10% of their time engaging in repetitive behaviour (see Table 13.5) gives rise to some concern.

It is important to emphasize that group housing may also have adverse welfare consequences through fighting. Not surprisingly, if two male dogs are confined in a pen with sufficient food for only one, raised cortisol levels may result (Knol, 1989). On a more practical note, it is often necessary to introduce unfamiliar dogs to pens already containing established groups; a procedure that is likely to be stressful or hazardous to the newcomer. Unfortunately, no published data are available on the degree of stress involved, although unpublished findings (Foster & Emmerson, personal communication) suggest that introducing dogs in pairs rather than singly does not reduce agonistic interactions, whereas agonistic interactions are reduced if a new animal is allowed to 'acclimatise' in an adjoining pen for 30 minutes prior to introduction.

Human social contacts and welfare

Provided that dogs have an opportunity to interact with humans during the primary socialization period, they will normally react positively to them in later life (Freedman, King & Elliot, 1961; Scott & Fuller, 1965). The time required to achieve this may be quite short. Wolfle (1990), for example, describes a programme that achieved adequate socialization with

Table 13.5. *Repetitive behaviour scores*

Site	% of dogs showing any repetitive behaviour	For all dogs, mean % of time spent in repetitive behaviour	% of dogs spending >10% of time in repetitive behaviour	Highest score for repetitive behaviour (% of time observed)
Woodgreen	42	4	14	47
Battersea	62	4	12	63
LGH	46	0.9	0	6
LSH	84	4.9	13	51

less than five minutes of human social contact per pup each week. More contact may, however, be beneficial. Fox & Bekoff (1975) suggested that the presence of a human is rewarding to dogs, and both Wolfle (1987, 1990) and Fox (1986) have argued that human social contact is important for dog welfare, possibly even more important than canine contact. Although experimental evidence concerning the benefits of human social contact is sparse, it is certainly true that the presence of humans can affect dogs' behaviour and physiology. Campbell *et al.* (1988) reported that the presence of people increased the activity levels of kennelled dogs, and Fox (1986) has suggested that withdrawal of regular contact with humans can result in dogs becoming 'people-shy'. Lynch & Gantt (1968) showed that handling can reduce laboratory dogs' heart rates, and Verga & Carenzi (1983) found that Rottweilers reared in human families show fewer fear reactions in a foraging test situation than those reared in kennels. This latter study, however, suffers from a number of potentially confounding influences. For example, in a family environment, dogs would presumably be exposed to a greater variety of novel situations. The familiarity of the person providing social contact may also be important, as may their gender. An intriguing study by Lore & Eisenberg (1986) showed that male – but not female – dogs were much less likely to approach an unfamiliar man in a kennel environment, although dogs and bitches were equally likely to approach an unfamiliar woman.

Despite the widely accepted benefits of human social contact, the majority of kennel-housed dogs probably receive little such contact. Hubrecht *et al.*'s (1992) study revealed that dogs at four different sites had little opportunity to interact with humans (from 2.52–0.34% of time observed), and spent even less of this time actually interacting (from 0.67–0.036% of time observed). It was apparent that many of the normal husbandry activities of kennel staff, such as the use of brooms, hoses or buckets of water, tended to discourage interactions. Management at both shelter and laboratory sites considered human contact to be beneficial, but the above results indicate that it may be necessary to provide more structured opportunities for human socialization.

Environmental enrichment and welfare

Exposure to a sufficiently stimulating environment during development is generally regarded as important as a means of avoiding the problem of excessively nervous or 'institutionalised' dogs (Fox & Spencer, 1969; Fox, 1971*a*). Some establishments provide toys for this purpose, but only one recent study has assessed their value as sources of environmental enrichment. DeLuca & Kranda (1992) looked at the responses of dogs, cats and pigs to a wide variety of novel toys or chews. Although hampered by observational difficulties, their study appeared to demonstrate breed and individual differences between dogs in their responses to the different items. Other forms of cage enrichment have not been widely used, and there is no published research available concerning their value.

As part of a study to determine whether various forms of enrichment early in life can improve dogs' ability to cope with later environmental changes, I have quantified the time spent playing with various toys and chews by ten-week-old beagle pups. Four litters of pups, aged between 5 and 14 weeks of age, were provided with either a plastic bucket and a

Gumabone tug toy[1], or a piece of rolled rawhide[2] and a dowel stick, all suspended from the ceiling by chains to avoid possessive aggression and monopolization by individual pups. This method had the additional advantage of keeping the items off the floor, out of drains and in the pen. The bucket and Gumabone tug toy, together with the rawhide and stick, were alternated weekly to provide some novelty. In addition, two of the litters were provided with a length of plastic piping large enough for the puppies to enter. At ten weeks (two weeks after weaning) the use of toys was examined by videotaping each litter over two days so that play with both sets of toys could be recorded. On each day, six hours of tape were recorded between 9.00 and 15.30 hrs. Use of a toy was defined as biting or chewing the toy, climbing on or into a toy, or interacting with the toy with other pups, for example in games of chase or ownership.

The dogs chewed all of the items but most of all the rawhide, Gumabone tug toy and stick. These were also used in tug-of-war play. The dogs climbed into the bucket and piping and used them for ownership play. If both ends of the pipe were clear of the pen walls, they also used it for chase play. The toys were generally well used; ranging from 28–94% with an average of 63.9% of the observed time. Total number of toys did not seem to be important, as increasing the number of toys by adding a length of pipe did not necessarily increase the overall amount of time spent playing with them. Novelty, however, appeared to be important since a particularly large amount of play was recorded on one occasion when the stick was pulled loose from its chain. It is interesting to note that deLuca & Kranda (1992) also found that beagles spent a great deal of time playing with noisy toys such as chains.

Calculation of the average percentage of total puppy time spent playing with each toy showed that the rawhide accounted for 12.01%, the stick 4.9% (or 1.5% after removal of time when stick came loose from chain), the pipe 4.67%, Gumabone 2.64% and bucket 1.91% of the pups' time. Fig. 13.9 shows the average amount of time per litter spent by different numbers of pups playing simultaneously with each

[1] Nylabone Ltd, PO Box 15, Waterlooville, PO7 6BQ, UK Nylabone Corp, Neptune, NJ 07753, USA

[2] Centaur House, Torbay Road, Castle Cary, Somerset, UK

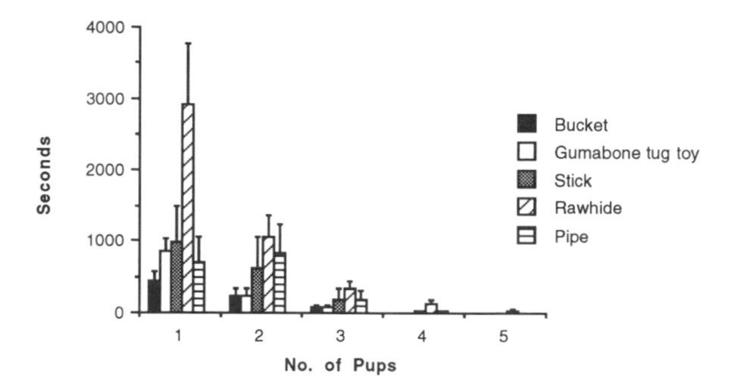

Fig. 13.9. Average amount of time spent playing with toys by varying numbers of pups.

of the toys. Items were most often used by only one or two pups at a time. As in DeLuca & Kranda's (1992) study, rawhide was the most favoured item, particularly when played with by single pups, and was perhaps perceived as food. Groups of two or three pups played with the stick, rawhide and tube slightly more than the bucket and Gumabone tug toy.

Although clearly specific to the particular toys, these results show that dogs of this age make extens ive use of enrichment items – particularly if there is novelty – and different items tend to be used in different ways. Hubrecht (1993) recently tested the responses of five 13-month-old beagles to suspended, chewable toys. The study showed that even after a period of two months, dogs were still spending an average of 24% of their time using the toys, and that frequent changes were not necessary to avoid habituation. The method of presenting toys suspended within the pens was of value in helping to prevent possessive aggression as well as avoiding some of the difficulties involved in cleaning out pens containing loose toys.

The need for further research

At the beginning of this chapter, it was suggested that we have a duty to provide captive dogs with an environment that keeps them in a state of good welfare. However, one has only to consider the excitement of a pet dog when taken for a walk to understand that, just as humans enjoy some activities or experiences more than others, the same is true of dogs. There is therefore a case that we should go beyond basic welfare and attempt to improve the quality of captive animal's lives as much as possible. Unfortunately, as this paper has demonstrated, there are many areas where even basic information concerning dog welfare is simply not available.

Many pet dogs, for example, appear to enjoy a ride in the car, although car sickness can be a problem. Transporting dogs on a commercial basis, where the opportunities to monitor the animals' health are limited, requires much more care. The IATA Live Animals Regulations (1990) provide detailed information concerning the transport of laboratory dogs. More general advice is given by Crittall (1986) and in the BVA (1978) leaflet 'Guidance for Local Authorities and their Veterinary Inspectors on the Breeding of Dogs Act 1973'. Again, these recommendations are based on little if any scientific evidence, and this author is aware of only one research article on the welfare aspects of transport in the domestic dog (Hanneman *et al.*, 1977). This ethically disturbing study reported that heat induced hyperthermia can be a major problem for dogs during air transport. To simulate this experience, the authors exposed dogs to temperatures up to 54.4 °C in the laboratory for 30 minutes. The result was permanent brain and tissue damage. Further research is needed, although not of this drastic nature.

The welfare implications of cosmetic mutilations are another area that remains unexplored. Even where anaesthetics are used, there is likely to be some post-operative suffering and reliable research findings might well hasten the decline of some of these practices. The welfare of working dogs has also received little attention. In the USA and increasingly in Europe, working service dogs are used to give independence to disabled people, and may perform such functions as pulling wheel chairs or helping the owner to stand up using the dog's back as a support. There is at least a potential here for welfare problems. In the United Kingdom the Guide Dogs for the Blind Association is currently carrying out research into methods of assessing stress in its dogs during training procedures and while in dog-housing (Vincent & Mitchell, 1992; Vincent, 1993).

In large kennelling establishments, noise can exceed industry safety levels for humans. Yet at high frequencies a dog's auditory acuity is substantially greater than that of a person (reviewed in Fox & Bekoff, 1975). We do not fully understand why dogs bark, but it has probably become exaggerated through artificial selection and is apparently not specific to particular contexts (Fox, 1971*b*, 1978). Individual dogs can produce barks of over 100 dB (Kay, 1972 cited in van der Heiden, 1992), and sound levels within the human hearing range can regularly reach values between 85 and 122 dB in both kennelling facilities (Ottewill, 1968; Peterson, 1980; Sales, Milligan & Khirykh, 1992) and veterinary hospitals (Senn & Lewin, 1975). Often, barking spreads through social facilitation following some stimulus, such as the imminent arrival of food (Fox, 1971*b*). There is little published information on the control of barking in establishments housing dogs, and its possible effects on canine welfare have not been considered. Limiting the number of dogs per room may help to control the spread of barking, and an architect's advice should be taken on the use of sound absorbent materials that are compatible with good hygiene. It is also possible that partition height could be a factor in stimulating or inhibiting barking.

A number of other questions relevant to the welfare of kennelled dogs should also be mentioned. For instance, what is the optimum number of dogs that should be housed together, and the most appropriate stocking density? Is human contact really more valuable than inter-specific contact, and if so, how much and how best should it be provided? Further work is also needed on the stress implications of different methods of introducing dogs into groups. In addition, we still do not know whether different dogs have different strategies for coping with particular environments. Scott & Fuller (1965) have demonstrated genetic differences in behaviour between breeds, but little is known about how the welfare requirements of different breeds might also vary.

To conclude, the design of appropriate housing that caters to the welfare needs of dogs is still more of an art than a science, and there is a considerable need for further research in this area.

Acknowledgements

The dog housing study was funded through a UFAW grant funded by Battersea Dogs' Home, Wood Green Animal Shelters, Fisons, Pfizer and The Douglas Turner Trust. My thanks are due to Drs T. B. Poole and J. Serpell for initiating the project and for helpful comments on earlier drafts. I am grateful to the managements and staff of the Wood Green

King's Bush Animal Shelter, Battersea Dogs' Home and Pfizer Central Research for provision of facilities and for permission to work at their sites.

References

Althaus, T. (1989). Die Beurteilung von Hundehaltungen. *Schweizer Archiv für Tierheilkunde*, **131**, 423–31.

Andersen, A. C. & Good, L. S. (1970). *The Beagle as an Experimental Animal*. Ames: Iowa State University Press.

Anderson, J. R. & Chamove, A. S. (1984). Allowing captive primates to forage. In *Standards in Laboratory Animals Management*, pp. 253–6. Potters Bar, Herts.: Universities Federation for Animal Welfare.

Anderson, R. S. (1973). Obesity in the dog and cat. *The Veterinary Annual*, **14**, 182–6.

Appleby, M. (1991). Frustration on the factory farm. *New Scientist*, 30 March 1991, pp. 34–6.

Arluke, A. (1994). Managing emotions in an animal shelter. In *Animals and Human Society: Changing Perspectives*, ed. A. Manning & J. A. Serpell, pp. 145–65. London: Routledge.

Bebak, J. & Beck, A. M. (1993). The effect of cage size on play and aggression between dogs in purpose-bred beagles. *Laboratory Animal Science*, **43**, 457–9.

Beck, A. M. (1975). The ecology of 'feral' and free roving dogs in Baltimore. In *The Wild Canids*, ed. M. W. Fox, pp. 380–90. New York: Van Nostrand Reinhold.

Berman, M. & Dunbar, I. (1983). The social behaviour of free-ranging surburban dogs. *Applied Animal Ethology*, **10**, 5–17.

Borchelt, P. L. (1983). Aggressive behavior of dogs kept as companion animals: classification and influence of sex, reproductive status and breed. *Applied Animal Ethology*, **10**, 45–61.

Bradshaw, J. W. S. & Brown, S. L. (1990). Behavioural adaptations of dogs to domestication. In *Pets, Benefits and Practice*, ed. I. H. Burger, pp. 18–24. London: British Veterinary Association.

Broom, D. M. (1983). Stereotypies as animal welfare indicators. In *Indicators Relevant to Farm Animal Welfare*, ed. D. Smidt, pp. 81–7. The Hague: Martinus Nijhoff.

Broom, D. M. (1988). The scientific assessment of animal welfare. *Applied Animal Behaviour Science*, **20**, 5–19.

British Veterinary Association (1978). Breeding of Dogs Act 1973: Guidance for Local Authorities and their Veterinary Inspectors. BVA Publications.

Campbell, S. A., Hughes, H. C., Griffin, H. E., Landi, M. S. & Mallon, F. M. (1988). Some effects of limited exercise on purpose-bred Beagles. *American Journal of Veterinary Research*, **49**, 1298–301.

Carpenter, E. (1980). *Animals and Ethics*, London: Watkins.

Carr, G. M. (1985). Behavioural ecology of feral domestic dogs (*Canis familiaris*) in Central Italy. *XIXth International Ethological Conference, Toulouse*, abstract 267.

Clark, J. D., Calpi, J. P. & Armstrong, R. B. (1991). Influence of type of enclosure on exercise fitness of dogs. *American Journal of Veterinary Research*, **52**, 1024–8.

Committee on Care and Use of Laboratory Animals, Institute of Laboratory Animal Resources (1985). *Guide for the Care and Use of Laboratory Animals*. Bethesda, MD: National Institutes of Health Publications, No. 85–23.

Council for Science and Society (1988). *Companion Animals in Society; Report of a Working Party*. Oxford: Oxford University Press.

Crittall, J. L. W. (1986). The transport of greyhounds and showdogs. In *The Welfare of Animals in Transit*, Proceedings of the Animal Welfare Foundation 3rd Symposium, ed. T. E. Gibson, pp. 98–103. BVA, Animal Welfare Foundation.

Crofts, W. (1984). *A Summary of the Statute Law Relating to Animal Welfare in England and Wales* (revised 1987). Potters Bar, Herts.: UFAW.

Daniels, T. J. (1983*a*). The social organisation of free-ranging urban dogs. I. Non-estrous social behavior. *Applied Animal Ethology*, **10**, 341–63.

Daniels, T. J. (1983*b*). The social organisation of free-ranging urban dogs. II. Estrous groups and the mating system. *Applied Animal Ethology*, **10**, 365–73.

Daniels, T. J. & Beckoff, M. (1989). Population and social biology of free-ranging dogs *Canis familiaris*. *Journal of Mammalogy*, **70**, 754–62.

Dantzer, R., Mormède, P. & Henry, J. P. (1983). Physiological assessment of adaptation in farm animals. In *Farm Animal Housing and Welfare*, ed. S. H. Baxter & J. A. D. MacCormack, pp. 8–19. The Hague: Martinus Nijhoff.

Dawkins, M. S. (1980). *Animal Suffering. The Science of Animal Welfare*. London: Chapman & Hall.

Dawkins, M. S. (1988). Behavioural deprivation: a central problem in animal welfare. *Applied Animal Behaviour Science*, **20**, 209–25.

DeLuca, A. M. & Kranda, K. C. (1992). Environmental enrichment in a large animal facility. *Laboratory Animals*, **21**: 38–44.

Duncan, I. J. H. (1983). Assessing the effect of housing on welfare. In *Farm Animal Housing and Welfare*, ed. S. H. Baxter & J. A. D. MacCormack, pp. 27–35. The Hague: Martinus Nijhoff.

Duncan, I. J. H. & Dawkins, M. S. (1983). The problem of assessing 'well-being' and 'suffering' in farm animals. In *Indicators Relevant to Farm Animal Welfare*, ed. D. Smidt, pp. 14–24. The Hague: Martinus Nijhoff.

Font, E. (1987). Spacing and social organisation: urban stray dogs revisited. *Applied Animal Behaviour Science*, **17**, 319–28.

Fox, M. W. (1965). Environmental factors influencing stereotyped and allelomimetic behaviour in animals. *Laboratory Animal Care*, **15**, 363–70.

Fox, M. W. (1968). Socialization, environmental factors, and abnormal behaviour development in animals. In *Abnormal Behaviour in Animals*, ed. M. W. Fox, pp. 332–55. Philadelphia, PA: W. B. Saunders.

Fox, M. W. (1971*a*). *Integrative Development of Brain and Behaviour in the Dog*. Chicago: University of Chicago Press.

Fox, M. W. (1971*b*). *Behavior of Wolves, Dogs and Related Canids*. London: Jonathan Cape.

Fox, M. W. (1978). *The Dog, Its Domestication and Behaviour*. New York: Garland STPM Press.

Fox, M. W. (1986). *Laboratory Animal Husbandry: Ethology, Welfare and Experimental Variables*. Albany: State University of New York Press.

Fox, M. W., Beck, A. M. & Blackman, E. (1975). Behaviour and ecology of a small group of urban dogs (*Canis familiaris*). *Applied Animal Ethology*, **1**, 119–37.

Fox, M. W. & Bekoff, M. (1975). The behaviour of dogs. In *The Behaviour of Domestic Animals*, 3rd edn, ed. E. S. E. Hafez, pp. 370–409. London: Baillière Tindall.

Fox, M. W. & Spencer, J. W. (1969). Exploratory behaviour in the dog: experimental or age dependent. *Developmental Psychobiology*, **2**, 68–74.

Fox, M. W. & Stelzner, D. (1967). The effects of early experience on the development of inter and intraspecies social relationships in the dog. *Animal Behaviour*, **15**, 377–86.

Frank, H. & Frank, M. G. (1982). On the effects of domestication on canine social development and behaviour. *Applied Animal Ethology*, **8**, 507–25.

Freedman, D. G., King, J. A. & Elliot, O. (1961). Critical periods in the social development of dogs. *Science*, **133**, 1016–17.

Gentry (1983). *When Dogs Run Wild*. Jefferson, NC: McFarland & Company.

Grant, M., (1989). *The Classical Greeks*. London: Weidenfeld & Nicholson.

Hanneman, G. D., Higgins, E. A., Price, G. T., Funkhouser, G. E., Grape, P. M. & Snyder, L. (1977). Transient and permanent effects of hyperthermia in dogs: a study of simulated air transport environmental stress. *American Journal of Veterinary Research*, **38**, 955–8.

Harden, R. H. (1985). The ecology of the dingo *Canis familiaris dingo* in northeastern New-South Wales, Australia. 1. Movements and home range. *Australian Wildlife Research*, **12**, 25–38.

Hart, B. L. (1977). Three disturbing behavioural disorders in dogs: ideopathic viciousness, hyperkinesis and flank sucking. *Canine Practice*, **6**, 10–16.

Hemmer, H. (1990). *Domestication: The Decline of Environmental Appreciation*. Cambridge: Cambridge University Press.

Hetts, S. (1991). Psychologic well-being: behavioral measures and implications for the dog. *Advances in Companion Animal Behavior*, **21**, 369–87.

Hetts, S., Clark, J. D., Calpin, J. P., Arnold, C. E. & Mateo, J. M. (1992). Influence of housing conditions on beagle behaviour. *Applied Animal Behaviour Science*, **34**, 137–55.

Hite, M., Hanson, H. M., Bohidar, N. R., Conti, P. A. & Mattis, P. A. (1977). Effect of cage size on patterns of activity and health of beagle dogs. *Laboratory Animal Science*, **27**, 60–4.

HMSO (1989). *Code of practice for the housing and care of animals used in scientific procedures*. Pursuant to Animals (Scientific Procedures) Act 1986. London: HMSO.

HMSO (1990). *Statistics of Scientific Procedures on Living Animals 1989*. London: HMSO.

HMSO (1992). *Statistics of Scientific Procedures on Living Animals 1990*. London: HMSO.

HMSO (1993). *Statistics of Scientific Procedures on Living Animals 1991*. London: HMSO.

Hubrecht, R. C. (1993). A comparison of social and environmental enrichment methods for laboratory housed dogs. *Applied Animal Behaviour Science*, **37**, 345–61.

Hubrecht, R. C., Serpell, J. A. & Poole, T. B. (1992). Correlates of pen size and housing conditions on the behaviour of kennelled dogs. *Applied Animal Behaviour Science*, **34**, 365–83.

Hughes, C. H., Campbell, S. & Kenney, C. (1989). The effects of cage size and pair housing on exercise of beagle dogs. *Laboratory Animal Science*, **39**, 302–5.

Hume, C. W. (1962). In praise of anthropomorphism. In *Man and Beast*. Potters Bar, Herts.: UFAW.

IATA. Live Animals Regulations (1990), 17th edn, ISBN 92-9035-259-0. London: IATA.

Jeppesen, L. L. & Pedersen, V. (1991). Effects of whole-year nest boxes on cortisol, circulating leucocytes, exploration and agonistic behaviour in silver foxes. *Behavioral Processes*. Special issue, December 1991.

Kelley, K. W. (1980). Stress and immune function: a review. *Annales de Recherche Vétérinaire*, **11**, 445–78.

Kennel Gazette (1992). London: Kennel Club, 8 December.

Kennel Gazette (1993). London: Kennel Club, 3 June.

Kleiman, D. G. (1967). Some aspects of social behaviour in the canidae. *American Zoologist*, **7**, 365–72.

Kleiman, D. G., & Brady, C. A. (1978). Coyote behavior in the context of recent canid research: problems and perspectives. In *Coyotes: Biology, Behavior and Management*, ed. M. Bekoff, pp. 163–88. New York: Academic Press.

Knol, B. W. (1989). Influence of Stress on the Motivation for Agonistic Behaviour in the Male Dog: Role of the Hypothalamus-Pituitary-Testis System. Ph.D. thesis, University of Utrecht.

Liang, B., Verrier, R. L., Melman, J. & Lown, B. (1979). Correlation between circulating catecholamine levels and ventricular vulnerability during psychological stress in conscious dogs. *Proceedings of The Society for Experimental Biology and Medicine*, **161**, 266–9.

Lore, R. K. & Eisenberg, F. B. (1986). Avoidance

reactions of dogs to unfamiliar male and female humans in a kennel setting, *Applied Animal Behaviour Science*, **15**, 261–6.
Luescher, U. A., McKeown, D. B. & Halip, J. (1991). Stereotypic or obsessive-compulsive disorders in dogs and cats. *Veterinary Clinics of North America: Small Animal Practice*, **21**, 401–13.
Lynch, J. J. & Gantt, W. (1968). The heart rate component of the social reflex in dogs: the conditional effects of petting and person. *Conditioned Reflex*, **3**, 69–80.
MacArthur, J. A. (1987). The Dog. In *The UFAW Handbook on the Care and Management of Laboratory Animals*, 6th edn, ed. T. Poole, pp. 456–75. Harlow, Essex: Longman.
Macdonald, D. W. (ed. advisor) (1985). *The Complete Book of the Dog*. London: Pelham Books.
Malcolm, J. R. (1983). Paternal care in canids. *American Zoologist*, **23**, 912.
Mason, G. J. (1991*a*). Stereotypies: a critical review. *Animal Behaviour*, **41**, 1015–37.
Mason, G. J. (1991*b*). Stereotypies and suffering. *Behavioural Processes*, **25**, 103–115.
Mench, J. A. & Krulisch, L. (1990). *Canine Research Environment*. Scientists Center for Animal Welfare, 4805 St Elmo Avenue, Bethesda, MD 20814, USA.
Mendl, M. (1991). Some problems with the concept of a cut-off point for determining when an animal's welfare is at risk. *Applied Animal Behaviour Science*, **31**, 139–46.
Meyer-Holzapfel, M. (1968). Abnormal behaviour in zoo animals. In *Abnormal Behaviour in Animals*, ed. M. W. Fox, pp. 476–503. Philadelphia, PA: W. B. Saunders.
Morris, D. M. (1970). The response of animals to a restricted environment (first published 1966). In *Patterns of Reproductive Behaviour*, pp. 490–511. London: Jonathan Cape.
Morton, D. B. & Griffiths, P. H. M. (1985). Guidelines on the recognition of pain, distress and discomfort in experimental animals and an hypothesis for assessment. *The Veterinary Record*, **116**, 431–6.
Mugford, R. A. (1984). Aggressive behaviour in the English cocker spaniel. In *The Veterinary Annual*, 24th issue, pp. 310–14. Bristol: John C. Wright.
Neamand, J., Sweeny, W. T., Creamer, A. A. & Conti, P. A. (1975). Cage Activity in the laboratory beagle: a preliminary study to evaluate a method of comparing cage size to physical activity. *Laboratory Animal Science*, **25**, 180–3.
Nesbitt, W. H. (1975). Ecology of a feral dog pack on a wildlife refuge. In *The Wild Canids*, ed. M. W. Fox, pp. 391–5. New York: Van Nostrand Reinhold.
Nowak, R. M. & Paradiso, J. L. (1983). *Walker's Mammals of the World*, 4th edn. Baltimore, MD: The John Hopkins University Press.
O'Farrell, V. (1989). *Problem Dog Behaviour and Misbehaviour*. London: Methuen.
Ottewill, D. (1968). Planning and design of accommodation for experimental dogs and cats. *Laboratory Animal Symposium*, **1**, 97–112.
Peterson, E. A. (1980). Noise and laboratory animals. *Laboratory Animal Care*, **13**, 340–50.
Pettijohn, T. F. (1978). Environment and agonistic behaviour in male telomian dogs. *Psychological Reports*, **42**, 1146.
Pettijohn, T. F., Davis, K. L., & Scott, J. P. (1980). Influence of living area space on agonistic interaction in telomian dogs. *Behavioral and Neural Biology*, **28**, 343–9.
PFMA Profile (1993). Pet Food Manufacturers Association: London.
Ritchie, Carson, I. A. (1981). *The British Dog. Its History from Earliest Times*. London: Robert Hale.
Rollin, B. E. (1986). Animal pain. In *Advances in Animal Welfare Science 1985*, ed. M. W. Fox & L. D. Mickley, pp. 910–1106. Boston: Martinus Nijhoff.
Royal Society/UFAW, (1987). *Guidelines on the Care of Laboratory Animals and Their Use for Scientific Purposes. I Housing and Care*. London: The Royal Society and the Universities Federation for Animal Welfare.
Rubin, H. D. & Beck, A. M. (1982). Ecological behaviour of free ranging urban pet dogs. *Applied Animal Ethology*, **8**, 161–8.
Ryder, R. D. (1989). *Animal Revolution: Changing Attitudes Towards Speciesism*. Oxford: Basil Blackwell.
Sales, G. D., Milligan, S. R. & Khirnykh, K. (1992). *The Acoustic Environment of Laboratory Animals*. A Report to the RSPCA. Universities Federation for Animal Welfare, Potters Bar.
Sandys-Winsch, G. (1984). *Your dog and the Law*. London: Shaw & Sons.
Schneiderman, N. & McCabe, P. M. (1985). Biobehavioral responses to stressors. In *Stress and Coping*, ed. T. M. Field, P. M. McCabe & N. Schneiderman, pp. 13–61. London; Lawrence Erlbaum Associates.
Scott, J. P. & Fuller, J. L. (1965). *Genetics and the Social Behaviour of the Dog*. Chicago: University of Chicago Press.
Scott, M. D. & Causey, K. (1973). Ecology of feral dogs in Alabama. *Journal of Wildlife Management*, **37**, 253–65.
Seligman, M. E. P. (1975). *Helplessness: On Depression, Development and Death*. San Francisco: Freeman.
Senn, C. L. & Lewin, J. D. (1975). Barking dogs as an environmental problem. *Journal of the American Veterinary Medical Association*, **166**, 1065–8.
Serpell, J. (1986). *In the Company of Animals*. Oxford: Basil Blackwell.
Solarz, A. K. (1970). Behaviour. In *The Beagle as an Experimental Animal*, ed. A. C. Anderson, pp. 453–68. Ames: Iowa State University Press.
Taylor, P. M. (1985). Clinical measurement of pain, distress and discomfort in dogs and cats. In *The Detection and Relief of Pain in Animals*, Proceedings of the Animal Welfare Foundation 2nd Symposium,

ed. T. E. Gibson, pp. 75–80. BVA, Animal Welfare Foundation.

Tester, J. R. (1987). Changes in daily activity rhythms of some free-ranging animals in Minnesota USA. *Canadian Field-Naturalist*, **101**, 13–21.

Thompson, W. R., Melzack, R. & Scott, T. H. (1956). 'Whirling behaviour' in dogs as related to early exposure. *Science*, **123**, 393.

van der Heiden, C. V. (1992). The problem of noise within kennels: what are its implications and how can it be reduced? *The Veterinary Nursing Journal*, **7**, 13–16.

Verga, M. & Carenzi, C. (1983). Behavioural tests to quantify adaption in domestic animals. In *Indicators Relevant to Farm Animal Welfare*, ed. D. Smidt, pp. 97–108. The Hague: Martinus Nijhoff.

Vincent, I. C. (1993). Is there a link between behaviour and blood pressure in dogs? *The Veterinary Nursing Journal*, **8**, 43–5.

Vincent, I. C. & Mitchell, A. R. (1992). Comparison of cortisol concentrations in saliva and plasma of dogs. *Research in Veterinary Science*, **53**, 342–5.

Wemelsfelder, F. (1985). Animal boredom: is a scientific study of the subjective experiences of animals possible? In *Advances in Animal Welfare Science 1984*, ed. M. W. Fox & L. D. Mickley, pp. 115–54. Boston: Martinus Nijhoff.

Willis, M. B. (1989). *Genetics of the Dog*. London: H. F. & G. Witherby.

Wolfensohn, S. (1981). The things we do to dogs. *New Scientist*, 14 May, pp. 404–7.

Wolfle, T. L. (1987). Control of stress using non-drug approaches, *Journal of the American Veterinary Medical Association*, **191**, 1219–21.

Wolfle, T. L. (1990). Policy, program and people: the three P's to well-being. In *Canine Research Environment*, ed. J. A. Mench & L. Krulisch, pp. 41–7. Scientists Center for Animal Welfare, 4805 St Elmo Avenue, Bethesda, MD 20814, USA.

Young, M. (1976). The mutilation of pet animals. In *The Mutilation of Animals*, Proceedings of the 2nd Symposium Sponsored by the Farm Livestock Committee RSPCA, pp. 109–113.

14 Variation in dog society: between resource dispersion and social flux

D. W. MACDONALD AND G. M. CARR

Introduction

Domestic dogs, *Canis familiaris*, live in various degrees of association with people. Extremes of this continuum range from the lap dog to those living on uninhabited islands (e.g. Kruuk & Snell, 1981). Dogs that are not strictly controlled by their owners might be expected to modify their behaviour to match their ecological circumstances according to the same principles that affect wild Carnivora (e.g. Kruuk, 1975; Macdonald, 1983) and free-ranging domestic ones such as cats, *Felis catus* (e.g. Macdonald *et al.*, 1987). Therefore, we would expect that free-ranging dogs studied under different conditions would behave differently, and that these differences would be adaptive (in so far as artificial selection had not compromised the traits in question). This prediction, however, has not been formally tested. Studies of feral dogs in urban areas have tended to conclude that they formed amorphous, and probably ephemeral associations (e.g. Beck, 1973). Furthermore, the questions of whether these associations are adapted to ecological circumstances and whether membership of groups has functional consequences have received little attention. Such questions have both theoretical and practical importance. The abundance, accessibility and potential manipulability of free-ranging dogs, combined with background knowledge of the genetic and physiological differences between breeds, make them strong candidates for testing ideas about canid socio-ecology. In addition, an understanding of dog behaviour should help the design of control programmes intended to counteract their potential as pests – whether as predators of stock or game, vectors of disease (e.g. rabies, distemper and *Echinococcus*; WHO, 1988), or as genetic contaminators of wild canid populations (see Boitani, 1983; Ginsberg & Macdonald, 1990). Finally, free-ranging dog behaviour is interesting because of the opportunity it affords for comparison between breeds and with the ancestral wolf, *Canis lupus*, and as part of the study of domestication (Scott & Fuller, 1965; Zimen, 1972).

We therefore compared the behaviour of free-ranging dogs in contrasting ecological conditions in order to observe whether their social organization differed, and to determine whether their behaviour matched predictions based on their ecology. Our approach was based on three general findings about the behaviour of the Carnivora.

First, in addition to obvious inter-specific variation (e.g. Gittleman, 1989), there is enormous intra-specific variation in the social organization of carnivores (Kruuk, 1975; Macdonald, 1983). A classic example is Kruuk's (1972) demonstration that populations of spotted hyaena, *Crocuta crocuta*, live in different sorts of societies depending on whether their prey are migratory or sedentary (see also Mills, 1990). A similar example, especially apposite to the case of the free-living dogs we studied, is Doncaster & Macdonald's (1991) finding that the spatial organization, group dynamics and reproductive behaviour of red foxes, *Vulpes vulpes*, varied between an urban and an adjoining rural habitat (see also Macdonald, 1981, 1987).

Second, intra-specific variation may be explained in terms of the pattern in which resources, such as food, are available. One idea, reviewed by Macdonald (1983), is called the Resource Dispersion Hypothesis (RDH) and stems from the proposition that groups may develop where resources are dispersed such that the smallest economically defensible territory for a minimum-sized social unit might also sustain additional animals. In a simple case, territory size would be determined by the dispersion of the minimum number of patches of food required to sustain an individual, whereas group size would be determined by the richness of those patches. A related concept is food security, which is the probability that a territorial occupant can satisfy its food requirements at any given time (Carr & Macdonald, 1986). The more variable the pattern in which food is available, the greater the resources that must be commandeered in a territory to achieve a given food security, and consequently the greater the opportunity for providing substantial food security to additional group members (Woodroffe & Macdonald, 1993 review this and related models).

Third, the functional consequences of group membership vary between populations and between individuals even within one group. Amongst the many advantages that have been proposed for group-living carnivores (see reviews in Gittleman, 1989; and a general account in Macdonald, 1992), obvious candidate benefits that might apply to free-living domestic dogs include cooperative hunting, communal care of young and corporate defence of food. Depending on the structure of groups, a vested interest in the survival of kin may be important, as elaborated in Lindstrom's (1986) Territory Inheritance Model. The marginal contribution made by each additional group member to the fitness of each member will be one of the factors that determines whether the benefits of larger groups compensate for any expansionism necessary to provide adequate food security for their members (Macdonald & Carr, 1989).

The study

It was with these three points in mind that we designed our study. The fieldwork was carried out in Italy, where Boitani & Fabbri (1983) estimated that 10% of 80 000 free-ranging dogs lived independently of people. The study area was in the mountainous Abruzzo region of central Italy (42° N, 13° 30′ E), which afforded the opportunity to compare dogs living in adjacent but different habitat types: small villages and open countryside. Preliminary observations suggested that the former depended on several food patches, each unpredictable in the time-table and abundance of food, whereas the latter depended mainly on a single very rich patch of food. (The RDH predicts that these differing patterns would have different consequences for group and territory sizes and social behaviour.)

There was wide phenotypic variation between individual dogs, but no observable systematic differences in phenotype between those living in the different habitats. The dominant phenotypes were the traditional Abruzzese (pastore maremmano), and the more recently introduced German shepherd. The Abruzzese is a large, predominantly white breed, similar to the Pyrenean. Both types were once used to guard flocks of sheep against the depredations of wolves. The virtual elimination of wolves from the area has, according to local people, coincided with the establishment of the free-living dog population in the countryside. The phenotypic similarity of the village and country populations, the youth of the latter, and the proximity of the two habitats, permit the hope that any differences in their social and spatial organization result from different ecological circumstances, rather than selective differences (whether natural or artificial) between the populations involved.

Three dog populations were observed intermittently between 1981 and 1983; two of these were located in villages, and the third in open countryside.

- Rovere (Fig. 14.1): This ancient village (population 225) occupies one side of a small hill (altitude 1413 m) and part of the surrounding plain. About half of the settlement is on the hill and consists of old terraced stone buildings and narrow streets. More modern houses, together with some larger farms, have spilled out onto the plain. These latter form a roughly triangular structure enclosing an open meadow.
- Sant'Iona (Fig. 14.2): Another ancient settlement, similar to Rovere (population 142), but at lower altitude (969 m). A large (600 m^2) open space is enclosed within the village, and there is an area of ruined buildings – the result of an earthquake in 1915. Sant'Iona has very little modern development, although some older houses have been renovated as holiday homes.
- Altopiano delle Rocche (Fig. 14.3): This small upland plain (1340 m) formed the focus of the third study area. It is approximately elliptical, with major and minor axes of 4 km and 1.5 km, and is surrounded by mountains that rise to more than 2400 m. The plain is mainly meadowland, while the surrounding mountains are covered by beech woods interspersed with herbaceous grassland. Cattle, sheep and horses are grazed in the area. Near the middle of the plain is a large dump that receives the domestic refuse of nearby settlements and is replenished on a daily basis. Between 6–10 wolves occurred in the vicinity.

The nomenclature of free-living dogs has become somewhat confused (Nesbitt, 1975; Daniels & Bekoff, 1989*a*). We shall use three terms: any dog not living under the close control of a human being may be referred to as 'free-living'; dogs living in the Altopiano, or similar habitats are 'sylvatic'; dogs living in villages will simply be known as 'village' dogs. We shall eschew the words 'feral', 'stray', etc.

Methods

Data on the sylvatic dogs were collected by a combination of observation (using image intensifying equipment at night) and radio-tracking. Candidates for radio-tracking were captured in baited cage traps and fitted with home-made radio-collars (173 KHz, following the construction described by Macdonald & Amlaner, 1980). Eight sylvatic individuals, weighing 20–30 kg, were collared during the course of the study, one of them twice. Radio-tracking was used to locate daytime resting places, and to follow movement at night. Fixes at night were taken *ad lib*., according to circumstances. Radio data were collected on 175 days (approximately 2000 radio sweeps), principally during an intensive study between April–August 1981, and during occasional monitoring thereafter. For the purposes of this account, the spatial organization of dogs is adequately represented by hand-drawn boundaries of home ranges based on combined radio-location and direct observation (Macdonald, Bale & Hough, 1980).

Village dogs, being habituated to humans, could be watched at close quarters. Focal animals were selected each day and their behaviour and exact location (based on a one metre grid) sampled once per minute, normally for a four hour observation period by day. Behaviour was divided into three broad categories: sedentary, active and translocatory (i.e. moving from one place to another), and subdivided within these categories to allow for more detailed analysis. Dogs not selected for focal studies were scan-sampled regularly to record their locations, and thus their ranges. Eighty hours of focal observations were made in Rovere in March 1981, and 350 hours in Sant'Iona, of which 100 directly comparable hours recorded in March 1983 are used in direct comparisons with Rovere.

Results

Demography

The initial census for Rovere (February 1981) gave 10 dogs (8 male, 2 female); for Sant'Iona (March 1982) 18 (13 male, 5 female). Both village populations were thus biased towards males (binomial test Rovere: $P = 0.055$; Sant'Iona: $P = 0.045$). The initial census of the Altopiano revealed 15 dogs (8 male, 7 female) living there. Tables 14.1 shows how the composition of one Altopiano pack changed over the course of the study. In addition to these animals, a group of around 8 sylvatic dogs lived to the north of the study area, apparently in conjunction with a second rubbish dump near the town of Rocca di Mezzo.

Village dog society

Casual observations in both villages suggested a rather amorphous society. Although certain dogs regularly kept one another company, there were no obvious packs. However, more detailed scrutiny identified discrete social groups. Tables 14.2(*a*) and 14.2(*b*) summarize the springtime patterns of amicable interaction (greeting, play, co-ordinated movement) between village dogs. Dogs varied in their amicability (percentage of observations in company varied between 5–70%) and often had preferred associates. But it is clear from the tables that they split up into mutually exclusive groups under this analysis. The converse is also true (Tables 14.3(*a*) and 14.3(*b*)): if social groups are provisionally erected on the basis of company keeping, hostile interactions are almost completely confined to encounters between dogs from different groups. Only about 2% of all hostile interactions were between group members. However, hostile encounters were not observed between all possible neighbouring dyads, so mere lack of hostility is not a sufficient indicator of group membership.

Despite their not routinely accompanying one another, and, indeed, often spending time alone, it thus appears that village dogs do belong to social groups.

These social groups also have a spatial dimension: they appear to be defending territories. The evidence for territoriality comes from two sources: their use of space, and the locations and nature of their hostile interactions with one another.

Figure 14.1 depicts the observed ranges of Roveresi dogs in spring 1981. The figure shows that the ranges of individual dogs formed three more or less exclusive clusters, which we shall refer to as the western, central and eastern (collective) ranges. Collective ranges averaged roughly 2 ha. Within two of these collective ranges, some dogs appeared to use the whole area, while others restricted most of their activities to a part of it (ten individual ranges are depicted on Fig. 14.1). However, no dog's individual range was observed to straddle the border of a collective range to any significant degree.

There were, as Fig. 14.1 indicates, two zones of overlap between collective ranges, of less than 5% of the range areas. These zones were the areas around two rubbish dumps at which the dogs fed.

A similar pattern of collective and largely exclusive ranges was found in Sant'Iona in spring 1983 (Fig. 14.2). The village was divided into four such areas, averaging 0.25 ha. Because of the larger canine population of Sant'Iona, not all of its dogs were the objects of extended focal studies, though all were subjected to scan sampling to determine their ranges. Figure 14.2 shows three collective ranges as established from scan and focal sampling.

Table 14.1. *Annual changes in the composition of the sylvatic dog pack (HIWG)*

	Summer 1981	Summer 1982	Summer 1983
Males			
Radio	P	D	D
Dark	P	D	D
Big White	P	P	P
Second White	P	P	P
Cream	P	P	P
Shaggy	P	D	D
Houdini[a]	–	P	P
Black[b]	–	–	P
Females			
1st Mama	P	P	P
2nd Mama	P	D	D
3rd Mama	P	P	P
Odd Mama	P	D	D
Noisy[b]	–	P	D
Pseudo Wolf[b]	–	–	P
Total	10	7	8
Sex Ratio	6 : 4	4 : 3	5 : 3

P, Present; D, Died or Disappeared.
[a]External recruit.
[b]Matured pup.

The overlap area of the Sant'Ionan collective ranges was somewhat larger than in Rovere but this overlap was mostly in the extensive open central area, which thus appeared to form a sort of 'no dog's land' with few geographical features, where range boundaries were not clear cut (although it was the object of considerable scent marking activity).

Figures 14.1 and 14.2 also show the locations of one month's hostile interactions in the two villages. These were largely confined (45 of 59 observations in 180 hours) to the putative borders of the collective ranges, and were particularly concentrated around important feeding sites, further confirming the idea that these ranges are territories in the sense of being exclusive, defended areas. Another confirmation of this hypothesis came from the rare occasions when a dog trespassed in a neighbouring range and was detected. In these cases, the trespasser would either leave immediately, or be chased out. On two occasions a trespasser was ejected by an individual it had itself ejected from its own range on a previous occasion, suggesting that location, rather than individual prowess was the deciding factor in such contests.

Village dogs were occasionally seen travelling, sometimes in company, to rubbish dumps in the vicinity and although we have no data on the frequency of these trips, Boitani *et al.* (Chapter 15) emphasize that such excursions provide an important opportunity to meet, and be recruited by, the sylvatic population.

Sylvatic dog society

Sylvatic dog society was different from that in the villages in one immediately noticeable respect. Sylvatic dogs were generally seen in company. It became clear during the course of the census that such associations regularly involved the same individuals, either pairs, or subsets of a single larger closed group. Four such units were identified at the beginning of the study, although, initially, one of these was actually a lone bitch, and two of the others were pairs.

There was a very strong tendency for these dogs to keep one another company, although nursing mothers

Table 14.2. *Pattern of amicable interactions between adult dogs in (a) Rovere and (b) Sant'Iona in March 1981 (80 hours) and 1983 (100 hours), respectively. The dogs are identified by two letter codes and sex (M/F), and are grouped on the tables in accordance with amicable ties as revealed by these data.*

(*a*) *Rovere*

	AU(M)	MA(M)	LI(F)	SU(M)	TI(M)	CA(M)	CR(M)	FA(M)	PO(M)
AU(M)	–								
MA(M)	78	–							
LI(F)	139	93	–						
SU(M)	0		0	–					
TI(M)	0	0	0	11	–				
CA(M)	0	0	0	0	0	–			
CR(M)	0	0	0	0	0	116	–		
FA(M)	0	0	0	0	0	37	15	–	
PO(M)	0	0	0	0	0	152	98	24	–
AG(F)	3	0	0	0	0	74	47	3	43

This table shows the pattern of amicable interaction between dogs in the village of Rovere. The demographic make-up altered considerably at the end of this period. Dogs have been organized into social groups on the basis of amicable interactions (greetings, play, etc. and coordinated non-aggressive movement). Eighty hours of observations.

(*b*) *Sant'Iona*

	WH(M)	NI(M)	MA(M)	BL(M)	OE(F)	OP(F)	TP(F)	EY(M)	CH(M)	BI(M)
WH(M)	–									
NI(M)	49	–								
MA(M)	101	37	–							
BL(M)	13	0	0	–						
OE(F)	2	0	0	121	–					
OP(F)	0	0	0	35	33	–				
TP(F)	0	0	0	42	27	113	–			
EY(M)	0	0	0	0	0	0	0	–		
CH(M)	0	0	0	0	0	0	0	41	–	
BI(M)	0	0	0	5	2	0	0	48	92	–
NE(F)	0	0	0	0	0	0	0	77	63	69

This table shows the pattern of amicable interaction between dogs in the village of Sant'Iona in a comparable month to that shown elsewhere for Rovere. The dogs have again been organized into social groups on the basis of amicable interactions (greetings, play, etc. and coordinated non-aggressive movement). One hundred hours of observations.

were a partial exception to this rule. In April 1981, no sylvatic dog (apart from the bitch, Nera, who was then the only animal in her range) was seen alone on more than 5% of observations. July 1981 presents a contrast. By July the bitches, First Mama, Second Mama and Third Mama, had all given birth (May 1981) and they were rarely seen in the company of their adult fellows.

Members of different sylvatic groups were never

Table 14.3. *Pattern of aggressive interactions between adult dogs in (a) Rovere and (b) Sant'Iona in social groups defined from the results presented in Table 14.2*
(*a*) *Rovere*

	AU(M)	MA(M)	LI(F)	SU(M)	TI(M)	CA(M)	CR(M)	FA(M)	PO(M)
AU(M)	–								
MA(M)	0	–							
LI(F)	0	0	–						
SU(M)	5	2	2	–					
TI(M)	1	0	0	0	–				
CA(M)	3	0	0	4	1	–			
CR(M)	1	2	0	4	0	0	–		
FA(M)	0	0	0	0	0	0	0	–	
PO(M)	2	0	0	5	1	1	0	0	–
AG(F)	1	0	0	0	0	0	0	0	0

This table shows the pattern of hostile interactions between dogs in the village of Rovere. Hostilities were observed only once between members of social groups as previously defined.
(*b*) *Sant'Iona*

	WH(M)	NI(M)	MA(M)	BL(M)	OE(F)	OP(F)	TP(F)	EY(M)	CH(M)	BI(M)
WH(M)	–									
NI(M)	0	–								
MA(M)	0	0	–							
BL(M)	0	3	0	–						
OE(F)	1	0	0	0	–					
OP(F)	0	0	1	0	0	–				
TP(F)	0	0	1	0	0	0	–			
EY(M)	3	5	0	3	0	0	0	–		
CH(M)	0	2	1	1	0	0	0	0	–	
BI(M)	1	2	0	0	0	0	0	0	0	–
NE(F)	0	0	0	0	0	0	0	0	0	0

This table shows the pattern of hostile interactions between dogs in the village of Sant'Iona for a month comparable to that shown elsewhere for Rovere. No hostilities were observed between members of social groups as previously defined.

seen amicably in one another's company. Meetings between such animals were characterized by avoidance or aggression, one indication that, like their village counterparts, they were territorial.

Figure 14.3 shows the patterns of observed range-use of each of the four sylvatic social units between the summers of 1981 and 1982. The general level of separation between the groups' ranges is immediately clear. Equally clear is that there is one principal area of overlap. This is the communal dump that all sylvatic dogs used as a feeding site on a regular basis. A second 'contested' area was the village of Rovere itself. This was regularly visited during the summer of 1981 by the largest group, known as HIWG, and also by dogs of the northern pair. Rovere had its own population of resident dogs (see above) and they fought the trespassing sylvatic dogs.

Fifteen aggressive incidents between members of

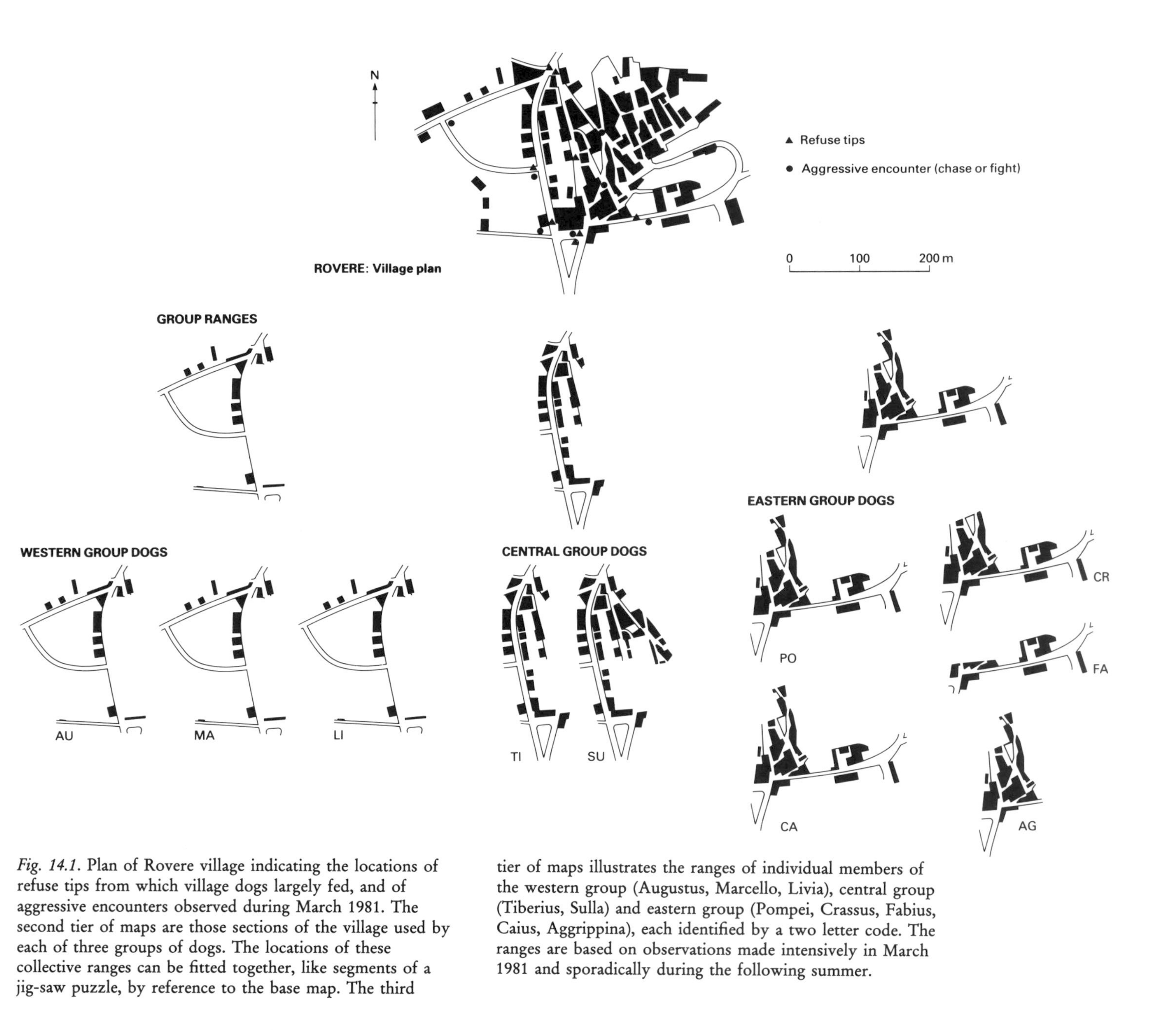

Fig. 14.1. Plan of Rovere village indicating the locations of refuse tips from which village dogs largely fed, and of aggressive encounters observed during March 1981. The second tier of maps are those sections of the village used by each of three groups of dogs. The locations of these collective ranges can be fitted together, like segments of a jig-saw puzzle, by reference to the base map. The third tier of maps illustrates the ranges of individual members of the western group (Augustus, Marcello, Livia), central group (Tiberius, Sulla) and eastern group (Pompei, Crassus, Fabius, Caius, Aggrippina), each identified by a two letter code. The ranges are based on observations made intensively in March 1981 and sporadically during the following summer.

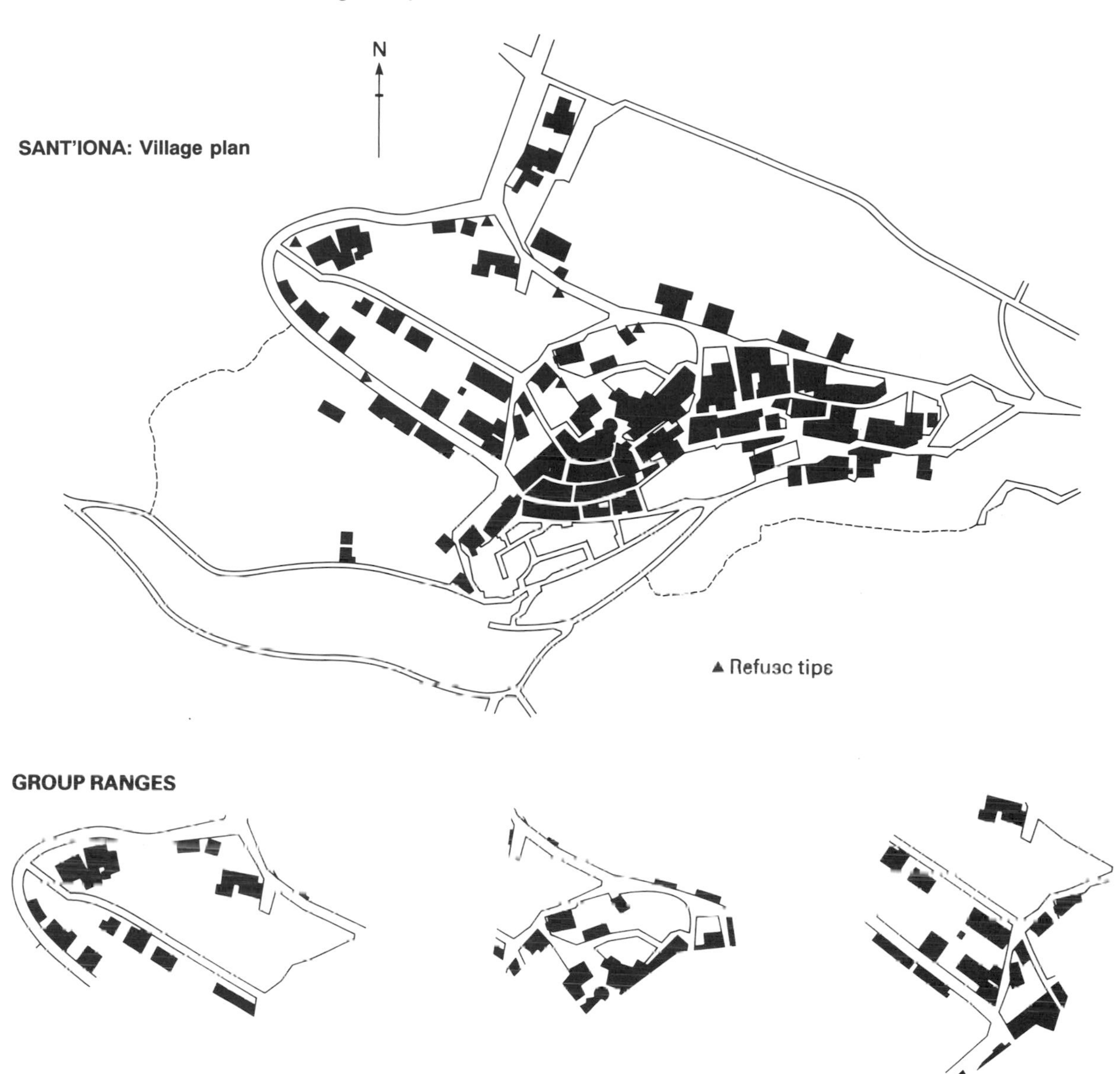

Fig. 14.2. Map of Sant'Iona village, showing the locations of refuse tips from which village dogs largely fed. The second tier of maps are those sections of the village used by each of three groups of dogs (displayed as explained in Fig. 14.1).

different sylvatic groups were observed during 70 days of observation between April–August 1981 (Fig. 14.3). They almost always occurred near the dump or village. Such aggression did not result in physical contact. In all 13 episodes involving the HIWG (2 vs Nera, 4 vs the southern pair and 7 vs the central pair) they were clearly victorious. When the large HIWG was involved, the other group always withdrew. In these cases, barking was often sufficient to drive off the opposition and, if this was not successful, the ensuing chase would do so. Indeed, aggression was rarely face to face: on all of seven occasions when the HIWG made a challenging bark from lying-up sites as far away as 1 km this was observed to clear the dump of rivals before any members of the group actually arrived. The remaining two encounters were between the northern and central pairs and the central and southern pairs. These encounters were confined to barking matches, and neither side forced a withdrawal by the other.

Food supply

Both sylvatic and village dogs depended heavily on human activity for their food. Village dogs fed mainly on domestic refuse, farmyard refuse and 'handouts' from people. The distribution of refuse tips within the villages is shown in Figs. 14.1 and 14.2. One dog (Livia; Rovere) was reported to have taken some live chickens. She was killed shortly afterwards by the farmer involved. Based on our observations of abundant garbage, there was no indication that the village dogs were limited by food availability. Only one dog (Video; Sant'Iona) was observed to cache food, although he did not do so on a regular basis.

In the countryside, dependence on human activity was almost as complete as in the villages. All the dogs of the Altopiano fed at the large refuse dump. Dogs belonging to HIWG scavenged from the carcasses of stock animals (two cattle and a horse) that had died in their range, although there is no evidence that they killed them (their observed interactions with horses, at least, were entirely playful). They also scavenged from Rovere and its surrounding farmyards, and broke into chicken coops on two occasions. Only once were sylvatic dogs in the study area seen to hunt a large animal. This was a red fox, and it successfully evaded them. However two dogs, Blackwhite and Clone (who were putative father and son), were observed digging for small mammals in the banks of drainage ditches on a total of 18 occasions during the summer months of 1982 and 1983. The apparent lack of hunting in this area may be atypical – perhaps encouraged by the reliable sources of immobile food. Nearby sylvatic dogs apparently did take livestock: a pack 25 km away was reliably reported to kill foals until (illegally) poisoned. The pack of eight, occasionally observed at Rocca di Mezzo, was alleged, on two occasions, to have killed large numbers of sheep.

The dogs of HIWG seemed to decide where they would feed on a given night during the early hours of the previous morning. They visited their food sources, the dump and the village of Rovere, with approximately equal frequency during the spring and summer of 1981 and then spent the day resting in the mountains close to one or other of them. Irrespective of where they fed the previous night, if they spent the day close to Rovere they fed there by night, whereas if they slept close to the dump they fed there. On all nights when they were observed during this summer, HIWG fed at one of the two sites. During and after the winter of 1981–2, visits to Rovere became considerably rarer, but the pattern of travelling to a lying-up site adjacent to the next night's foraging site remained.

Reproductive behaviour

A total of 17 litters whelped by nine different sylvatic bitches were recorded during the study period. Of these, 15 were born to HIWG and one each in the central and southern ranges.

Village dog reproduction was subjected to considerable human interference. Only one litter was observed in Rovere (born to the bitch Agrippina). This was destroyed almost immediately. Of three litters noted in Sant'Iona, one litter of five pups was allowed to live. Table 14.4 summarizes the details of sylvatic litters. Sylvatic bitches are generally dioestrus. Summer litters were born in May or June, winter litters in November or December. The two bitches that survived for the whole study period followed different patterns: Third Mama produced large litters in alternate breeding seasons and small litters or none at all in the intervals; First Mama produced median litters every season. Unfortunately, there are too few data to generalize from this observation.

Pregnant sylvatic females usually selected a rocky cleft or a tangle of exposed tree roots as a den for their pups (although, on one occasion, an underground burrow of unknown provenance was chosen). HIWG females chose sites more than 100 m above the level of the plain. For the first six weeks, mothers remained with their pups unless disturbed or left for relatively short periods while feeding.

In the two cases of non-HIWG sylvatic litters, the male and female of the pair stayed closely associated before and after the birth of the pups. The male slept in close proximity to the bitch and her litter and, when the pups became mobile, played with them, although in neither case was he observed to bring food to the den.

Pregnant HIWG females remained associated with the rest of the pack until they whelped. But once they had given birth, they led a semi-detached exist-

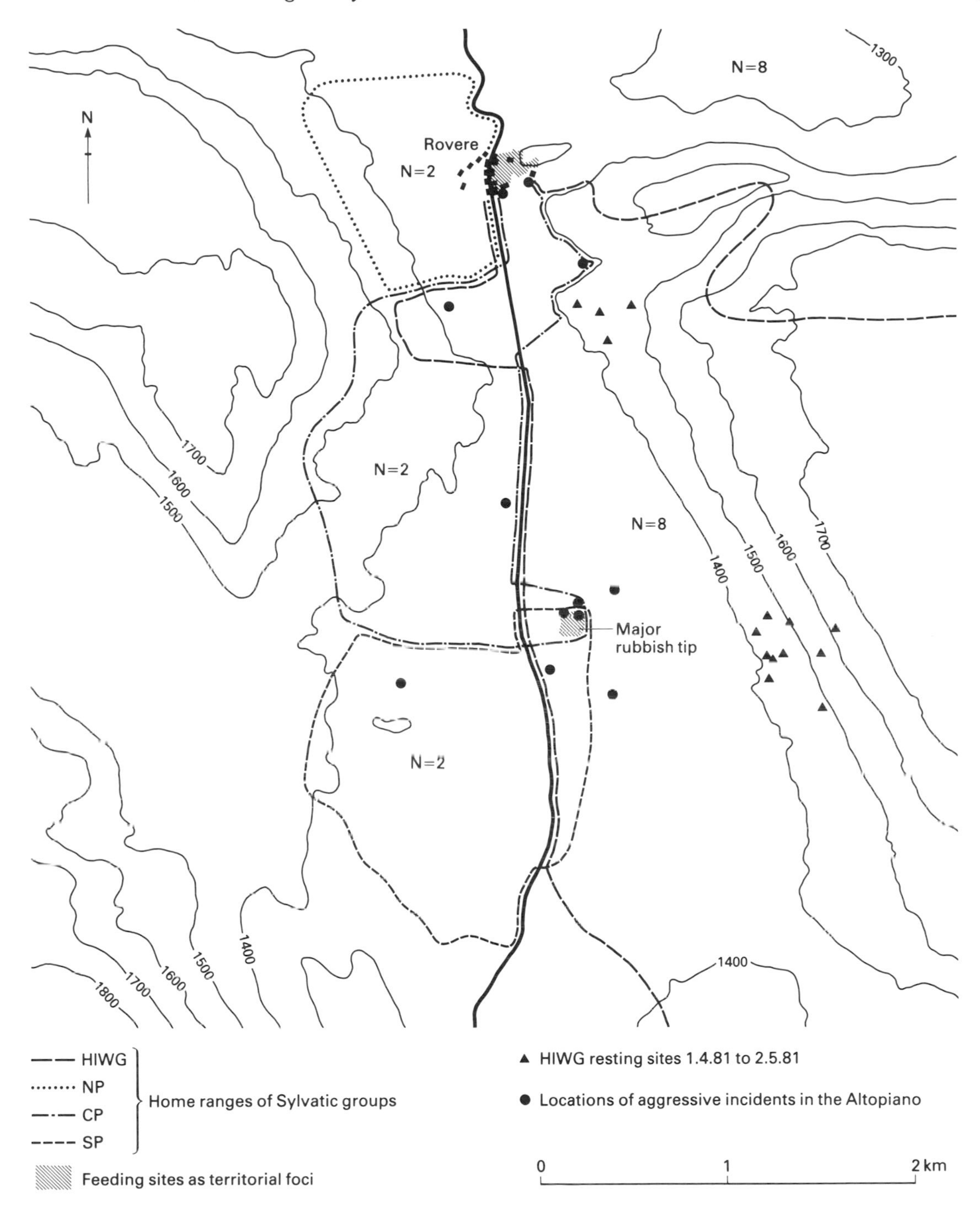

Fig. 14.3. Map of the Altopiano delle Rocche showing the location of the village of Rovere and the major communal rubbish tip. The hand-drawn borders of home ranges are shown for northern, central and southern pairs of sylvatic dogs (group sizes indicated by: $N = 2$), and for the western edge of the HIWG's range. To the north ($N = 8$) no borders are plotted for the group that operated in the vicinity of the rubbish tip at Rocca di Mezzo. The HIWG (Hole in the Wall gang, named after their site of entry to Rovere village) rested in beech woodland on the valley walls, either south of Rovere or east of the rubbish tip (day-time resting sites of radio-tagged dogs indicated for spring 1981). The sites of aggressive encounters between resident sylvatic dogs and between them and unaccompanied non-residents are also shown, clustered particularly around the two important feeding sites.

Table 14.4. *The breeding pattern of HIWG sylvatic females during summers and winters 1981–3. The data are observed litter sizes*

Bitch	Season				
	S 81	W 81	S 82	W 82	S 83
1st Mama	5	5	5	?2	5
2nd Mama	9	0	D	D	D
3rd Mama	9	3	7	?2	6
Odd Mama	1	D	D	D	D
Noisy	–	9	0	D	D
New Mama	–	3	D	D	D
Pseudo Wolf	–	–	–	–	5
Total	24	20	12	?4	16
Mean pups/litter: 5.5					

D, indicates that the female died; – indicates that she had not entered the group (New Mama) or reached adult size (Pseudo Wolf and Noisy).

ence for several months. No dog other than the mother was seen to visit any HIWG litter, although bitches sometimes whelped in close proximity (within 100 m of each other on two occasions). For the first six weeks, or so, the only contact HIWG mothers appeared to have with other adult group members was when they met at feeding sites.

If the den site was approached while the mother was present, she would sometimes engage in what appeared to be a distraction display – exiting noisily from the site, and then circling quietly back to it. Such clearly detected approaches to dens sometimes resulted in a change of den site, while physical disturbance of the den, always did so. Den site changes were a significant cause of pup mortality (see below). However, two litters were apparently destroyed in their dens shortly after birth (by unknown agents – see below), so relocation may have been a less expensive option than staying put.

By the age of about ten weeks, pups were relatively autonomous behaviourally and highly mobile. At this stage HIWG litters would move to a new site, usually still on the mountainside, but closer to a feeding site, though on two occasions such 'base camps' were established in hedgerows on the plain. The two non-HIWG sylvatic litters stayed put.

From 10 to 20 weeks mothers spent an increasing proportion of time away from their litters, and from about 14 weeks, surviving pups began to socialize with other adults from the group, and with any other surviving pups. Until the age of about 20 weeks, however, they continued to sleep at different sites from the other adults.

The mortality rate was high. Only 16% of 76 pups born survived to the age of 20 weeks. Identified causes of death were: adverse weather, getting left behind during den changes, getting lost when first mobile and predation (two episodes of predation involved circumstantial evidence that the pups had been killed by a large canid, possibly a wolf or a dog). An additional hazard was the removal of pups by people seeking pets.

Of 12 pups surviving to 20 weeks over three years, 4 were observed to be recruited into their natal groups: three into HIWG (one male, two female) and one (male) into the central range. The fate of the remainder is unknown, but they did not disperse into other monitored packs.

Non-resident dogs and external recruitment

Accompanied dogs (sheepdogs with shepherds and their flocks, gundogs with hunters, or pets with tourists) passed routinely through the area. More interesting was the arrival of unaccompanied dogs. These sometimes arrived by human agency (e.g. the reportedly common practice of releasing unwanted pets) but were, when first detected, free ranging.

A dog was counted as an unaccompanied non-resident (UNR) if it was detected on two different days. Four such dogs were noted in Rovere, three in Sant'Iona and twenty in the Altopiano. Two of the Roveresi dogs were, when first detected, a mixed-sex pair. The Altopiano dogs included two groups, one of five individuals, and one of nine, and also a pair.

UNRs suffered one of four fates: some established themselves as residents, either by incorporating themselves into existing social groups, or establishing new ones. Some were killed by existing residents, and some, after remaining in the area for a few days, disappeared to an unknown fate.

Three of the four observed UNRs established themselves successfully in Rovere; two were a pair, who moved into the vacated western range six months after its occupants had been killed in April

1981. The third, a male, joined the eastern group a year after the beginning of the study. However, a fourth dog, also male, was killed shortly after his arrival in a nocturnal fight with members of HIWG on the edge of the village. Three male UNRs arrived in Sant'Iona in the summer of 1982.

The most notable UNRs in the Altopiano were the two large groups. These appeared in successive summers (1981, 1982), in each case first being spotted in the south of the area. The first group consisted of five individuals (three male, two female). They had two observed hostile encounters with HIWG, both lasting for around 15 minutes and involving barking and chasing, but no physical contact. In both of these encounters, the intruders were driven off by the residents. The last contact with this group was five days after the first. The second group consisted of nine beagle-like hounds. No interactions between this group and any residents were observed, and the period between first and last sightings was seven days.

Five UNRs were observed to establish themselves in the area. Two were a pair who took over the northern range four months after the last observation of the bitch Nera. A third was a female who joined the pair occupying the central range, and the fourth a male who succeeded in becoming a member of HIWG. The fifth, the bitch New Mama joined the HIWG in late summer of 1981, whelped three pups in autumn of which none survived, and herself died before the following spring. One UNR in the Altopiano was almost certainly killed by resident dogs of HIWG although the fight was not observed.

All UNRs appeared during spring and summer. This reinforces Boitani *et al.'s* (Chapter 15) suspicion that dispersal from villages and recruitment to the sylvatic population is greatest during the mating season, perhaps involving the attraction of male village dogs to sylvatic bitches on heat.

Discussion

In summary, free-living dogs in this part of Italy appear to adjust their behaviour to their local circumstances. In the villages, they adopt a 'fox-like' existence in which individuals live in loose social groups that defend a territory, but pursue a relatively solitary existence. In the countryside, sylvatic dogs adopt a more wolf-like lifestyle, generally operating as a pack, although individuals sometimes travel alone.

Both village and sylvatic dogs were largely dependent for food on human activity and, in the villages at least, were subjected to significant human interference, a fact reflected in their sex ratios and the culled litters. That males are more tolerated by villagers than females was confirmed in conversations.

During 1981–3 population levels appeared stable, although recruitment from outside was important. No dog was observed to change group or to disperse. The groups of dogs clearly had social integrity: relationships within groups were largely amicable, those between groups were hostile, and fights could be fatal between neighbouring groups and when resident dogs encountered itinerants.

Many of our observations agree with those of other free-ranging dogs: male-biased sex ratios are common in free-ranging dogs (e.g. Daniels, 1983*a*, *b*), and probably result from human interference with the population from which feral dogs are drawn. The home range sizes we recorded are of the same order as those noted elsewhere, varying between 2–10 ha for urban dogs (e.g. Berman & Dunbar, 1983) and 4–10 km^2 (e.g. Scott & Causey, 1973) for rural dogs. From Alabama to Illinois, packs of 'feral' dogs number 2–6 adults (e.g. Scott & Causey, 1973; Nesbitt, 1975) – compared with which the HIWG was rather large. We have few data on the breeding system of our dogs, but the association between pairs even within the HIWG is in accord with other observations (Daniels, 1983*b*). Multiple pairs breeding within larger packs of wild canids are uncommon, but have been recorded for coyotes (Camenzind, 1978), wolves (van Ballenberghe, 1983) and, circumstantially, golden jackals (Macdonald, 1979). The behaviour of Abruzzese free-ranging dogs is reviewed more fully by Boitani *et al.* (Chapter 15), and our observations fit within the general pattern of previous observations.

We will conclude by readdressing our three initial questions. First, how does the behaviour of these free-living dogs relate to observations on other canids? Second, can the contrasting behaviour of village and sylvatic dogs be interpreted within the framework of ideas on ecological pressures affecting carnivore society? Third, is it appropriate to seek adaptive explanations for the behaviour of these dogs? In exploring these questions we have the great advantage of comparing our results with those gathered subsequently on the same population by our colleagues Boitani *et al.* (Chapter 15).

Inter-specific comparisons

While this is not the place for a comprehensive review of canid social systems, we note certain striking parallels between the behaviour of these dogs and that of some other species.

The sylvatic dogs formed packs that have much in common with those described for wolves in many parts of their range (see review by Harrington *et al.*, 1982). Like wolves, these dogs apparently benefited from strength of numbers and, on occasion, killed rival conspecifics. Unlike wolves, they showed no evidence of co-operative care of young or of adoption, although these traits are widespread in wild canids (reviewed in Macdonald & Moehlman, 1982). Nevertheless, these dogs were sympatric with the remnant wolf population of the Abruzzo, and actually travelled much smaller territories and behaved as a more coherent pack than is common amongst the local wolves (Boitani, 1983). In contrast, the village dogs displayed a social system that was strikingly similar to that of red foxes (Macdonald, 1981) although their territories did not drift as do those of some urban foxes (Doncaster & Macdonald, 1991).

Out of this comparison the obvious question is why do foxes not form packs when feeding at superabundant rubbish tips, such as on the Altopiano. Indeed, red foxes did feed from that same tip, and from similar ones in the Abruzzo region (Macdonald, Barasso & Boitani, 1980). In these cases even if a group used the tip they never moved as a cohesive pack, although members of larger groups monopolized access and defeated members of smaller groups and occasionally joined forces to repel intruders (e.g. Macdonald, 1987, pp. 80–2). The ecological circumstances of tip-feeding foxes are so similar to those of the HIWG that the contrast is puzzling. Of course, in such cases, the foxes have larger competitors around (e.g. dogs and wolves), but so too the sylvatic dogs (though larger than foxes) were obliged to compete with wolves. Although it is an untestable hypothesis there seems little choice but to resort to a phylogenetic argument: packing is a behaviour never observed amongst vulpine foxes, but frequently observed in lupine canids.

Ecological pressures acting on dog society

The refuse on which both village and sylvatic dogs depend is clearly defendable. Furthermore, a dog arriving at a rubbish site in the wake of an intruder could find the food gone. Territoriality is the general solution to safeguarding economically defendable resources. Range configurations and aggression between neighbours both indicated that village and sylvatic dogs were territorial.

The refuse tips, farmyard scraps and food handouts on which the village dogs depended were unpredictable. Householders dumped food on an unpredictable schedule, in variable quantities, and it was collected erratically each week. It was obvious that no single dump would provide adequate food security for even one dog, and so we conclude that it was necessary for a village dog to configure its territory so as to encompass several potential feeding sites, thereby hedging its bets. Within such a territory, however, several dumps might be fruitful on a particular day, and sometimes a particular dump might be supplied with enough edible scraps to satisfy more than one dog. These are the circumstances that Macdonald (1981) proposed would lead to red foxes living in social groups, and there is much in common between the behaviour of the village dogs and that of foxes. If group size is limited by food (more probably it is limited, in this case, by human interference), the simplest prediction of the RDH would be a correlation between patch richness during periods of minimum food availability (bottleneck periods) and group size. Our data on food availability are inadequate to test this.

Assessment of the availability of food to village dogs is complicated because they had the opportunity to make excursions to nearby communal rubbish dumps. Such excursions led to a relatively predictable and abundant food source, but may have involved some danger in encounters with sylvatic dogs.

The sylvatic dogs depended on food scavenged from the Altopiano tip and throughout Rovere, sites separated by a distance of roughly 2.7 km. There is thus an association between patch dispersion and territory size, but this is complicated by the fact that sylvatic dogs also required access to safe refuges in the forest, and they may have visited villages in search of resources other than food (e.g. bitches).

Why did group size in the villages vary between 2–5, whereas in the countryside it was between 1–10? We propose that the mean group size of 3.5 in seven village groups reflects the modest richness of the unpredictable dumps from which they fed. In contrast, food at the Altopiano dump was very abundant and possibly superabundant. The high maximum group size (10 adults, HIWG 1981) accords with this high patch richness. The great variance in group size of sylvatic dogs is explicable because all the other groups in the vicinity were subjugated by the HIWG, and thus had effectively reduced access to food. The eight adults using the Rocca di Mezzo dump were in analogous circumstances to those of the HIWG.

Why did the sylvatic dogs form packs? We found no evidence of co-operative hunting, none of communal pup care (nor adoption), but ample evidence of the benefits of strength of numbers. The success of the HIWG in driving smaller groups of sylvatic dogs (and itinerants) from the Altopiano dump, and from Rovere during raids, was conspicuous. Furthermore, it directly parallels observations on other canids in which larger packs prevail (e.g. coyotes; Bekoff & Wells, 1980). It also parallels interspecific clashes, e.g. the capacity of larger groups of spotted hyaena to withstand piratical lions and vice versa (Kruuk, 1972). Such advantages of co-operative defence are particularly marked amongst competing species of canids (Lamprecht, 1978), making it interesting to know whether larger packs of sylvatic dogs fare better in encounters with wolves. It seems likely that wolves were a major factor affecting the sylvatic dogs both as predators and competitors. Competition with coyotes affects the spatial organization of red foxes (Voigt & Earle, 1983), inter-specific competition may result in character displacement in canids (Dayan *et al.*, 1989), and behavioural domination of smaller canids by larger sympatric species may be widespread (Hersteinsson & Macdonald, 1992). Certainly, similarities in the niches of wolf and sylvatic dog in Italy indicate the potential for strong competition between them (Boitani, 1983). The movement patterns of sylvatic dogs and their tendency to travel as a pack may both have been affected by the likelihood of encounters with wolves. Indeed, the configuration of their territory may have tessellated between wolf ranges in just the way that red fox ranges squeeze between coyote territories (Voigt & Earle, 1983). The cohesiveness of the HIWG was forcefully illustrated on the occasion when one of them was snared and the remainder of the pack stayed at hand barking ferociously even when closely approached, in contrast to their normal timorousness. The advantages of strength of numbers would be predicted to lead to an arms race to recruit a larger group and perhaps even to expansionism (*sensu* Kruuk & Macdonald, 1985).

It seems unsurprising that the sylvatic dogs packed together, but why then did the village dogs not do so as well? One possibility is that they did not need to: the small diameter (the largest, in Rovere, were less than 200 m across) of village territories meant that members of a collective group were invariably within earshot of each other. If the food available at any site was often likely to be insufficient for more than one dog then travelling in company could be disadvantageous. Equally, travelling singly might be the best way to intercept intruders, knowing that a bark can bring help within seconds. According to this scenario, greater patch dispersion, combined with large patch size, might favour packing to ensure that assistance was on hand.

The adaptive significance of dog behaviour

The interpretation of the behaviour of feral domestic animals is complicated by effects of artificial selection. Insofar as behaviour patterns may have been directly selected by humans, or may be pleiotropic corollaries of selection for other traits (e.g. see Belaev & Trut, 1975), it may be inappropriate to assume that domestic animals behave adaptively. This misgiving has been widely voiced in terms of the interpretation of the behaviour of free-ranging cats. However, Macdonald *et al.* (1987) and Kerby & Macdonald (1988) have argued that farm cats are sufficiently independent of humans that their behaviour may plausibly be interpreted in functional terms. Nonetheless, Macdonald (unpublished) has recently radio-tracked the ancestral *Felis silvestris lybica* and found that it persists in feeding solitarily around rubbish dumps at which free-ranging domestic cats form groups.

Is it sensible to consider free-living dogs as wolves, of which they are, by any genetical standard, merely a subspecies? There are some substantial differences, of which the wolf's 25% larger relative brain size is one (Coppinger & Schneider, Chapter 3). Furthermore, there are behavioural anomalies. Kleiman & Malcolm's (1981) observation that domestic dogs are the only canids in which males show no paternal care may indicate either an adaptive departure, an artifact of domestication, or a dearth of observation. In comparison to domestic cats, it is far harder to argue that dogs have continued to be subject to natural selection throughout their association with humans. The anatomical consequences of selective breeding are obvious in the differences between domestic breeds, and behavioural differences are equally marked (Zimen, 1972). Coppinger & Schneider (Chapter 3) have shown intriguing ontogenetic differences in the behaviour of puppies from livestock-guarding and herding breeds of dog. The quality of play in these breeds is almost totally different from the age of ten weeks. Breed differences have tangible physiological bases: the concentration of dopamine (a precursor of adrenal biosynthesis) in the brains of herding border collies is substantially greater than in the brains of the guarding maremmas from which our study population partly descended. The minimal exposure of most domestic dogs to natural selection, the extreme pressure of artificial selection, the scope for pleiotropic genetic effects, and the potential impact that breed differences might have on the behaviour of a group of free-living dogs, all combine to sound a note of extreme caution in seeking too precise an adaptive interpretation to their behaviour.

On the other hand, the speed of selective change is sometimes remarkable. Although the example is back to front, Belaev and Trut (1975) managed to breed placid, sociable foxes from hysterically nervous wild-types in just 20 generations (and this change coincided, pleiotropically, with the appearance of attractively piebald coat colours and dioestrous reproductive cycles). Furthermore, there is one overwhelmingly compelling line of evidence that our free-living dogs were behaving adaptively: as predicted on adaptationist arguments, individuals from the same population behaved quite differently when exposed to contrasting ecological circumstances.

The interpretation of village dog society is complicated, more than is that of sylvatic dogs, by the active involvement of people. Although the link may be tenuous, most of these dogs were notionally owned by somebody, and we cannot exclude the influence of this ownership during ontogeny as a factor shaping adult dog behaviour and movements (as argued elsewhere by Daniels & Bekoff, 1989*b*). Furthermore, villagers may have affected group size directly, as they manifestly affected sex ratio and breeding success. Nonetheless, we were impressed by the autonomy of the village dogs and judge that it is fruitful to consider seriously the proposition that at least many aspects of their behaviour have functional explanations.

The role of chance: the social flux model

Both sylvatic and village communities were subject to drastic density independent mortality at various ages. Pups were routinely killed *en masse* and bitches were selectively culled. In Rovere the seemingly stable organization described in 1981 was shattered by the violent deaths of several dogs (shot, hit by cars, etc.), and the composition of groups was affected by recruitment of immigrants into existing groups and by immigrants occupying vacant territories. The sylvatic dogs were shot and poisoned, their pups were stolen as pets, killed by storms and by other canids. Between 1981–3 the HIWG recruited three pups born into the pack, and two apparently unrelated immigrants. We conclude that groups of free-living dogs suffer frequent changes in composition and in kinship structure. Stochastic events may determine the balance of kin versus outside recruits in the pack at any given time.

This state of affairs leads to two observations. First, immigration of unrelated outsiders into groups where individuals of the same sex are already present occurs in some wild carnivore societies (e.g. Rood, 1987) including wolves (Lehman *et al.*, 1992). Therefore, the commonly reported presence of immigrants in dog groups (e.g. Scott & Causey, 1973; Daniels & Bekoff, 1989*b*) does not make them qualitatively different from other carnivore groups, although they may be unusual in the high frequency of immigration. Recruitment into sylvatic groups may be associated with some social turmoil, particularly if it tends to occur in the mating season. Second, many canid societies are largely based on kin groups, and kin selection is frequently invoked as a contributor to their behaviour (e.g. Macdonald & Moehlman, 1982). Furthermore, high rates of density independent mortality have been linked in urban red foxes with drifting territoriality, absence of social suppression of reproduction and an hypothesized disruption of the social hierarchy (Doncaster & Macdonald, 1991). In summary, frequent perturbation of dog group structure might be expected to alter the nature of social relationships, and we predict that these relationships will be qualitatively different in periods of stability where kinship bonds develop, as distinct from periods of high mortality when unrelated recruits may predominate.

In this context, it is noteworthy that between 1981–3 we found only two unrelated immigrants in the HIWG. In contrast, Boitani *et al.* (Chapter 15) studied the same group between 1984–7 during which time, although the total group size remained unchanged, there was a continual flux of immigrants such that only one survivor of the 1981–3 period was present in 1987.

We conclude that 1981–3 was, by chance, a less perturbed period for the HIWG, and was characterized by a move towards longstanding social ties and kinship ties, whereas the opposite circumstances prevailed in Boitani's 1984–7 study. If this hypothesis has any validity it could be that corollaries of the shift are: (i) synchronous reproduction in 1981–3 but not in 1984–7; and (ii) stability of territorial configuration. Boitani *et al.* report that at approximately six month intervals the home range borders shifted so that over three years the total range of the HIWG was roughly 30 km^2, whereas it was only about 10 km^2 at any given time. In red foxes perturbation is associated with shifting prevailing ranges (Doncaster & Macdonald, 1991). Range shifts in our sylvatic dog groups may arise because resources shift, wolf ranges shift, a change in composition of the pack alters the balance of strength, or because new recruits blend part of their previous range with that of the existing group.

Acknowledgements

This work was funded by a grant to DWM from the Waltham Centre for Pet Nutrition. We warmly acknowledge the friendship and collaboration of our colleague, Professor Luigi Boitani, and we thank him, along with Drs John Bradshaw and James Serpell for their forebearance. The manuscript benefited greatly due to comments from L. Boitani and S. Creel.

References

Beck, A. M. (1973). *The Ecology of Stray Dogs: a Study of Free-Ranging Urban Animals.* Baltimore MD: York Press.

Bekoff, M. (1979). Scent-marking by free-ranging domestic dogs. *Biology of Behaviour*, **4**, 123–39.

Bekoff, M. & Wells, M. C. (1980). The social ecology of coyotes. *Scientific American*, **242**, 130–51.

Belaev, D. K. & Trut, L. N. (1975). Some genetic and endocrine effects of selection for domestication in silver foxes. In *The Wild Canids*, ed. M. W. Fox, pp. 416–26. New York: Van Nostrand Reinhold.

Berman, M. & Dunbar, I. (1983). The social behaviour of free-ranging suburban dogs. *Applied Animal Ethology*, **10**, 5–17.

Boitani, L. (1983). Wolf and dog competition in Italy. *Acta Zoologica Fennica*, **174**, 259–264.

Boitani, L. & Fabbri, M. L. (1983). Censimento dei cani in Italia con particulari reguardo al fenomeno del randagismo. *Ricerche di Biologia della Selvaggina* (INBS, Bologna), **73**, 1–51.

Camenzind, F. (1978). Behavioural ecology of the coyote in the national Elk Refuge, Jackson, Wyoming. In *Coyotes: Biology, Behavior and Management*, ed M. Bekoff, pp. 267–94. New York: Academic Press.

Carr, G. M. & Macdonald, D. W. (1986). The sociality of solitary foragers: a model based on resource dispersion. *Animal Behaviour*, **35**, 1540–9.

Daniels, T. J. (1983*a*). The social organization of free-ranging urban dogs. I: Non-estrous social behaviour. *Applied Animal Ethology*, **10**, 341–63.

Daniels, T. J. (1983*b*). The social organization of free-ranging urban dogs. II: Estrous groups and the mating system. *Applied Animal Ethology*, **10**, 365–73.

Daniels, T. J. & Bekoff, M. (1989*a*). Spatial and temporal resource use by feral and abandoned dogs. *Ethology*, **81**, 300–12.

Daniels, T. J. & Bekoff, M. (1989*b*). Population and social biology of free-ranging dogs, *Canis familiaris*. *Journal of Mammalogy*, **70**, 754–62.

Dayan, T., Tchernov, E., Yom-Tov, Y. & Simberloff, D.(1989). Ecological character displacement in Saharo-Arabian *Vulpes*: outfoxing Bergmann's rule. *Oikos*, **55**, 263–72.

Doncaster, C. P. & Macdonald, D. W. (1991). Drifting territoriality in the red fox, *Vulpes vulpes*. *Journal of Animal Ecology*, **60**, 423–39.

Ginsberg, J. & Macdonald, D. W. (1990). *Foxes, Wolves, Jackals and Dogs: Action Plan for the Conservation of Canids.* Gland, Switzerland: IUCN Publications.

Gittleman, J. L. (1989). Carnivore group living: comparative trends. In *Carnivore Behaviour, Ecology, and Evolution*, ed. J. L. Gittleman, pp. 183–207. London: Chapman & Hall.

Harrington, F. H., Paquet, P. C., Ryon, J. & Fentress, J. C. (1982). Monogamy in wolves: a review of the evidence. In *Wolves of the World, Perspectives of Behavior, Ecology and Conservation*, ed. F. H. Harrington & P. C. Paquet, pp. 209–22. Park Ridge, NJ: Noyes Publications.

Hersteinsson, P. & Macdonald, D. W. (1992). Interspecific competition and the geographic distribution of red and arctic foxes (*Vulpes vulpes* and *Alopex lagopus*). *Oikos*, **58**, 505–15.

Kerby, G. & Macdonald, D. W. (1988). Social behaviour of farm cats. In *The Domestic Cat: The Biology of its Behaviour*, ed. D. Turner & P. Bateson, pp. 67–81. Cambridge: Cambridge University Press.

Kleiman, D. G. & Malcom, J. R. (1981). The evolution of male parental investment in mammals. In *Parental Care in Mammals*, ed. D. J. Gubernik & P. H. Klopfer, pp. 347–87. New York: Plenum.

Kruuk, H. (1972). *The Spotted Hyaena: a Study of Predation and Social Behaviour*. Chicago: University of Chicago Press.

Kruuk, H. (1975). Functional aspects of social hunting by carnivores. In *Function and Evolution in Behaviour*, ed. G. Baerends, C. Beer & A. Manning, pp. 119–41. Oxford: Clarendon Press.

Kruuk, H. & Macdonald, D. W. (1985). Group territories of carnivores: empires and enclaves. In *Behavioural Ecology. Ecological Consequences of Adaptive Behaviour*, ed. R. Sibley & R. Smith, pp. 521–36. Oxford: Blackwell Scientific Publications.

Kruuk, H. & Snell, H. (1981). Prey selection by feral dogs from a population of marine iguanas (*Amblyrhynchus cristatus*). *Journal of Applied Ecology*, **18**, 197–204.

Lamprecht, J. (1978). The relationship between food competition and foraging group size in some larger carnivores. *Zeitschrift für Tierpsychologie*, **46**, 337–43.

Lehman, N., Clarkson, P., Mech, L. D., Meier, T. H. & Wayne, R. (1992). A study of genetic relationships within and among wolf packs using DNA fingerprinting and mitochondrial DNA. *Behavioural Ecology & Sociobiology*, **30**, 83–94.

Lindstrom, E. (1986). Territory inheritance and the evolution of group living in carnivores. *Animal Behaviour*, **34**, 1825–1835.

Macdonald, D. W. (1979). Flexibility of the social organization of the golden jackal, *Canis aureus. Behavioral Ecology and Sociobiology*, **5**, 17–38.

Macdonald, D. W. (1981). Resource dispersion and the social organisation of the red fox (*Vulpes vulpes*). In *Worldwide Furbearer Conference Proceedings*, Vol. 2, ed. J. Chapman & D. Pursley, pp. 918–49. Frostburg, MD: R. R. Donnelley.

Macdonald, D. W. (1983). The ecology of carnivore social behaviour. *Nature*, **301**, 379–84.

Macdonald, D. W. (1987). *Running with the Fox*. London: Unwin Hyman.

Macdonald, D. W. (1992). *The Velvet Claw: a Natural History of the Carnivores*. London: BBC Books.

Macdonald, D. W. & Amlaner, C. J. (1980). A practical guide to radio tracking. In *A Handbook on Biotelemetry and Radio Tracking*, ed. C. J. Amlaner Jr & D. W. Macdonald, pp. 143–59. Oxford: Pergamon Press.

Macdonald, D. W., Apps, P. J., Carr, G. & Kerby, G. (1987). Social behaviour, nursing coalitions and infanticide in a colony of farm cats. *Advances in Ethology*, **28**, 1–66.

Macdonald, D. W., Ball, F. & Hough, N. G. (1980). The evaluation of home range size and configuration using radio-tracking data. In *A Handbook on Biotelemetry and Radio Tracking*, ed. C. J. Amlaner Jr & D. W. Macdonald, pp. 405–24. Oxford: Pergamon Press.

Macdonald, D. W., Barasso, P. & Boitani, L. (1980). Foxes, wolves and conservation in the Abruzzo Mts., Italy. In *The Red Fox, Behaviour and Ecology*, ed. E. Ziman, pp. 223–35. The Hague: W. Junk.

Macdonald, D. W. & Carr, G. M. (1989). Food security and the rewards of tolerance. In *Comparative Socioecology: the Behavioural Ecology of Humans and other Mammals*, ed. V. Standen & R. A. Foley, pp. 75–99. Oxford: Blackwell Scientific Publications.

Macdonald, D. W. & Moehlman, P. D. (1982). Cooperation, altruism and restraint in the reproduction of carnivores. In *Perspectives in Ethology*, Vol. 5, ed. P. P. G. Bateson & P. Klopfer, pp. 443–67. New York: Plenum Press.

Mills, M. G. L. (1990). *Kalahari Hyaenas: the Comparative Behavioural Ecology of Two Species*. London: Unwin Hyman.

Nesbitt, W. H. (1975). Ecology of a feral dog pack on a wildlife refuge. In *The Wild Canids*, ed. M. W. Fox, pp. 391–5. New York: Van Nostrand Reinhold.

Rood, J. (1987). Dispersal and intergroup transfer in the dwarf mongoose. In *Mammalian Dispersal Patterns: the Effects of Social Structure on Population Genetics*, ed. B. D. Chepko-Sade & Z. T. Halpin, pp. 85–103. Chicago: University of Chicago Press.

Scott, M. D. & Causey, K. (1973). Ecology of feral dogs in Alabama. *Journal of Wildlife Management*, **37**, 253–65.

Scott, J. P. & Fuller, J. L. (1965). *Genetics and the Social Behaviour of the Dog*. Chicago: University of Chicago Press.

van Ballenberghe, V. (1983). Two litters raised in one year by a wolf pack. *Journal of Mammalogy*, **64**, 171–3.

Voigt, D. R. & Earle, B. D. (1983). Avoidance of coyotes by red fox families. *Journal of Wildlife Management*, **47**, 852–7.

Woodroffe, R. & Macdonald, D. W. (1993). Badger sociality – models of spatial grouping. *Proceedings of the Symposia of the Zoological Society of London*, **65**, 145–69.

World Health Organization. (1988). *Report of WHO Consultation on Dog Ecology Studies Related to Rabies Control*. WHO/Rabies Research 88.25.

Zimen, E. (1972). *Wolfe und Königspudel – Vergleichende Verhaltensbeobachtungen*. Munich: R. Piper Verlag.

15 Population biology and ecology of feral dogs in central Italy

L. BOITANI, F. FRANCISCI, P. CIUCCI
AND G. ANDREOLI

Introduction

In Italy, free-ranging dogs (*Cani familaris*) are a familiar sight to anyone visiting the countryside. A nationwide census conducted in 1981 (Boitani & Fabbri, 1983*a*) revealed that almost every region reported free-ranging dog populations. The national overall estimate was about 800 000 free-ranging dogs, mostly distributed in the central and southern regions (Boitani & Fabbri, 1983*a*). Feral dogs, i.e. those domestic dogs that live without any direct contact with, or dependence on, humans (Boitani & Fabbri, 1983*a*), were estimated to represent about 10% of the total (they are only a component of the more numerous population of free-ranging dogs, as these also include stray dogs and all those left by their owners to move freely in and out of villages and into the surrounding wild areas).

Despite their significant impact on human and natural environments, free-ranging dogs have rarely been investigated and few studies on their ecology have been reported (Gipson, 1972; Beck, 1973; Scott & Causey, 1973; Nesbitt, 1975; Causey & Cude, 1980; Barnett & Rudd, 1983; Daniels, 1983*a*, *b*; Daniels & Bekoff, 1989*a*, *b*). In this paper we present data on group composition, life stories, recruitment, home range, movements and activity patterns of a group of feral dogs in the mountain region of Abruzzo, central Italy.

One of the critical difficulties encountered in feral dog research is the determination of the true status of the dogs being investigated, and previous researchers have employed a variety of different definitions (see Causey & Cude, 1980; Boitani & Fabbri, 1983*a*; Daniels & Bekoff, 1989*a*, *b*). Our study was concerned only with dogs we found to be living in a completely wild and free state with no direct food or shelter intentionally supplied by humans (Causey & Cude, 1980). The dogs also showed no evidence of socialization to humans (Daniels & Bekoff, 1989*a*), and tended to display a strong and continuous avoidance of direct human contact.

Methods

Study Area

The 250 km^2 study area is located in the Velino–Sirente mountain group, one of the Apennine ridges that crosses the Abruzzo region. The central part of the study area is a flat carst highland at 1300 m elevation, surrounded by mountains up to 2490 m. The highland is crossed by the main paved road that connects the villages of Ovindoli, Rovere, Rocca di Mezzo, Rocca di Cambio, and Secinaro to the east. Several dirt roads reach side valleys and higher altitude pastures, and they are mostly used in the summertime by shepherds and tourists.

The mean annual temperature is 7.6 °C with a minimum in January (–1.4 °C) and maximum in August (16.2 °C). Extreme low temperatures in the order of –20 °C are not unusual. Annual precipitation averages 90 cm for a total of 100 rain days, a third of them in the autumn. Snow depth is at its maximum in February–March, averaging 100 cm during the years 1982–84 at 1800 m altitude.

Thirty-two per cent of the area is covered by beech (*Fagus sylvatica*) forests, mostly in pure stands, with other species (*Pinus nigra*, *Fraxinus excelsior* and *Quercus cerris*) occurring only at lower altitudes. The bottom of the valley is covered by abandoned fields, pastures and fields cultivated annually with potatoes and cereals. At 1800 m the beech forests give way to alpine meadows that have been severely degraded by centuries of grazing. Large mammals were almost totally exterminated before the end of last century; today roe-deer (*Capreolus capreolus*) are very rare, and wild boar (*Sus scrofa*) are increasing following a recent introduction. Hares (*Lepus europaeus*), squirrels (*Sciurus vulgaris*) and foxes (*Vulpes vulpes*) are the only common mammals, while 6–10 wolves (*Canis lupus*) are estimated to be permanently in the area (Boitani & Fabbri, 1983*b*). In summer, the area is used by about 7000 sheep – mostly in the alpine pastures. At night, they are kept in stables or enclosures, often heavily guarded by shepherd dogs. Roughly 1400 cows and 200 horses are also kept in the most productive pastures of the valley.

Feral dogs were studied from February 1984 through May 1987. However, the same group of dogs had been observed repeatedly, though not intensively, since 1981, in parallel with a study reported by Macdonald & Carr (Chapter 14).

Capture and handling

Dogs were captured alive using a variety of traps depending on different trapping conditions, e.g.

snow, ground texture, presence of human activities, and so on (see Boitani & Fabbri, 1983*b*). Baiting stations and lures were used, but most traps were set along known trails. A blow-pipe (Telinject) was used to inject the dogs intramuscularly with a mixture of xylazine hydrochloride (Rompun, Bayer) (3 mg/kg body weight) and ketamine hydrochloride (Ketalar, Parke Davis) (4 mg/kg of body weight). This mixture made it possible to reduce the amount of liquid injected to 1.6 ml per 30 kg dog. After 15–20 minutes the dogs were sufficiently sedated to be approached and handled safely, and the anaesthetic's effect lasted for about 20–30 minutes. All immobilized dogs recovered physical control quickly and apparently completely.

Classification of dog types

Visual and radiotracking observations were used to distinguish feral dogs from other free-ranging dogs. The amount of aggression shown by trapped dogs – as used by Scott & Causey (1973) and Daniels & Beckoff (1989*a*, *b*) – was not found to be consistent with the known status of dogs and was therefore discarded as a means of classification. Some feral dogs were never trapped and they were classified as feral because they consistently associated with other marked feral dogs.

Sex of unmarked feral dogs was easily determined by observing the animals and their behaviour, especially their urination postures (see Bekoff, 1979). Visual observation was also used to classify the dogs into broad age classes: pups (up to 3 months), juveniles (3 months–1 year), adults (1–5 or 6 years), and old (more than 5 or 6 years). Trapped animals were aged according to the eruption and wear patterns of their dentition (Kirk, 1977), up to the limit of these techniques.

Morphology

All trapped dogs were weighed and their body measurements, coat colour and pattern, and breed type recorded. These external morphological characters were used to help identify possible breed combinations, as all the dogs were crossbred.

Radiotelemetry

Dogs were marked with plastic, numbered eartags (Rototag, Dalton, England) and the nine captured adults were fitted with radiotransmitter packages[1]. Radio-collared dogs were normally monitored five times a day, rotating the time when each dog was located. In addition, at least once a week, a single dog was monitored every 10 minutes for 24 hours. Several periods of continuous observation were also dedicated to particular situations. For example, during spring 1985 and spring 1986, two den sites were monitored continuously for the time they were attended, or when unexpected movements occurred. Telemetry was also used to locate and approach animals so that visual observations could be made on group composition, behaviour and habitat utilization.

Snowtracking, sightings, howling, scat collection, as well as telemetry on one animal, were used to monitor the home ranges of two wolf packs living in the area.

Life histories and group composition

The study area was continuously searched for signs of dogs, and all free-living animals were identified and located by radiotelemetry or visual observation. Frequent radio locations and snowtracking made it possible to monitor the dogs' activities very closely, while group composition and evolution was monitored by direct observation. Mating times, breeding, litter sizes and mortality causes were all checked by direct observations. (Litter sizes were based on the first sighting of pups, and this was usually possible within a couple of weeks after birth.) Pup survival was therefore estimated from the difference between numbers at first sightings and numbers observed over the following months.

Stray dogs from the villages were also monitored so that they could be identified when found outside their normal urban environment.

[1] Radio-collars were assembled with an AVM SB-2 transmitter cast in dental acrylic and packed in a PVC pipe with a lithium battery and transmitting antenna; the total weight of these collars was about 300 g. Transmission was in the 150–151 MHz frequency range. Signals were received by vehicle roof-mounted and hand-held directional antennas (4 and 5 elements yagi). Location of dogs was accomplished by triangulation (Mech, 1983). Locations were plotted on topographic maps (1 : 25 000) and referred to a grid system where the unit was a 250×250 m^2 to account for the maximum tested error obtained in using the radio-location technique in that mountain area and on the dogs.

Home range and spatial patterns

The size, configuration and seasonal dynamics of the dogs' home ranges were estimated by an extension of the Harmonic Mean method (Dixon & Chapman, 1980), utilizing the HOME RANGE program (made available from E. O. Garton, Moscow, Idaho). The Harmonic Mean method allows home range to be represented as a series of contours (isopleths) as illustrated in Fig. 15.3 below. Although isopleths have a strict mathematical meaning, their interpretation is made easier when extended to area contours including different percentages of all observations. Habitat use was analyzed by integrating home range contours with a computerized land use map[2].

Activity and temporal patterns

Radio fixes were classified as either 'resting', 'active' or 'travelling' according to signal reception patterns. 'Travelling' involved a change in location of the subject animal during the observation period. These activity classes were established by testing radio-collars on domestic dogs. Disproportionate use of the home range was tested for statistical significance using a one-tailed Kolmogorov-Smirnov test (Samuel, Pierce & Garton, 1985). Counts and proportion data were analyzed with chi-square tests.

Results

A total of 9 adults (4 males and 5 females) were captured and radio-collared. Four individuals were subsequently recaptured a second time. No apparent injuries to dogs occurred, apart from some molar tooth wear in female 05. Forty pups were born during the study period, of which two (coded 14 and 25 in Table 15.1) reached adulthood.

[2] A subsample of the original data set was randomly selected to eliminate the bias due to sampling procedures (i.e. autocorrelation of locations and sampling efforts not equally distributed through time) and inclusion of extreme locations (i.e. outliers). We chose 95% Harmonic Mean isopleths to represent home-range boundaries, and 50% isopleths to delineate core areas (Spencer & Barret, 1984).
To assess the internal anatomy of the home range, we used a GIS (ARC-INFO)/HOME RANGE integration (Boitani *et al.*, 1989). Habitat topology included vegetative cover (woodland, open forest/shrubland complex, grassland complex, open field and farmed area), elevation, roads, villages and other areas of human activity. On a smaller scale, habitat features characterizing the feral dogs' refuge areas were identified by analyzing the portion of home range included within 25% isopleths.

From July 1984 to March 1987, a total of 7956 locations were recorded, of which 1387 (17%) were visual observations of radio-collared individuals, and 883 (11%) involved observations of uncollared individuals belonging to the study group. Fifty-six per cent of the total records were made during the daytime, while the rest were recorded during the evening or after dark. From the original data set, a random subsample of 1618 radio-locations was utilized for the analysis of group ranging behaviour, representing an average of 1.6 locations/day. Two animals were observed for longer periods, – 712 and 649 days, respectively – while the shortest observation period per dog was 55 days (see Table 15.1).

All dogs were crossbred: the predominant breed appeared to be the 'Abruzzo' shepherd dog, together with some German shepherd and hound type animals. All were medium to large in size (weight ranging from 17–31 kg, mean = 22.8 kg), and small sized dogs were never observed among feral or other free-living dogs. Colours varied from almost solid brownish yellow to the more common combination of white with black or brown patches.

Group composition and life histories

A group of feral dogs was living in the same area since at least 1980 and in summer 1982 numbered at least nine adults. This group, however, was decimated by poison baits set out by local people in the autumn of 1983, and the only surviving animal (male 04) became part of the present study group until the end of the study.

Intensive search for tracks was conducted in areas within and outside the known range of the group in order to investigate the possible presence of other feral dogs. No tracks were found apart from those of dogs living within the villages. This indicated that the population of feral dogs we studied was relatively isolated. Searches for wolf tracks were also carried out to investigate wolf distribution and movements, along with supporting data from one radio-collared wolf.

Figure 15.1 summarizes the group's history from February 1984 to the end of 1987, showing arrivals of new animals, and deaths and births that changed the pack composition during the study period. The group ranged in size from a minimum of 3 animals, in the fall of 1984 and spring of 1986, to a maximum of 15 dogs in the summer of 1984 when two litters,

Table 15.1. *Feral dogs (pups excluded) of the studied group*

Animal code	Sex	Age	Origin	Cause of Death	Date of Death	Radio-days no.
01	f	A	R/F	Poisoned	Dec. 85	128
02	f	A	R/F	Poisoned	Dec. 84	137
03	m	A	R/F	?	June 86	117
04	m	A	F	Poisoned	Apr. 88	712
05	f	A	R	Poisoned	Apr. 88	649
06	f	J	R	Shot	Dec. 85	112
14	x	J	F	Wolves	Apr. 86	–
17	m	A	R	Poisoned	Oct. 86	55
18	f	A	R	Poisoned	Apr. 88	177
25	f	J–A	F	Poisoned	Apr. 88	–
30	m	A	R	Captured	Nov. 87	117

f, female; m, male; x, unsexed; A, adult; J, juvenile; F, feral; R, recruited.

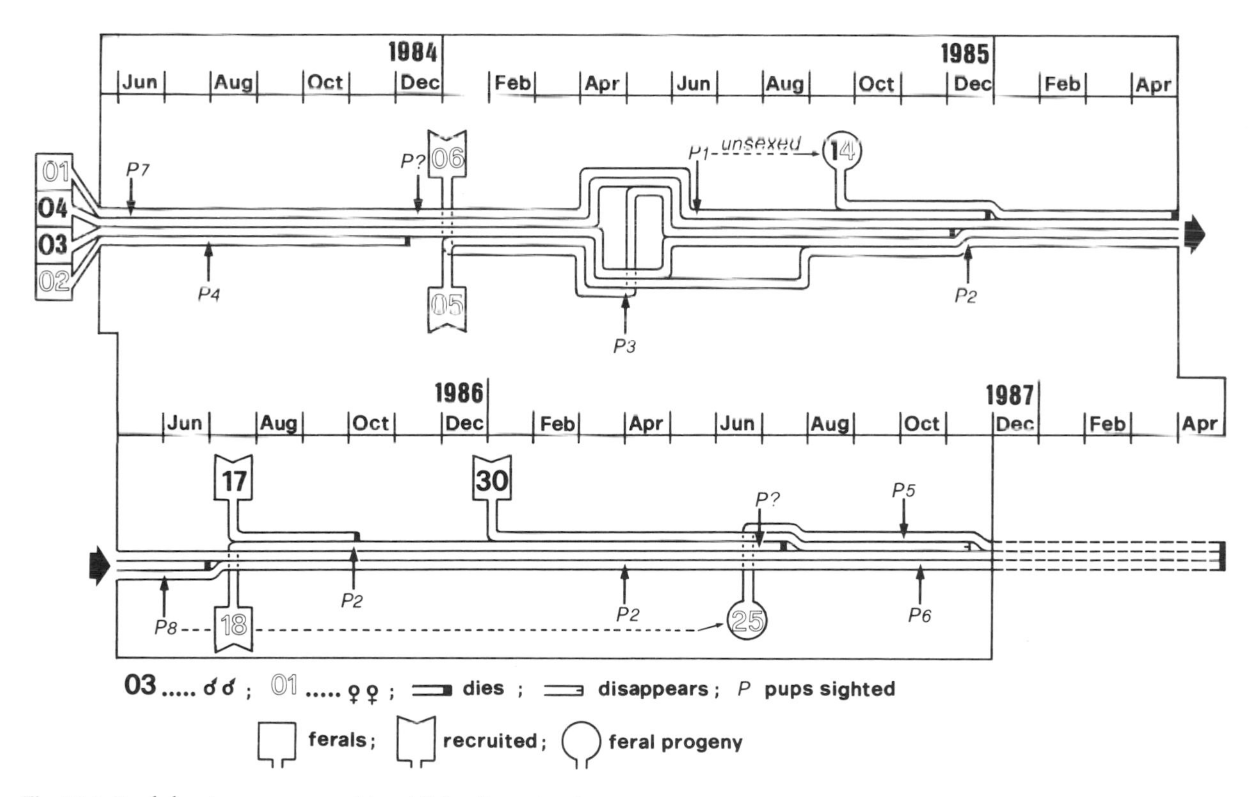

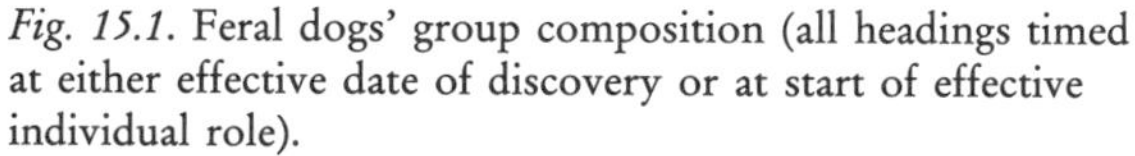

Fig. 15.1. Feral dogs' group composition (all headings timed at either effective date of discovery or at start of effective individual role).

comprising 7 and 4 pups respectively, joined the 4 adults. Overall, the group consisted of a core of two pairs, and when a partner died or went missing, he or she was usually replaced by free-ranging animals from the villages. The initial pairs (dogs 01 & 04 and 02 & 03) maintained their relationships until the death of one or other partner. New pairs were then formed with newcomers (in three cases) or with a young animal from a previous litter (one case), and these new relationships were then in turn maintained until the death or disappearance of one partner. During the study period three cases of new recruits were observed (dogs 05, 18, 30) and their arrival was always associated with females being in oestrus. Adult female 18 arrived accompanied by an adult male 17, who remained with the group for two months before his death. Adult female 05 arrived accompanied by her daughter (female 06) who was accepted and integrated into the group. Later, an adult male, phenotypically identical to female 06 and showing great confidence with her (probably her brother), tried repeatedly to approach and enter the group. However, he was aggressively rejected by the rest of the group, apart from females 05 and 06 (his probable mother and sister).

Group members always maintained very close contact, even when the den sites of the females were located far from the usual resting sites and required long journeys to visit. The two pairs split apart for a short time in 1985 when female 05 gave birth in an area about 15 km away from the usual range. Male 03 (her partner) stayed with her for about two months, although with frequent visits to the rest of the group, while the young female 06 (her daughter) stayed only ten days before leaving and rejoining the group. Female 05 then rejoined the group almost three months later, without encountering any apparent difficulty reintegrating.

Dog 30 was a persistent loner until he joined the group during the oestrus period of female 05 in December 1986 (05 had lost her previous partner in June 1986). Soon after this he became more closely associated with female 18.

In general, the group was relatively stable, basically rotating around two pairs and their offspring. Male 04 tended to act as leader of the group; he took the lead in aggressive behaviour toward intruders, and he was usually last to retreat after such encounters. The group did not show competition or aggression toward unknown dogs found around villages or at the dumps or around sheep yards, but they usually showed strong aggression toward intruders trespassing in their core areas. In one extreme case, a shepherd dog from a nearby sheep yard crossed the feral group's core area and was physically attacked by the group in broad daylight. Most clashes with intruders, however, ended with furious barking and holding the ground until the intruder slowly retreated.

Temporal associations with transient dogs were observed during oestrus periods, when aggressive tendencies seemed less pronounced. In the winter of 1986, for example, a male with a female on heat, both stray animals from the village of Rocca di Mezzo, joined the group (which consisted of five individuals at that time, but only one adult female) for a month. Similarly, in the winter of 1987 a group of nine stray dogs from Ovindoli met the feral dog group at the dump for several nights, sometimes spending the whole night within the group's home range, although without any apparent interaction with them.

We found no evidence that exclusive mating occurred within the group's breeding pairs, nor that the males in each pair were the fathers of the females' offspring. Figure 15.2 summarizes the occurrence of oestrus for all females during the study period. A relatively regular 6–7 month interval is evident for most females, but there is no sign of any synchronization of oestrus cycles among females. Out of 12 oestrus periods, 6 (50%) were in the spring (February to May) while the rest were scattered through most of the other months.

The earliest recorded oestrus and successful parturition occurred in a wild born female at 13 months of age. This female was the daughter of female 05 and she was the only wild born pup observed to reach sexual maturity and to reproduce.

Dens were not the centre of activities for the rest of the group, although they were often close to the more familiar resting sites (within 100–200 m, with the exception of female 05 in spring 1985). Other

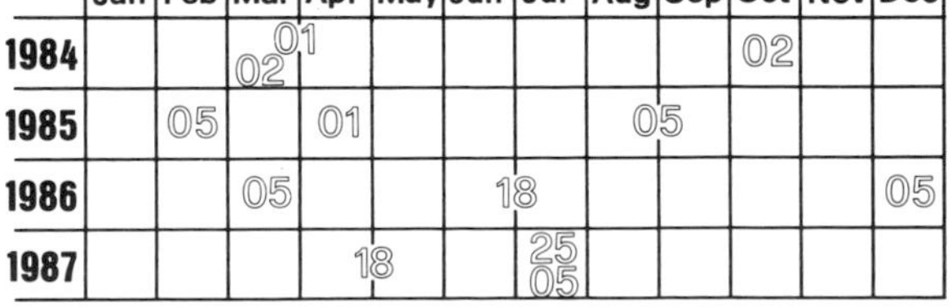

	Jan	Feb	Mar	Apr	May	Jun	Jul	Aug	Sep	Oct	Nov	Dec
1984			01 02							02		
1985		05		01				05				
1986			05				18					05
1987					18		25 05					

Fig. 15.2. Occurrence of oestrus in the group's females (numbers refer to animal code).

members of the group only occasionally visited den sites or females with pups, and never made any significant contribution to the feeding and/or care of the mother and her young. Dens were all in natural cavities, among boulders or under large rocks.

Eleven litters were observed, involving a total of 40 pups. In addition, postmortem examination of female 02 revealed 6 fetuses. Out of a total of 42 pups or fetuses that were sexed, 32 (76%) were males, with extreme cases of male : female ratios of 5 : 0 and 5 : 1 per litter. The mean litter size at first sighting was 3.63 ($N = 11$, range 1–8 pups).

Pups were difficult to observe and their carcasses were seldom found, limiting our ability to estimate mortality rates and causes. Pups disappeared even though they had been relatively easy to observe in the previous days. Foxes, crows and other small predators were fairly common in the area, and they may have been responsible for the rapid disappearance of any carcasses. Out of 40 pups born alive, 28 (70%) died within 70 days from birth, 9 (22.5%) within 120 days, and 1 (2.5%) within one year. Only 2 pups (5%) survived beyond the age of one year. Most deaths appeared to occur when pups started moving from the den site at 2–3 months of age. However, direct evidence of death is available for only a few pups; three 70-day old pups were killed at the den by a fox. When a pup survived to the juvenile age, it was always the only surviving member of its litter.

Seven adults were killed by poison baits, a one-year-old juvenile was killed by wolves, an 18-month-old female was shot, one was captured by local people, and we do not know the fate of one adult (see Table 15.1). Male 04 was the longest living individual (6 full years in the wild), followed by female 05 (3.4 years).

Recruitment of new and viable members of the group appeared to depend on village free-ranging dogs joining the group at intervals. At least four adults joined the group successfully at different stages of the group's history, and not a single pup born to the group survived to produce enough offspring to replace the losses due to mortality (see Fig. 5.10).

Home range and spatial patterns

From July 1984 to May 1987 the overall home-range size was 57.85 km^2, as shown within the 95% Harmonic Mean contour (see Fig. 15.3). Its mean elevation was 1500 m a.s.l. ranging from 1260–1950 m. The home range was not used uniformly; rather, different areas were consistently more frequented than others. This uneven use of the range is represented visually in Fig. 15.3 through contours delineating portions of the total utilization area (HOME RANGE software).

Most of the home range lies east of the paved road connecting three villages located at the lowered altitude of the highland. Ovindoli and Rocca di Mezzo are medium size villages with a permanent high level of human activity, especially during the summer and winter tourist seasons: these villages are mostly contained within the outer circle of the home range. Rovere is a smaller village, partly abandoned, where human presence is at a very low level for several months a year; this village is included in the core area of the range and is located close (about 1 km) to the 1984–87 Harmonic Mean Centre of activity.

Two garbage dumps are located in the open fields near the two main villages, and these sites are neither fenced nor protected in any other way from animal utilization. Both dumps lie at the edges of 50% contour zones (i.e. the core areas). Water is scarce but fairly homogeneously distributed throughout the area and easily accessible. About 76% of the range is utilized from spring to fall by free-ranging livestock, mostly sheep and, to a lesser extent, horses and cattle.

To evaluate the extent of road influences within the range, a buffer zone of 100 m was calculated along each paved road with more intensive traffic, and a buffer of 50 m along each less intensively used road. The total buffer area within each contour zone was then compared to each contour area (see Table 15.2). Road area as a proportion of total area declines from the 95–25% contour zones, indicating the tendency of the dogs to avoid roads. Habitat types were also calculated for each contour area (see Table 15.3). Although the proportions of the different habitat types within the 95% contour range did not differ significantly from those typical of the overall study area, there were significant differences between the inner and outer contour zones. Forest cover tended to increase, and open fields tended to decrease, toward the core areas.

Home-range size changed over the years, but without any seasonal fluctuation (see Fig. 15.4). Average size was 11.27 km^2 (±6.23 km^2, $N = 10$), with a minimum size of 2.24 km^2 in the summer of 1986 and a maximum of 21.15 km^2 in the summer of 1985. All

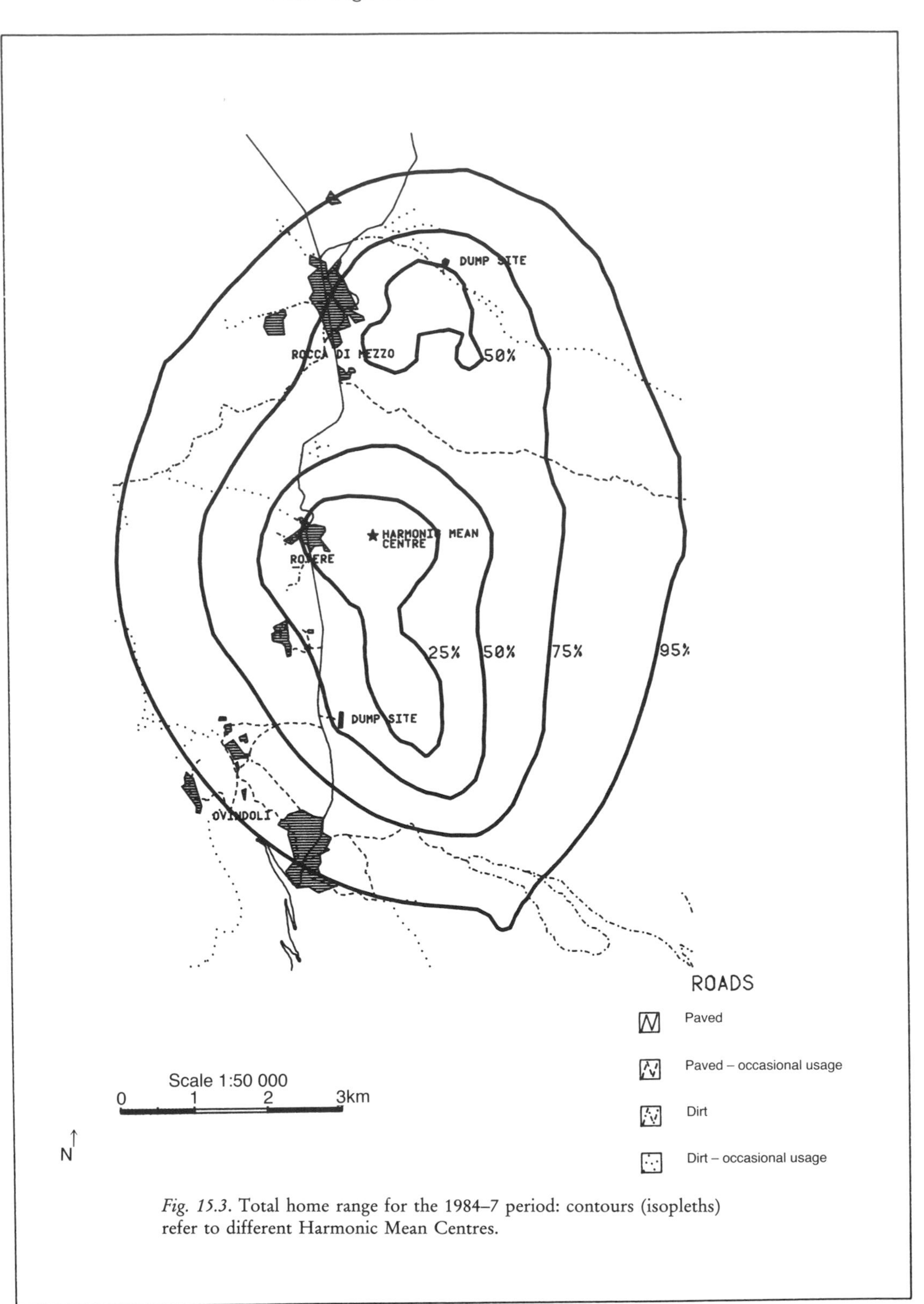

Fig. 15.3. Total home range for the 1984–7 period: contours (isopleths) refer to different Harmonic Mean Centres.

Table 15.2. *Road distribution within the home range*

Home-range isopleth (%)	Home range area (km^2)	Road area (km^2)	Road area (%)
95	57.8	6.6	11.4
75	29.0	2.5	8.5
50	11.9	0.6	5.6
25	3.5	0.1	3.5

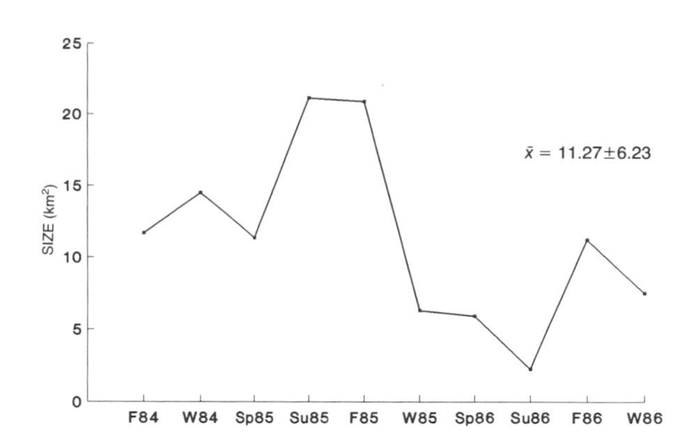

Fig. 15.4. Seasonal variation in home range size (fall 1984 through winter 1986–7).

seasonal ranges were in the same general area (see Fig. 15.5) and any variations in sizes and/or utilization patterns of the area were generally due to various occasional and unpredictable factors, such as disturbance by humans (hunters, shepherds, tourists), interference by other dogs or wolves, presence of substantial amounts of carrion (dead cattle or horses), previous territory knowledge of newly-recruited dogs (e.g. female 05 in 1985), or disturbance at the dumps.

Figure 15.5 summarizes the sequence of the seasonal home range changes. The large size in the summer and fall of 1985 is partly due to the arrival of two new group members who were used to roaming more widely around Rovere and visiting the Rocca di Mezzo dump before they joined the group. The particularly small range size in the summer of 1986 resulted from the combined effects of female 05's denning activities, and the low number of radio-tracked group members.

Individual home ranges showed remarkable variations. An extreme case is that of female 05 during her denning period of about four months in the spring/summer of 1986. When compared to her annual home range for the same year (37.5 km^2), the denning period range represented only 4.3% (1.6 km^2), and the 25% contour area was reduced to an even greater extent (from 0.98 km^2 for the whole year to only 0.00001 km^2 for the denning period). The den was located on a rocky and inaccessible slope at the edge of Rovere, fully protected from human disturbance and yet with easy access to garbage disposed of by the village households. The female had to move only a few hundred metres to find all she needed to feed herself and her pups.

Only three major excursions outside the home range occurred in the winter and spring of 1985 and they eventually led to the denning by female 05 in an area 15 km away from the usual range. These excursions, excluded from the overall home range calculations, were not apparently due to any particular factor, although in one case a visit to a nearby village

Table 15.3. *Habitat type distribution within study area and home range*

	Area (m^2)	Wood (%)	Prairie/Rocks (%)	Shrubland (%)	Open field (%)
Study area	232.0	34.2	38.3	16.2	11.3
95% isopleth[a]	57.8	34.0	33.9	11.0	21.1
75% isopleth	29.0	28.3	39.6	8.4	23.7
50% isopleth[b]	11.9	29.3	44.6	2.5	23.6
25% isopleth[c]	3.5	43.7	46.7	–	9.6

$^{a\text{–}c}P < 0.001$; $^{b\text{–}c}P < 0.01$.

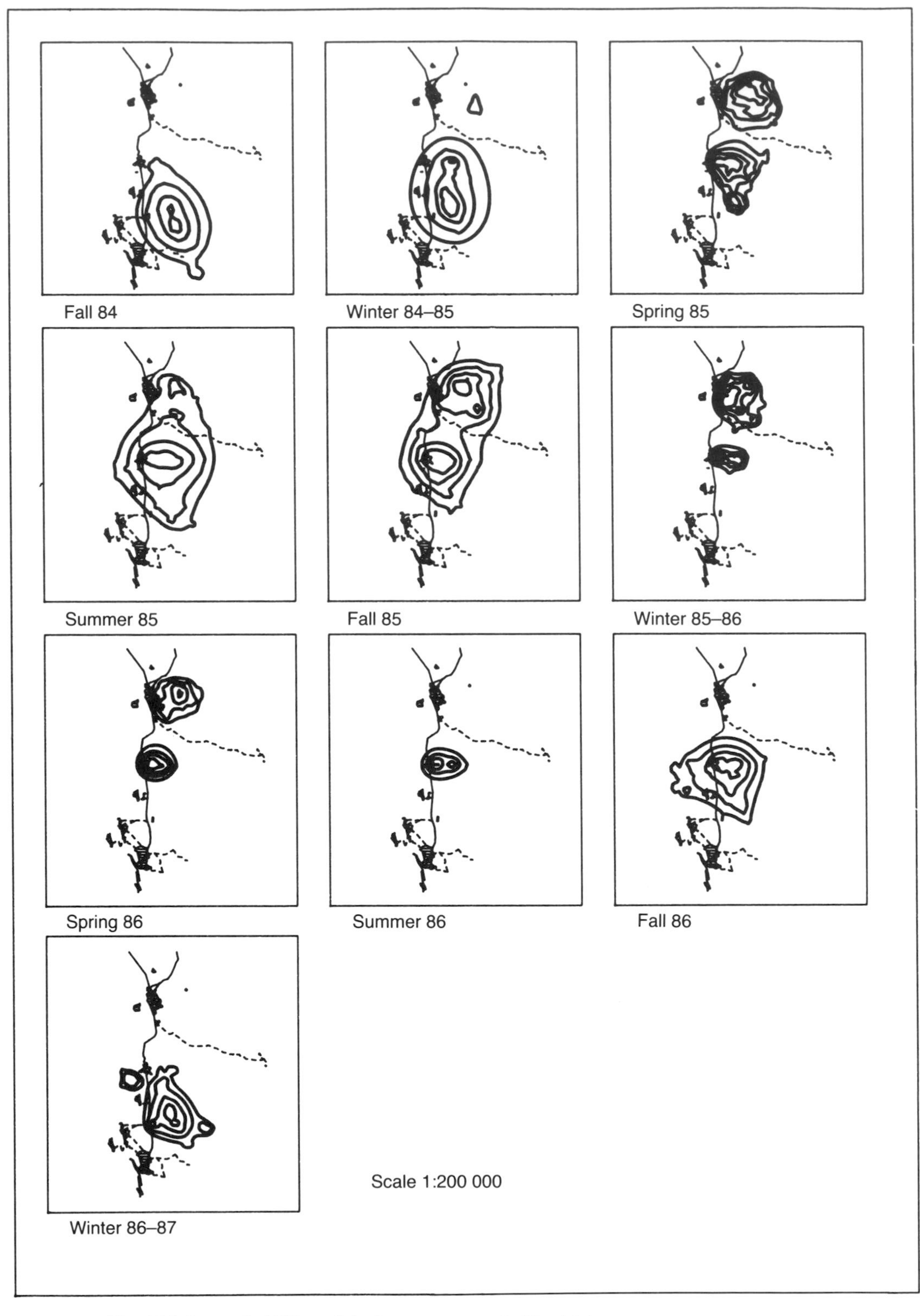

Fig. 15.5. Seasonal shifting of the home range: as in Fig. 15.3. Only the three villages and the connecting road are shown.

Table 15.4. *Seasonal core areas (25% isopleth)*

	Fall 1984	Wint. 1984–5	Spr. 1985	Summ. 1985	Fall 1985	Wint. 1985–6	Spr. 1986	Summ. 1986	Fall 1986	Wint. 1986–7	mean ± SE
Area (km²)	0.35	0.69	1.21	1.04	1.66	0.65	0.25	0.002	0.77	0.37	0.69 ± 0.49
% of H.R.	3.0	4.0	10.6	5.0	7.9	10.5	4.2	0.1	6.9	4.9	5.7 ± 3.3

dump may have provided the incentive. It is probably no coincidence that on the day of the first excursion to the dump a very strong wind, which may have carried attractive odours, was blowing from that direction. Although the new dump contained plenty of potential food, the dogs returned to their usual range after these few visits.

The 50% contour of the total home range defined a core area of about 11.97 km² (20.7% of the total). Apart from Rovere, the core area avoided villages, was crossed by a small portion of roads and bordered two garbage dumps. The 25% contours stress these features, and the Harmonic Mean Centre is almost precisely the same distance from the two dumps, i.e. the two main food sources. The seasonal 25% core areas show a significant variability in size (see Table 15.4) and, moreover, they show an interesting variability as percentages of their respective 95% contours. This confirms different patterns of range utilization during different periods, although without seasonal recurrence, and mostly due to occasional, unpredictable events or circumstances. Average core area size is 5.71% of the total home range, suggesting a high level of attachment to particular sites where dogs tend to stay most of their time and where they return after excursions to other parts of the range. Twenty-five per cent core areas include dens, resting sites and retreat sites, and a highly significant proportion of the nonactive radiofixes were found within their limits (see Table 15.5).

Shifts in core areas appeared to be largely random, although the key environmental features within them remained remarkably consistent (see Fig. 15.6). When seasonal centres of activity are plotted, three preferred areas can be identified: two by the two major dumps, and one by the village of Rovere, itself a safe food and shelter source with little human interference. All seasonal centres of activity lie away from roads, other villages and areas of human activity. Over the years, core areas shifted back and forth between two dumps and to the east of the main road, regardless of the number of group members or their former attachment to the area.

Core areas also have distinct territorial significance. Aggressive behaviour, as indicated by barking, chasing or approaching intruders aggressively, was more frequent and more intense when encounters were closer to the core areas. Aggression toward shepherd dogs or other free-ranging dogs, as well as to our own research working dogs, was often observed within the core areas, but rarely occurred in other parts of the range. Although there is no numerical evidence for this, levels of aggression were apparently lower during oestrus periods. As a consequence, these periods provided the only opportunities for the successful recruitment of new group members.

Interesting differences exist in the utilization patterns of villages and dumps. Day/night visits to the villages (see Table 15.6) show: (a) a significant avoidance of Ovindoli, the largest of the villages with enough human activity to scare the dogs away at any

Table 15.5. *Activity distribution within the home range*

	95% Iso	75% Iso	50% Iso	25% Iso
Activity				
Obs.	751	700	559	379[a]
(%)	(100)	(93.2)	(74.4)	(50.5)
No activity				
Obs.	837	803	691	489[b]
(%)	(100)	(95.9)	(85.5)	(58.4)

[a–b] $P < 0.001$.

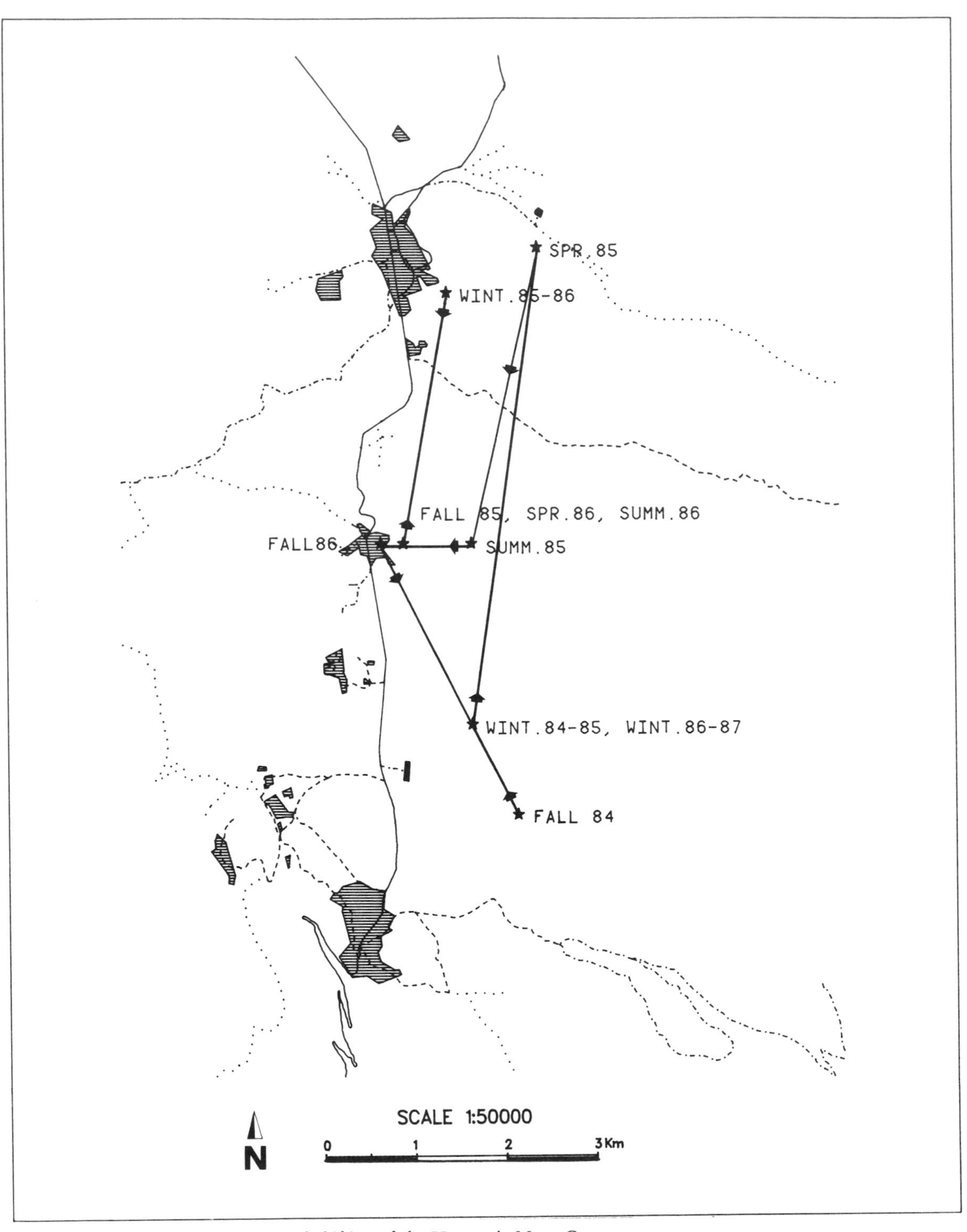

Fig. 15.6. Seasonal shifting of the Harmonic Mean Centres (stars).

Table 15.6. *Nocturnal and diurnal presence at villages*

	Day[a]	Night[b]
Ovindoli		
Obs.		
(%)	–	–
Rovere		
Obs.	34	123
(%)	(21.6)	(78.4)
Rocca di Mezzo		
Obs.		17
(%)	–	(100)

[a–b]$P < 0.05$.

time; (b) a very limited use of Rocca di Mezzo at night, when streets are empty; and (c) a significant use of Rovere. A ruined castle sits at the top of the hill where Rovere is located. Although these ruins are actually within the village boundary, they provided a safe place to retreat, a reliable shelter from storms and a place from which to venture into the streets. Hence the village could be used safely at night and, to some extent, during the day, especially in winter.

Dump utilization also reflects different levels of disturbance (see Table 15.7). Secinaro was visited during the excursions outside the usual home range, and female 05 made her den in close proximity to this dump. The Ovindoli dump (the largest and richest of all) is located near a major road and in the middle of a large, open field often frequented by people during

Table 15.7. *Nocturnal and diurnal presence at village dumps*

	Day[a]	Night[b]
Ovindoli		
Obs.	44	106
(%)	(29.3)	(70.6)
Secinaro		
Obs.	19	2
(%)	(90.5)	(9.5)
Rocca di Mezzo		
Obs.	39	18
(%)	(68.4)	(31.6)

[a–b]$P < 0.001$.

daylight hours; it is mostly visited by the dogs at night (70.6% of observations). On the contrary the Secinaro dump is located in a secluded and quiet, small valley where nobody ever goes but the little truck carrying the garbage there every morning. The dogs quickly learned this pattern and visited the dump immediately after the truck had left (90.5% of all fixes). The Rocca di Mezzo dump has intermediate characteristics: it is large and often disturbed by human presence, but it is located in a protected side valley surrounded by forest with plenty of potential retreats. It was visited significantly more often during the daytime (68.4%).

The dumps offered an unlimited source of food to the dogs: all kinds of refuse was thrown there in large quantities, including slaughterhouse leftovers. Large bones could be found all along the trails from the dumps to the dogs' resting sites. Dogs were observed feeding at the dumps for periods of up to an hour or more, and usually arriving in groups or pairs rather than singly. They were also tolerant of each other as they skillfully ripped open the plastic sacs delivered freshly every morning. The garbage cans in the streets of Rovere provided another food source; the dogs had learned to tip over the cans and spread the garbage on the ground.

During the years before our field study, local people had made several claims for damages to livestock caused by feral dogs. During the study period we saw no evidence of any killing of, or damage to, either domestic or wild animals by the feral dogs. A few chases of hares and squirrels were observed but they seemed to arise from playfulness rather than predatory behaviour. The dogs were monitored feeding on a wild boar carcass, but the animal had almost certainly died of natural causes. Occasional dead livestock had an important impact on the dogs' activities and movements, and the group sometimes moved its centre of activities to take complete and continuous advantage of such sudden food sources.

During the study period, wolf numbers and activities in the area were monitored by snowtracking, scat collections, elicited howlings, and also by one captured and radio-collared wolf. Two main packs of wolves were identified, whose home ranges and movements partially overlapped those of the feral dogs. The first pack comprised five to seven individuals and had its range north and east of the dogs' range for a total of about 300 km^2. It partially overlapped with the northern portion (about 30%) of the dogs' home range, since the wolves also utilized the Rocca di Mezzo dump as a food source. The second pack averaged about five individuals. Its entire home range was never fully defined, since it extended well south of the study area, but it overlapped with the southern part of the dogs' range (about 35%). This pack utilized the Ovindoli dump as a food source. While no evidence of direct contact between wolves and dogs was obtained, it is possible to infer some interactions from movement patterns and behavioural observations. For instance, excursions and exploratory journeys by the dogs into the northern pack's territory coincided with the oestrus periods of the wolves, and in April 1986 a juvenile feral dog was killed by the wolves in the proximity of the Rocca di Mezzo dump. In addition, our own imitations of wolf howling when the dogs were at the Rocca di Mezzo dump elicited immediate and strong aggressive reactions. The two wolfpacks' territories never came in contact with each other, and the central part of the dogs' home range (Rovere area) was located in the zone between the wolves' territories.

Activity and temporal patterns

A total of 7073 fixes of radio-collared animals was utilized for the analysis of activity patterns. Feral dogs 'rested' for 48% of the time, while they were 'active' (40%) or 'travelling' (12%) for the rest of the time. Activity patterns by season and by sex (see Table 15.8) show significant differences for both sexes among seasons and also, for some seasons, between sexes. Females in general travelled less and rested for longer, and this sex difference was more marked in the spring and, to a lesser extent, the fall when denning and pup-rearing kept them at the den sites. Males were more active and travelled for longer in the spring and summer. Winter activity showed a generally similar pattern for both sexes.

Female 02 during her autumn 1984 pregnancy period provided a good example of activity pattern changes: she spent only 3% of her time travelling, as this was limited to a single daily trip from the resting site to the dump and back. Most of the rest of the time she was resting (66%). Female 05 also rested most of the time (69%) during her denning period in the spring of 1985, and only 8% of her time was dedicated to her twice daily trips to the dump. During the same period, male 04 showed the highest recorded percentage of travelling (23%) as he was often checking the den and patrolling the rest of his range. His higher travelling percentage reduces the resting percentage, since the 'active' category remains largely unchanged.

Adding all 'active' and 'travelling' data results in a daily activity pattern involving two peaks indicating a significant preference for activity at dawn and dusk (see Fig. 15.7). Seasonal analysis of overall activity indicates, as expected, the two activity peaks shifting to adjust to dawn and sunset times. This pattern is best shown when diurnal activity is divided into six daily periods defined in relation to dawn and sunrise

Table 15.8. *Activity patterns by season and by sex*

		% No activity	% Activity	% Travelling	*N*	*P*
Winter	(m)[a]	44	44	12	800	0.08
	(f)[e]	47	40	13	949	
Spring	(m)[b]	46	36	18	621	<0.001
	(f)[f]	66	26	8	313	
Summer	(m)[c]	43	42	15	750	0.003
	(f)[g]	48	41	11	1015	
Fall	(m)[d]	47	44	9	849	0.01
	(f)[h]	53	38	9	1616	

f, female; m, male.
[a–b–c–d] $P < 0.001$; [e–f–g–h] $P < 0.001$.

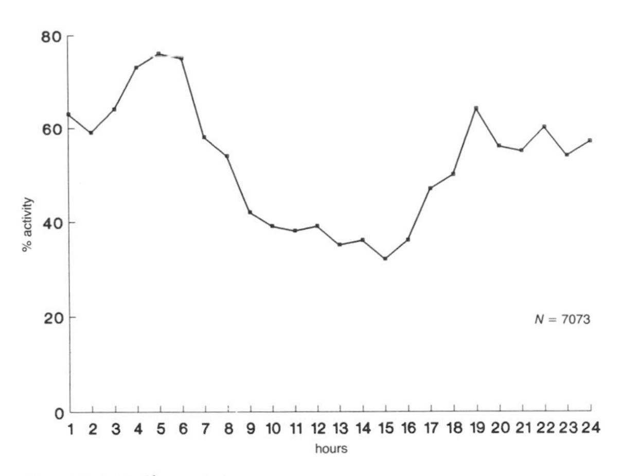

Fig. 15.7. Daily activity patterns.

(see Fig. 15.8). Travelling maintains its two peaks, but is evenly distributed through the other periods, including the daytime. 'Active' is significantly higher during the hours of darkness. The dogs show a basic pattern of becoming active at their resting sites around sunset, travelling to the dumps, staying active around the dump areas and at lower altitudes, and then travelling back at dawn to their resting sites.

In spring, breeding activities modify the basic activity pattern and a third significant peak of activity is evident during the central part of the day. Mating, pregnancy, denning and visits to the den sites by males influence both sexes' activities at this time. Intensive radiotracking during breeding times

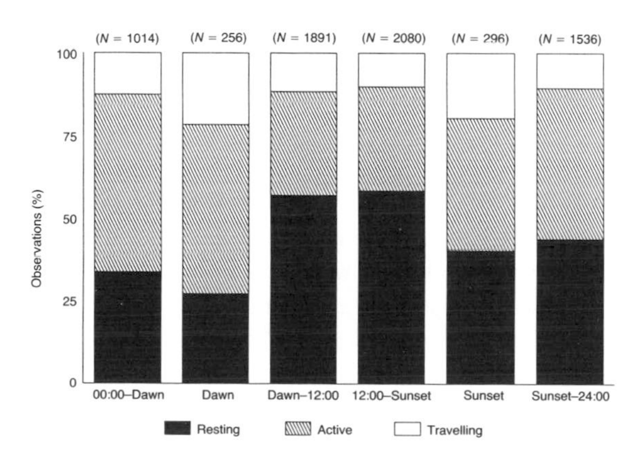

Fig. 15.8. Activity distribution in different periods of the day.

showed that female activities were strongly affected. Female 02, during her pregnancy in October–November 1984, did not show any significant activity peaks throughout the 24 hours. Female 05 was closely monitored during three separate breeding periods in very different denning conditions. In the spring of 1985 when she denned by the Secinaro dump, a quiet place she could visit at any time of the day, her activity showed no significant peaks. In the fall of 1985 she denned in an abandoned stable at the edge of Rovere, and the village's garbage cans were her main source of food. Here her activity was mainly nocturnal (although, statistically-speaking, not highly significantly so). In the summer of 1986 she denned in a cave outside Rovere, and again her activity was strictly (and statistically highly significantly) nocturnal.

Activity patterns were also highly correlated with elevation (see Table 15.9). Lower altitude areas, where the dumps and villages are located, showed the higher activity percentages as compared with the resting activities at higher altitudes (core areas). Correlation between activity and vegetation cover (see Table 15.10) is also highly significant: daytime resting was mostly confined to wooded or mixed habitat, as was 'active' behaviour, though to a lesser extent. During the hours of darkness, activity was more likely to occur in areas with no vegetation cover, since the dogs were travelling to the more exposed areas around dumps and villages.

Activity patterns were also significantly associated with moonlight ($P < 0.001$), though this is difficult to explain. 'Active' and 'travelling' behaviour occurred at a higher rate under moonlit conditions, although only significantly so during the second half of the night (midnight–dawn). The first half (sunset–midnight) shows no significant effect of moonlight on activity patterns.

Discussion

Origin of feral dogs

The distinction between feral, stray and free-ranging dogs is sometimes a matter of degree (Nesbitt, 1975). Moreover, the category of 'abandoned' has recently been introduced (Daniels, 1987; Daniels & Bekoff, 1989*a*) and treated as a separate entity. Dogs have been classified on the basis of behavioural or ecological traits (Scott & Causey, 1973; Causey & Cude,

Table 15.9. *Activity distribution at different elevations (m)*

	<1200	1200–1400	1400–1600	>1600
% No activity[a] ($N = 3492$)	2.5	28.4	56.2	12.9
% Activity[b] ($N = 3519$)	1.4	41.8	47.0	9.8

$^{a–b}P < 0.001$.

Table 15.10. *Diurnal and nocturnal activity distribution within habitat types*

	Wood	Prairie/ Shrubland	Open field
Day			
% Activity[a]	26.7	41.3	32.0
% No activity[b]	32.2	45.5	22.2
Night			
% Activity[c]	15.0	41.9	43.1
% No activity[d]	21.7	46.4	31.9

$^{a–b}P < 0.001$; $^{c–d}P < 0.001$; $^{a–c}P < 0.001$; $^{b–d}P < 0.001$.

1989), their origins (Daniels & Bekoff, 1989*a*, *b*), their main type of range – i.e. rural vs. urban (Berman & Dunbar, 1983) and degree of access to public areas (Beck, 1973) – and their kind and level of dependency on, and control by, humans (WHO, 1988). This diversity of definitions contributes to the difficulty of comparing results from different studies, and the great variety of urban, rural and 'wild' habitats involved means that such comparisons may only yield a confirmation of the high ecological and behavioural flexibility of dogs. Also, more theoretical analyses of canid evolutionary strategies (see e.g. Bekoff, Daniels & Gittleman, 1984) are of limited value when carried out on individuals living under the pressure of human rather than natural selection.

The terms 'stray' and 'feral' appear to describe robust categories, at least with respect to the social dimension of the human–dog relationship. Stray dogs maintain social bonds with humans, and when they do not have an obvious owner, they still look for one. Feral dogs live successfully without any contact with humans and their social bonds, if any, are with other dogs. It may also be useful to maintain 'free-ranging', 'urban', 'suburban' and 'rural' as a second tier of description to be added to the basic categories as required.

Most authors agree that 'owned', 'stray' and 'feral' are not closed categories and that a dog may change its status during its lifetime (Scott & Causey, 1973; Nesbitt, 1975; Hitara, Okuzaki & Obara, 1987, Daniels, 1988; Daniels & Bekoff, 1989*a*). Only 3 out of the 11 adults in the present study were definitely born in the wild, while the others were recruited from village populations, shifting from a stray condition to a feral one. Only one dog apparently went back to his original condition, after being captured and locked in an enclosure. Changes in status may depend upon several natural and artificial factors (see Fig. 15.9). A dog, for example, may become stray by escaping human control, by being abandoned or simply by being born to a stray mother (Beck, 1975). The way back is only likely to occur when a stray is readopted by humans. A stray dog can become a feral one when forced out of a human environment or when accepted by a feral group existing nearby (Daniels, 1988; Daniels & Bekoff, 1989*a*), as was the case with the majority of our group members.

Group composition

Among canids, packs are defined as social units that hunt, rear young and protect a communal territory as a stable group (Mech, 1970). Pack members are usually related (Bekoff *et al.*, 1984). The feral dogs in our study area showed these characters only to a limited extent and they were not fully related, as is the case with most stray and feral dogs studied elsewhere (see Scott & Causey, 1973; Nesbitt, 1975; Causey & Cude, 1980; Berman & Dunbar, 1983; Daniels & Bekoff, 1989*a*, *b*). The kinds of associations and social bonds formed among feral dogs do not follow the precise rules of pack living, as

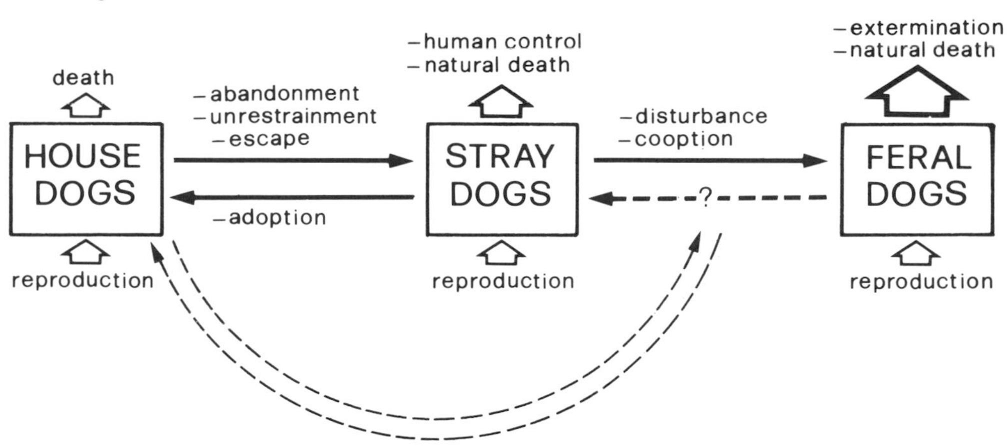

Fig. 15.9. The feralization model (size of arrows is proportional to relative dimension of processes).

described for the other canids (see Kleiman & Eisenberg, 1973; Bekoff *et al.*, 1984; Gittleman, 1989), and the term 'group' seems more appropriate than pack.

Overall dog density for the Abruzzo region, where the study area was located, is 2.59 individuals per square kilometre: owned but uncontrolled dogs account for the bulk of animals (1.38 dogs/km^2), followed by strays (0.89 dogs/km^2) and feral dogs (0.32 dogs/km^2) (Boitani & Fabbri, 1983*a*). Feral dogs are not evenly distributed over the entire region; they show a disjunct pattern of small, partially-isolated populations, and we were unable to find evidence of other feral dogs inhabiting the areas adjacent to our studied group.

Extrapolating from our group's total home-range size gives a mean density of 0.2–0.05 dogs per square kilometre for the area as a whole, depending upon group composition. These densities would tend to confirm previous indications that the majority of feral dogs in central Italy live at lower altitudes in more densely populated areas where food resources are richer, and opportunities for exchanges with village-based, stray dog populations are higher (Boitani & Fabbri, 1983*a*; Boitani, 1983).

Feral dog group sizes tend to be similar across a wide range of situations. Two studies in Alabama reported group sizes of 2–5 and 2–6 individuals, respectively (Scott & Causey, 1973; Causey & Cude, 1980). Daniels & Bekoff (1989*b*) reported 2–4 animals per group in a feral population in Arizona, and Nesbitt (1975) found a mean group size of 5–6 animals, in his 5-year study of feral dogs in Illinois. Although Boitani & Racana (1984) observed feral dogs in Basilicata (southern Italy) mostly in pairs, the group size of 3–6 adults found in the present study (Fig. 15.10) tends to agree with those reported previously. Studies of urban free-ranging dogs report mainly solitary animals or pairs (Beck, 1975; Berman & Dunbar, 1983; Daniels, 1983*a*, *b*; Hirata, Okuzaki & Obara, 1986; Daniels & Bekoff, 1989*b*; Macdonald & Carr, Chapter 14), and there has been some debate about whether smaller group sizes in urban areas are a response to scarce (Beck, 1973; Daniels & Bekoff, 1989*b*) or plentiful (Berman & Dunbar, 1983) food resources. Unfortunately, neither hypothesis is supported by accurate estimates of food resource distribution in either spatial or temporal terms. Daniels & Bekoff (1989*b*) recognize that patterns of social organization at urban and rural sites are based largely on dog-ownership practices. They stress the level of care and food provided by the owner as a primary reason for the lack of sociality in urban dogs; in other words, that the existing social bonds with owners and other humans tend to lower the dogs' motivation to form other social contacts. The higher levels of sociability among feral dogs, and the relatively strong bonds we observed among the members of our group, seem to be maintained despite the presence of abundant and easily accessible food resources.

The presence of wolves in areas partly overlapping the dogs' range may have provided an important pressure for group living. Group-living would have resulted in improved vigilance and improved protection from potential predators (see also Macdonald & Carr, Chapter 14).

Group composition (excluding pups) was relatively stable during the study period (see Fig. 15.10), although there was little evidence for any functional mechanism(s) regulating group size. Although the study lasted more than three years, all of the events that resulted in a reduction or increase in group numbers appeared to occur unpredictably. All deaths were accidental or caused by human interference, while newborns from feral parents contributed almost nothing to the group's long-term stability. The process of recruitment of new group members from the village stray population appeared to be the most powerful force maintaining group size. At the end of the study, all but one dog in the group originated from strays (see Fig. 15.10). The presence of breeding pairs appears essential to trigger the process of recruitment: new adults were accepted into the group only when a resident adult was left alone during the breeding period. Breeding periods in canids are accompanied by extensive social interactions that, in turn, may contribute to stronger pair bonds (Kleiman & Eisenberg, 1973). Increased social interactions may facilitate the acceptance of outsiders (most of the temporal association with dogs from the villages were observed during these periods), while the formation of strong pair bonds may be the major factor preventing further recruitment.

Although these speculations provide a promising hypothesis, they do not fully explain how a stable group size is maintained, and further data are needed on the behavioural responses of individual dogs to attempts by outsiders to approach and join the group. Macdonald & Carr (Chapter 14) propose that recruitment tends to occur in association with social disturbance. Our data support this view, if mating times are treated as periods of increased social stress.

Quantity and distribution of food resources are often cited as primary causes of social groups and determinants of group size (see Macdonald, 1983; von Schantz, 1984; Macdonald & Carr, 1989 and Chapter 14). The Resource Dispersion Hypothesis suggests that 'groups may develop in an environment where resources are dispersed such that, under certain circumstances, the smallest economically defensible territory for a pair can also sustain additional animals' (Macdonald & Carr, 1989). The theory also predicts that group size will be determined by richness of food resource patches during periods of minimum food availability. The garbage dumps in our study area provided a superabundant food supply during all seasons, and food did not appear to be a limiting resource. In such cases, group size is likely to be related more to social factors than to ecological ones. It is interesting to note that the marked philopatry of our dogs meets the general premises of the Territory Inheritance Hypothesis (Lindstrom, 1986). This hypothesis on the evolution of group living in carnivores gives greater importance to the attachment of individuals to their parents' territories, and predicts an optimal group size which agrees with that observed in our study. However, further and more detailed research on feral dog ecology is needed in order to test the strength of these various alternative hypotheses.

Group-splitting was observed only in conjunction with denning and pup-rearing by female 05, and it lasted more than five months. During this period her male partner maintained close contact with the rest of the group, travelling back and forth between the den and the group's usual home range. Daniels & Bekoff (1989*b*) have suggested that group-splitting may serve an adaptive function for pack-living canids as a means of reducing both the burden of alloparental care on the pack, and the threat of infanticide by the dominant female. Conversely, group participation in pup-rearing is adaptive for precisely the opposite reasons, i.e. it relieves the female from the burden of caring alone for her pups, and it provides more protection for the young from predators (Kleiman & Eisenberg, 1973). In our study group, splitting appeared to be more an accidental event, linked to the finding of a new, abundant and relatively undisturbed food source (the dump at Secinaro). In all other observed cases of denning, females always reared their pups without any assistance or threats from other group members. In the absence of any adaptive advantage to group-splitting, the result is a positive pressure against it. In other words, females are better off denning within the group's core areas so as to increase protection from intruders and potential predators.

Reproduction and life histories

Denning and rearing pups apart from the group has been reported by Daniels (1988) and Daniels & Bekoff (1988*b*), although actual distances are not

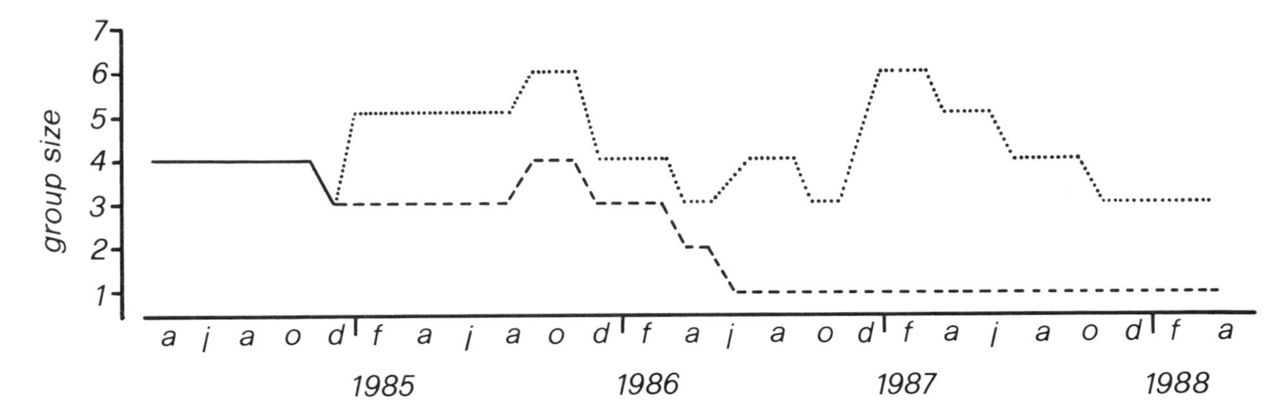

Fig. 15.10. Group size dynamics of original members and their progeny (solid and dashed line), and total number including recruited stray dogs (dotted line).

given. With the exception of female 05 in the spring of 1985, all other dens were within short distances from the usual resting sites. Denning females spent most of the time at the dens, where they were often visited by other group members, and there was no evidence that females were actually separated from the group. Again, the presence of potential predators (including humans) may have played a role in keeping dens within the group core areas.

Rearing pups without male assistance may be an artifact of the domestication process. Domestic dogs stand alone, among all canids, in their total lack of paternal care (Kleiman & Malcolm, 1981). Domestic dogs usually breed twice a year, although artificial selection for faster reproductive rates seems to have disrupted any seasonal patterns. Feral dogs maintain the pattern with an average of 7.3 months (range 6.5–10 months) between successive oestrus periods (female 01 may have bred late in 1984, but we failed to gather direct evidence for this). Since 50% of the breedings occurred during February–May (33% in April–May), they indicate a seasonal increase in reproduction in the spring. Although this spring concentration is statistically significant, it has not been possible to demonstrate any real synchrony of breeding among females. Macdonald & Carr (Chapter 14) report a much higher level of synchrony of breeding in their dogs and relate this to a period of group stability. This hypothesis would also fit our data, and would merit further study. Increases in breeding frequency in the spring and fall were reported by Gipson (1972) and suggested as likely by Daniels & Bekoff (1989*b*). In terms of probability of pup survival, the time of the year when breeding occurs is critical. Wild canids in Italy give birth in April (*Vulpes vulpes*) or May (*Canis lupus*) (Boitani, 1981), and the pattern shown by feral dogs may indicate a converging strategy.

In wolves, breeding is generally restricted to a single dominant female (Mech, 1970), although there are reported cases of two litters being raised successfully within the same pack (Van Ballenberghe, 1983). In feral dogs we saw no indication of any attempts to control reproduction of any (subordinate) adult. All females reproduced, giving the pack its full potential for demographic increase.

Domestic dogs are known to have litter sizes of up to 17 pups, although ten is the more usual upper limit (Kleiman, 1968; Kleiman & Eisenberg, 1973). A litter of five and a total of eight for two other litters have been reported for feral dogs by Nesbitt (1975), while Daniels & Bekoff (1989*b*) report ten pups from two litters. Most estimates of litter size rely on numbers obtained at the first sighting of pups, which rarely occurs before they are mobile (at 3–4 weeks). Earlier mortality may therefore contribute to the smaller litter sizes of feral animals. Our mean litter size of 3.63 is lower than previously reported, and also smaller than the 5.5 pups/litter obtained by Macdonald & Carr in the same area (Chapter 14). Variability in age-related fecundity in different breeds may help to explain these differences.

Pup survival rates were also low compared with figures quoted elsewhere. For example: Nesbitt (1975) reports 3 pups surviving out of 8 (37% survival); Scott & Causey (1973) describe 33% survival to 4 months and 22% survival to 1 year; Daniels & Bekoff (1989*b*) give a figure of 34% survival to 4 months of age for 5 litters; and Macdonald & Carr (Chapter 14) report 16% survival up to 5 months. In the present study we obtained lower survival rates and high early mortality (70% mortality within 70 days). Two factors may contribute to increased mortality at the time when the pups becomes independent: (a) the tendency of juveniles to explore the range without adult supervision; and (b) the fact that the mother is entering a new oestrus and is likely to lower her interest in previous offspring. Bekoff (1977) has proposed that a female dog entering a new oestrus would benefit energetically from weaning pups from the previous litter early. The very low survival rate at four months of age would suggest that the majority of mortality occurs during this period of enforced early independence, although further study of life histories and mortality in pups and juveniles is needed to confirm this. Low reproductive efficiency in feral dogs can be attributed to two main causes: (a) oestrus onsets are irregular and many litters are born at bad times of the year; and (b) mothers suffer the entire burden of parental care. In such conditions, it is perhaps not surprising that reproductive failures and high mortality rates are common.

With pup survival rates of only 5% after one year, it becomes easier to understand why free-ranging dogs have such difficulty maintaining their population levels. The same problem has already been noted but not explained for urban dogs (Beck, 1973; Daniels, 1983*b*) and for feral dogs in Arizona (Daniels & Bekoff, 1989*b*), and our study suggests that, without continuous recruitment of new group members from outside sources, feral groups cannot maintain their own population levels (see Table 15.10).

A critical factor contributing to negative demographic balance is the skewed sex ratio in our feral group. Most urban and rural/suburban free-ranging dog populations show sex ratios skewed toward males. Beck (1973), for example, reported a ratio of 1.8 : 1 in favour of males in Baltimore, Maryland; Daniels (1983*b*) obtained a ratio of 3 : 1 in favour of males in three study areas in Newark, New Jersey; Boitani & Racana (1984) found a ratio of 5 : 1 in favour of males in Bella, a village of southern Italy, and a ratio of 4 : 1 in the surrounding rural areas; Daniels & Bekoff (1989*b*) found ratios of 1.6 : 1 and 2 : 1 in favour of males in urban areas, and ratios of 4 : 1 and 3 : 1 in rural areas; three independent studies in Tunisia, Sri Lanka and Ecuador found male-biased sex ratios (60–65%) (WHO, 1988) and, finally, Macdonald & Carr (Chapter 14) found ratios of 4 : 1 and 2.6 : 1 in the villages of our study areas. Possible reasons for such findings have already been discussed (see Beck, 1973; Daniels & Bekoff, 1989*b*): the unbalanced sex ratios of urban dogs result from direct selection of males as pets and from the selective removal of females from the population, either temporarily to avoid unwanted pregnancies, or permanently by killing them as newborn puppies. Differential mortality rates for the two sexes are unlikely to occur in the absence of human interference. In one case, Daniels & Bekoff (1989*b*) noted a sex ratio of 1 : 3.5 in favour of females in a feral dog population in Arizona, in an area adjacent to an urban area. They attributed this result to the differential abandonment of female dogs by urban owners, since they could find no evidence of skewed sex ratios among puppies or of differential survival rates between the sexes.

When considering only the adult members in our group, a sex ratio of from 1 : 2 to 1 : 1.5 (group composition at various stages) in favour of females was obtained. From an initial 1 : 1 ratio, three females and two males were recruited from outside at various stages. Such small numbers, however, would not support the higher female abandonment hypothesis and they would not justify, in our opinion, any generalizations concerning the origins of skewed sex ratios. Two other animals that joined the group (a female and an unsexed individual) were the progeny of the feral group and their addition skewed the ratio more significantly in favour of females. This result is even more anomalous given that overall litter composition was highly skewed (3.2 : 1) in favour of males. Higher female survival rates would appear to be the only explanation for the observed adult sex ratio. Male-biased litter sex ratios have been reported in other canids, e.g. wolves (Mech, 1975), and wolves and hunting dogs (*Lycaon pictus*) (International Zoo Yearbook, vols. 5–11), but this sort of reproductive strategy would appear to have little adaptive value given the biological context of feral dogs. Further research is needed on different philopatric tendencies between the sexes, and on whether philopatry increases survival. A further interesting and open question is whether pups of the two sexes receive differential parental care.

Home range

Home range estimates can yield very different results as a consequence of the methods adopted for their measurement (Laundre & Keller, 1984), and any comparison between estimates obtained by different methods should be very carefully evaluated (Macdonald, Ball & Hough, 1980). Several environmental factors determine home range size, and human activities can be among the most powerful variables affecting the behaviour of any given species. This is especially true for canids (Kleiman & Brady, 1978), and for dogs in particular.

Of all the quantitative methods available for the estimation of home range, the Harmonic Mean (HM) method proved to be the most appropriate in view of the dogs' highly flexible and diverse spatial patterns, both at the individual and group level. The HM method, in conjunction with a GIS based habitat analysis, allowed a more accurate study of the home range's internal anatomy. The more internal contours reveal lower densities near roads and higher densities in wooded areas, and they show the strategic locations of core areas with respect to garbage dumps and villages. We know of no previous study that has provided such detailed definition of the environmental and activity patterns within feral dogs' home ranges.

Within the overall home range (95% contour, 57.8 km^2) the group used smaller portions at any one time, shifting its core area in response to several factors, such as the finding of new food resources (i.e. a large livestock carcass), disturbance by humans, denning activities, previous spatial use patterns of newly-recruited dogs (i.e. when females 05 and 06 joined the group), unpredictable fluctuations in food availability at dumps, and possible interference from wolves. These factors had no seasonal predictability (Fig. 15.4), and they appeared to occur as random events in the group's history. Daniels & Bekoff (1989*a*) found seasonal variations in home range related to the presence of dependent pups in one group they studied, while another showed no such changes. Differential energetic requirements were suggested as a possible reason for the two groups' behaviour (Daniels & Bekoff, 1989*a*), one group being slightly larger and having less food available. Scott & Causey (1973) also found a shifting of core areas depending on the presence or absence of pups.

Drifting home ranges have been recorded for urban foxes as a consequence of social instability following abrupt changes in population structure and food availability (Doncaster & Macdonald, 1991). In our case, we suggest that drifting of seasonal ranges reflects not only direct environmental changes, but also, and perhaps more importantly, the influence of previous knowledge of the area by new members of the group. In our study, the random shifting of core areas (Fig. 15.6) is essentially maintained within three main alternative areas (the dumps of Ovindoli and Rocca di Mezzo, and the village of Rovere), and within the same long-term boundaries. This may suggest that a tradition of area use inhibits more random movements or dispersal, and may perhaps increase the group members' fitness through the inheritance of territory (see Lindstrom, 1986). The presence and optimal distribution of a number of key environmental resources, including dumps, retreat areas and safe denning sites, could also be responsible for keeping the dogs within the same overall boundaries. Other possible explanations, such as unsuitable environmental parameters in surrounding areas, competition with adjacent groups, or limitations in movement capability, do not seem to work as well on our group.

Seasonal home ranges, though only a portion of the total, show the same key environmental features, and their core areas are similarly located with respect to roads, dumps, forests, villages, etc. They are more interesting, however, in relation to territorial behaviour. The seasonal home range sizes we obtained (average 11.27 km^2 ± 6.23, range 2.24–21.15) are comparable with those found in previous feral dog studies. Nesbitt (1975), for example, recorded a home range of 28.5 km^2 in his five-year study. Gipson (1983) reported a 70 km^2 home range for feral dogs in Alaska but he did not describe any temporal variations. It should also be noted that he obtained his estimate using a different method (Minimum Convex Polygon) – when computed by that method, the home range size of our group was 91.7 km^2. Scott & Causey (19733) found home range sizes of from 4.44–10.50 km^2 for three groups of dogs in Alabama.

One solitary individual had a smaller home range of 2.83 km^2, but it also used to travel up to 1.6 km away from its usual range. Causey & Cude (1980) found a minimum home range of 18.72 km^2 for another group of dogs in Alabama. Boitani & Racana (1984) found a home range of 6 km^2 for a group of feral dogs in southern Italy, but the estimate was based only on sightings and snowtracking data. Telemetry techniques can yield different results from those obtained by other indirect techniques of animal location. Daniels & Bekoff (1989*a*) reported a mean home range size of 1.62 km^2 for five feral dogs after the pups reached independence (but only 0.14 km^2 when the pups were still dependent). Their estimate, however, is based on the average of four animals located by sightings, and only one followed by radio-tracking. It is not surprising that the average for the four animals' home range was only 0.59 km^2, since the fifth had by far the largest home range of the entire study (5.08 km^2).

In the present study the relative distances of dens, dumps and other resting areas determined the final home range sizes, to a large extent independently of group size. According to the Resource Dispersion Hypothesis (Macdonald & Carr, 1989), territory size should be determined by the dispersion of food patches: in our study, if food were the only limiting resource, we would expect a territory size reduced to a minimum, as small as that allowed by social factors. If we expand the resource concept to include also dens and retreat areas, however, then our data fit the general predictions of the hypothesis. Urban and suburban dogs are reported to have much smaller home ranges, of the order of 2–11 ha (Beck, 1973; Berman & Dunbar, 1983; Daniels, 1983*a*, Santamaria, Passannanti & Di Franza, 1990), although up to 61 ha has been reported by Fox, Beck & Blackman (1975). Food availability patterns, small group sizes and reduced social contacts are probably the main determinants of such ranging behaviour, and this tends to confirm the importance of resource dispersion as a determinant of feral dog ranging behaviour.

Territorial aggression was observed more consistently in our dogs than reported previously (Scott & Fuller, 1965; Bekoff, 1979; Daniels & Bekoff, 1989*a*; Berman & Dunbar, 1983; Boitani & Racana, 1984), as it was displayed not only in the vicinity of den sites, but also within the entire core areas and at any time of the year. Similar observations have also been made by Macdonald & Carr (Chapter 14) in the same area. This high frequency of territorial behaviour may be related to the higher level of social integration within the group, the higher degree of isolation from other dogs and/or the fact that food resources were concentrated in the dumps. The partial overlap with two wolf pack ranges may also have increased the dogs' overall wariness and defensiveness.

Although we have little evidence of direct competition between feral dogs and wolves (apart from dog 14 being killed by wolves), the partial overlap of territories and the almost identical niche they share in central Italy (Boitani, 1983) make competition for food and for space highly likely. The presence of wolves may, therefore, be an important factor shaping the dogs' home range and determining its location. The central part of the dogs' home range lies close to Rovere, where human presence may help to keep wolves away, although equally the opposite might be the case, i.e. that wolves avoided Rovere because of the consistent presence of feral dogs. Only the experimental removal of the dogs would be able to determine which of these two alternatives is more likely.

No dispersal movements were observed, and only few brief excursions outside the usual home range were recorded. We have the impression that the dogs moved as if suddenly attracted by a scent: they went to check out the origin and possibly the nature and consistency of the signal. This impression was reinforced when the dogs went into the northern wolf pack territory at a time wolves are usually in oestrus. The dogs ran into and out of the area without stopping or slowing down, as if aware of the risks of being caught intruding in a wolf area.

Patterns of dump utilization (daytime/night) and of visits to the villages illustrated the flexibility of the group's behaviour. At the individual level, the same animal was able to adopt different strategies to suit local conditions and minimize its risks. This overall adaptability should make us wary of drawing functional generalizations on the dogs' behaviour and ecology based on few data or short-term studies.

Activity patterns

The tendency of dogs to show nocturnal and crepuscular activity patterns was first reported by Beck

(1973) for urban dogs. During the summer, activity was mainly restricted to two periods, 7–10 p.m. and 5–8 a.m., and similar bimodal activity peaks were found by Berman & Dunbar (1983) among the dogs of Berkeley, California. Hirata *et al.* (1986) reported that the dogs of several Japanese towns were most active from midnight to 6 a.m., with a peak just before and around 6 a.m. A prominent dawn peak of activity has also been observed in free-ranging rural dogs in Virginia (Perry & Giles, 1971), while bimodal, dawn and dusk activity have been reported in several studies of feral dogs (Scott & Causey, 1973; Causey & Cude, 1980; Boitani & Racana, 1984; Daniels & Bekoff, 1989*a*). Nesbitt (1975) found similar temporal patterns, although he suggested that feral dogs could be active and travel all day, and that they tended to restrict themselves to the nocturnal and crepuscular hours in an attempt to avoid human contact. The movements of female 05 in three different but comparable denning situations seem to confirm Nesbitt's suggestion: when human presence was low, she moved mostly during the daytime, while she later resumed nocturnal habits when visiting the more disturbed dump of Rocca di Mezzo and the village of Rovere. Avoiding humans may provide an explanation for nocturnal activity, but it does not explain the bimodal pattern found for all dogs during all seasons. Nine out of 17 canid species are strictly nocturnal (see Bekoff, Diamond & Mitton, 1981), and bimodal activity regimes are known for a great variety of carnivores, including foxes (*Vulpes fulvus*) (Ables, 1975), hyaenas (*Crocuta crocuta*) (Kruuk, 1972), and maned wolves (*Chrysocyon brachyurus*) (Dietz, 1984). Aschoff (1966) called it the 'bigeminus pattern' and suggested that it was an innate behavioural trait, independent from any environmental pressure. Achoff (1966) also pointed out that the second (dawn) peak of activity is usually less intense, although this was not the case in the present study where maximum activity levels were always observed at dawn.

Wolves show seasonal activity changes, being more nocturnal in summer and both nocturnal and diurnal in winter (Mech, 1970). The activity patterns of feral dogs may reflect this ancestral flexibility. Seasonal variation has been reported previously by Scott & Causey (1973) and Beck (1973) who suggested that on hot summer days dogs preferred lying around and resting in shaded areas, resulting in the concentration of foraging activities during the night. We know of no previous study that has quantified activity variations during breeding periods.

Activity patterns along the elevation gradient and as related to vegetation cover provide further evidence of the differential use of the home range and the role of core areas as the main resting and retreat sites. An association between activity and moonlight has not been reported before, and may have been due to the effects of improved visibility. If so, however, this would suggest that the dogs' nocturnal habits are an adaptation to local conditions rather than an innate trait.

Food sources and predation

Possible predation on wildlife and livestock has been the main impetus for feral dog studies. In North America, feral dogs have long been accused of predation on deer by the popular press, though on the basis of little evidence. In Italy too, the press and the conservation movement pointed to stray and feral dogs as primary predators of livestock and competitors of wild wolves, again with little supporting evidence (see Boitani, 1983). During our study we did not find any evidence of predation of livestock. The remains of wild boar (*Sus scrofa*), the only large ungulate in the area, were rarely found in dogs' scats and we never saw any evidence of predation on a live wild boar. Most previous studies of free-ranging dogs' feeding ecology have produced similar results. Perry & Giles (1971) studied owned radio-collared dogs; Causey & Cude (1980), Scott & Causey (1973) and Gipson & Sealander (1977) studied feral dogs, and all of them concluded that the dogs in their study areas were merely a nuisance, and had little overall impact on wildlife populations. Sweeney, Marchinton & Sweeney (1971) recorded 65 experimental chases of radio-marked deer by hounds, without a single deer suffering any injury. Progulske & Baskett (1958), Corbett *et al.* (1971) and Olson (1974) experimented with trained dogs chasing deer and were unable to document a single successful hunt. A small percentage (7%) of successful chases was reported by Hawkins, Kilmstra & Antry (1970) in Illinois, and, in Idaho, Lowry & MacArthur (1978) reported 12 deer killed out of 39 chases. Also Denney (1974) in Colorado and Gavitt, Downing & McGinnes (1974) in Virginia reported cases of deer being killed by feral dogs. These apparently contradictory results are probably best explained by local conditions, and whether dogs had adequate alternative sources of food. It is also likely that some individual dogs or groups of dogs acquire the ability to chase and kill deer and maintain the habit through cultural transmission to new group members. Learning and cultural transmission are known to be important in the acquisition of hunting skills in mammalian predators.

Several authors have observed feral dogs hunting and feeding on rodents, rabbits and other small game, but details are not reported. However, detailed accounts of a special predatory situation have been reported for Galapagos Islands' feral dogs feeding on marine iguanas (*Amblyrhynchus cristatus*) (Kruuk & Snell, 1981; Barnett & Rudd, 1983), and for feral and semi-feral dogs predating capybaras (*Hydrochoerus hydrochaeris*) in Venzuela's Llanos (Macdonald, 1981).

Absence of livestock predation was reported by Scott & Causey (1973), and Nesbitt (1975) was unable to document a single case of livestock depradation in his five year study. Similarly, in our study area, where cattle were free-ranging over most of the area, no interference with livestock was ever observed. In contrast, Nesbitt (1975) reported that free-ranging *pet* dogs killed three calves in his study area, and one of us (Boitani, unpublished data) was able to document severe damage by free-ranging owned dogs on livestock in other areas of Italy. Thus, it appears that free-ranging owned and stray dogs may be the primary agents of livestock predation, although the blame tends to fall on feral dogs and wolves. This attitude is deeply rooted in the traditional perception humans have of the role of dogs as friends and of wolves as enemies (see also Serpell, Chapter 16), and is difficult to counteract on the basis of often 'fragile' biological evidence.

Management and conservation implications

Feral dog populations, as represented in the results of our study, are not the intimidating problem they seemed to be when considered on purely theoretical grounds. The idea of a population of potential predators, with plenty of food resources to rely on, and with the reproductive potential given by all females producing litters twice a year, had all the makings of an ecologist's nightmare. However, we have shown that feral dogs had no impact on livestock during our three-year study, and our findings are the same as those obtained from all other studied populations. Predation on, and disturbance to, wildlife is also not a significant threat in any known context.

Nevertheless, feral dogs do have an impact on the human and natural environment. They can be effective carriers of potentially harmful diseases (rabies, echinococcus, toxocara, parvovirus, etc.) and they can enhance parasite transmission (Biocca *et al.*, 1984). Thus, in the absence of any conceivable positive role for feral dogs, it would be preferable to eradicate them from natural areas. Any eradication plan aimed directly at the dogs living in the wild would be very costly and difficult to implement, since the shooting or trapping of all feral dogs would be a formidable task, and would not provide a permanent solution to the problem, as long as stray dogs continue to proliferate in the villages. Direct control measures are rendered even more difficult by the need to be highly selective to avoid harming wolves, an endangered species throughout most of Europe (Boitani, 1983).

One particularly important and encouraging result from our study was the discovery that feral dogs are unable to maintain their population levels without the continuous recruitment of new individuals from the stray reservoir. This evidence would indicate that the easiest and most effective way of controlling feral dogs would be by emphasizing the control of the stray population, combined with direct control of feral dogs to accelerate the process of eradication. Nesbitt's (1975) study on feral dogs followed a massive removal effort of more than 100 dogs carried out a few years earlier by Hawkins *et al.* (1970). Yet, clearly, healthy groups of feral dogs still remained in the area despite these measures.

Stray dog control is an old problem that many countries around the world are tackling with varying degrees of success. Direct control of stray animals is technically feasible and, in fact, is being implemented by public authorities in many towns and districts. However, unless the source of stray dogs is eliminated, an ultimate solution will never be reached. In the end, the most effective way of solving the stray dog problem is by influencing people's overall attitudes toward dog-keeping. Great Britain is the home of over seven million dogs, and yet it is relatively hard to find a stray dog anywhere in either the towns or the countryside: it provides one of the best demonstrations that it is people's attitudes toward dogs, rather than any technical knowledge of control measures, which prevents the bulk of the dog population from being out of strict human control.

In conclusion, the feral dog problem should only marginally be managed by direct impact on the dogs themselves. Rather the problem needs to be approached by an intensive and effective campaign of public education. This should include a general information and education campaign, but it should also be focused more specifically on those human groups who are most responsible for the problem: hunters, shepherds, farmers and tourists who release, abandon or simply take inadequate care of their dogs.

An indirect way to affect stray and feral dog populations is to reduce their access to garbage dumps: these should be enclosed by effective fences, covered with earth or chemicals, or – the most radical and nature-friendly solution – they should not exist at all since refuse can be processed and recycled in more appropriate ways. Different solutions will suit local financial and technical capabilities, but it must be stressed that an effective answer (the installation of dog-proof fencing around dumps) should be possible under any circumstances. Whether these relatively simple management guidelines will ever be effectively implemented is questionable, however, and a good deal of pessimism comes from the evidence that stray and feral dogs are today as numerous as ever.

Feral dogs are old companions of human history. They probably existed in Eurasia soon after the domestication of the dog, as a result of the high integration of Mesolithic human cultures and the natural environment, and of the many opportunities the dogs had to move in and out of human settlements. Also, on the North American continent, feral dogs are believed to have originated as a consequence of dog domestication by Amerindian tribes long before the arrival of European colonists (McKnight, 1964). Although it is difficult to find reliable historical accounts of feral as opposed to stray dogs (the differences being so hard to define), there are at least two notable examples showing that the feralization process was already going on several millenia ago: the dingo in Australia, and its likely ancestor the pariah dog of southern Eurasia (Zeuner, 1963). More recently, in the eighteenth century, stray and feral dogs roamed many of the large cities of the Mediterranean basin (Istanbul, Alexandria) and were regarded, in some cases, as almost a separate subspecies (see Brehms, 1893). Mediterranean lifestyles and environmental conditions (the combination of relatively warm climate and *laissez faire* attitudes, and the presence of small game, free-ranging livestock and garbage dumps) appear particularly favourable to stray and feral dog populations (Boitani & Fabbri, 1983*a*).

A comprehensive account of the negative impact of feral dogs on the human and natural environment has never been attempted, and the separate evaluation of the different areas of conflict (i.e. health, and economic and ecological factors) rarely justifies the costs of a radical control effort. Human and veterinary health problems caused by stray dogs have been a major cause of concern for public health departments and international organizations, and they have generated an extensive literature (though little on truly feral animals). Economic considerations arise from the cost estimates of direct and indirect damages caused by free-ranging dogs, and include the costs of health problems, control measures and livestock predation.

The ecological problem has a number of different dimensions, including the role of the dog as a predator, as a disease carrier and as a competitor with biologically similar wild species. In Italy, recent evaluation of the feral dog's negative impact on wolf survival has provided new evidence supporting the need for feral dog eradication (Boitani & Fabbri, 1983*a*, *b*; Boitani, 1983). Potential conflicts between wolves and dogs occur at four major levels (Boitani, 1983). The first conflict arises over food resources, since both wolves and dogs feed mostly on garbage they find at village dumps. Quantity of food is not so much a limiting factor as its availability: dumps are usually close to villages and these locations are easier for dogs to exploit than for wolves. Once a dump is 'occupied' by a large and healthy group of dogs, wolves may find it difficult to gain access to the resource. The second area of conflict involves the space available for territorial ranges. Wolves in Italy mainly live in small packs or solitarily (Boitani & Fabbri, 1983*b*), and dispersing animals play a critical role in securing the species' survival. Wolf packs tend to become highly unstable as human disturbance displaces them, and competition with stable and strong feral dog groups may prevent the wolves from establishing new pairs and new territories.

Third, problems arise when wolves and dogs interbreed. Although they belong to the same biological species, the two gene pools have been separated for many thousands of years by domestication. Evidence of 'hybridization' was reported by Boitani & Fabbri (1983*b*), and it is currently being investigated with modern DNA techniques in order to obtain a better assessment of the frequency of its occurrence. Wolf/dog interbreeding has probably been going on for centuries, at least in Italy where shepherd dogs and wolves have been closely associated in the same habitat. However, hybridization may represent a greater danger to the wolf in the modern context, simply because of the large number of dogs and the high probability of wolf/dog encounters. More data are needed on the efficiency of behavioural barriers between wolves and dogs in nature in order to assess the potential threat posed by hybridization.

The final area of conflict concerns the attitudes of humans towards wolves and dogs. In the mountain regions of central Italy, dogs are traditionally regarded as companions and people tend to underestimate their potential impact as predators. Wolves, on the contrary, are perceived as vermin and any damage supposedly perpetrated by them is considered unacceptable. Shepherds and farmers consistently blame wolves for livestock predation, in order to get financial compensation, and as a result wolves suffer much heavier condemnation than they deserve. At present, free-ranging dogs are the major threat to wolf survival and conservation in Italy, and this fact adds additional weight to arguments in favour of their eradication.

Acknowledgements

We thank M. L. Fabbri, J. Geppert, E. Schoenfeld, D. Talarico, and all the others who volunteered with field work assistance. F. Corsi helped with GIS and computer analysis. The project was funded by the Istituto Nazionale di Biologia della Selvaggina (INBS), Bologna, and by an Earthwatch programme; their support is gratefully acknowledged. D. W. Macdonald is a long-time friend and there is no way to acknowledge properly his collaboration. An anonymous referee helped with improvements to the manuscript, and J. Serpell edited it and made it readable.

References

Ables, E. D. (1975). Ecology of the red fox in North America. In *The Wild Canids*, ed. M. W. Fox, pp. 216–36. New York: Van Nostrand Reinhold.

Aschoff, J. (1966). Circadian activity patterns with two peaks. *Ecology*, **47**, 657–702.

Barnett, B. D. & Rudd, R. L. (1983). Feral dogs of the Galapagos Islands: impact and control. *International Journal for the Study of Animal Problems*, **4**, 44–58.

Beck, A. M. (1973). *The ecology of stray dogs: a study of free-ranging urban animals*. Baltimore, MD: York Press.

Beck, A. M. (1975). The ecology of 'feral' and free-roving dogs in Baltimore. In *The Wild Canids*, ed. M. W. Fox, pp. 380–90. New York: Van Nostrand Reinhold.

Bekoff, M. (1977). Mammalian dispersal and the ontogeny of individual behavioural phenotypes. *American Naturalist*, **111**, 715–32.

Bekoff, M. (1979). Scent-marking by free-ranging domestic dogs. *Biology of Behaviour*, **4**, 123–39.

Bekoff, M., Daniels, T. J. & Gittleman, J. L. (1984). Life history patterns and the comparative social ecology of carnivores. *Annual Review of Ecological Systematics*, **15**, 191–232.

Bekoff, M., Diamond, J. & Mitton, J. B. (1981). Life history patterns and sociality in canids: body size, reproduction, and behaviour. *Oecologia*, **50**, 388–90.

Berman, M. & Dunbar, I. (1983). The social behaviour of free-ranging suburban dogs. *Applied Animal Ethology*, **10**, 5–17.

Biocca, M., Giovannini, A. Gradoni, L., Gramiccia, M., Mantovani, A., Pozio, E., Procicchiani, L. & Mantovani, A. (1984). Problemi di sanita' pubblica legati ai cani randagi e inselvatichiti. *Annali Dell' Istituto di Sanita*, **20**, 275–86.

Boitani, L. (1981). Lupo, *Canis lupus*. In *Distrubuzione e Biologia di 22 Specie di Mammiferi in Italia*, ed. M. Pavan, pp. 61–7. Roma: CNR.

Boitani, L. (1983). Wolf and dog competition in Italy. *Acta Zoologica Fennica*, **174**, 259–64.

Boitani, L., Ciucci, P., Corsi, F. & Fabbri, M. L. (1989). A geographic information system (GIS) application to analyze the internal anatomy of the home range: feral dogs in the Appenines (Italy). In Abstracts of the Fifth International Theriological Congress, pp. 890–91. Rome.

Boitani, L. & Fabbri, M. L. (1983*a*). Censimento dei cani in Italia con particolare riguardo al fenomeno del randagismo. *Ricerche di Biologia della Selvaggina* (INBS, Bologna), **73**, 1–51.

Boitani, L. & Fabbri, M. L. (1983*b*). Strategia nazionale di conservazione per il lupo (*Canis lupus*). *Ricerche di Biologia della Selvaggina* (INBS, Bologna), **72**, 1–31.

Boitani, L. & Racana, A. (1984). Indagine eco-etologica sulla popolazione di cani domestici e randagi di due comuni della Basilicata. *Silva Lucana* (Bari), **3/84**, 186.

Brehm, A. (1983). *Tierleben*, 4 vols. Liepzig-Wien.
Causey, M. K. & Cude, C. A. (1980). Feral dog and white-tailed deer interactions in Alabama. *Journal of Wildlife Management*, **44**, 481–4.
Corbett, R. L., Marchinton, R. L. & Hill, C. L. (1971). Preliminary study of the effects of dogs on radio-equipped deer in mountainous habitat. *Proceedings of the Annual Conference of the Southeastern Association State Game and Fish Commissioners*, **25**, 69–77.
Daniels, T. J. (1983*a*). The social organization of free-ranging urban dogs: I. Nonestrous social behaviour. *Applied Animal Ethology*, **10**, 341–63.
Daniels, T. J. (1983*b*). The social organization of free-ranging urban dogs: II. Estrous groups and the mating system. *Applied Animal Ethology*, **10**, 365–73.
Daniels, T. J. (1987). The Social Ecology and Behaviour of Free-ranging Dogs, *Canis familiaris*. Unpublished Ph.D. dissertation, University of Colorado, Boulder.
Daniels, T. J. (1988). Down in the dumps. *Natural History*, **97**, 8–12.
Daniels, T. J. & Bekoff, M. (1989*b*). Spatial and temporal resource use by feral and abandoned dogs. *Ethology*, **81**, 300–12.
Daniels, T. J. & Bekoff, M. (1989*b*). Population and social biology of free-ranging dogs, *Canis familiaris*. *Journal of Mammalogy*, **70**, 754–62.
Denney, R. N. (1974). Impact of uncontrolled dogs on wildlife and livestock. *Transactions of the North American Wildlife and Natural Resources Conference*, **39**, 257–91.
Dietz, J. M. (1984). Ecology and social organization of the maned wolf (*Chrysocyon brachiurus*). *Smithsonian Contributions to Zoology*, **392**, 1–51.
Dixon, K. R. & Chapman, J. A. (1980). Harmonic mean measure of animal activity areas. *Ecology*. **61**, 1040–4.
Doncaster, C. P. & Macdonald, D. W. (1991). Drifting territoriality in the red fox *Vulpes vulpes*. *Journal of Animal Ecology*, **60**, 423–39.
Fox, M. W., Beck, A. M. & Blackman, E. (1975). Behaviour and ecology of a small group of urban dogs (*Canis familiaris*). *Applied Animal Ethology*, **1**, 119–37.
Gavitt, J. D., Downing, R. L. & McGinnes, B. S. (1974). Effects of dogs on deer reproduction in Virginia. *Proceedings of the Annual Conference of the Southeastern Association State Game and Fish Commissioners*, **28**, 532–9.
Gipson, P. S. (1972). The Taxonomy, Reproductive Biology, Food Habits, and Range of Wild *Canis* (Canidae) in Arkansas. Unpublished Ph.D. dissertation, University of Arkansas, Fayetteville.
Gipson, P. S. (1983). Evaluation and control implications of behaviour of feral dogs in Interior Alaska. In *Vertebrate Pest Control and Management Materials: 4th Symposium*, ed. D. E. Kaukeinen. Philadelphia, PA: ASTM Special Technical Publication.
Gipson, P. S. & Sealander, J. A. (1977). Ecological relationship of white-tailed deer and dogs in Arkansas. In *Proceedings of the 1975 Predator Symposium, Montana Forest Conservation Experimental Station*, ed. R. L. Philips & C. Jonkel, pp. 3–16. Missoula: University of Montana.
Gittleman, J. L. (ed.) (1989). *Carnivore Behaviour, Ecology and Evolution*. London: Chapman & Hall.
Hawkins, R. E., Klimstra, W. D. & Antry, D. C. (1970). Significant mortality factors of deer in Crab Orchard National Wildlife Refuge. *Transactions of the Illinois State Academy of Sciences*, **63**, 202–6.
Hirata, H., Okuzaki, M. & Obara, H. (1986). Characteristics of urban dogs and cats. In *Integrated Studies in Urban Ecosystems as the Basis of Urban Planning* I., ed. H. Obara. Special Research Project on Environmental Science (B276-R15-3), pp. 163–75. Tokyo: Ministry of Education.
Hirata, H., Okuzaki, M. & Obara, H. (1987). Relationships between men and dogs in urban ecosystem. In *Integrated Studies in Urban Ecosystems as the Basis of Urban Planning* II., ed. H. Obara. Special Research Project on Environmental Science (B334R15-3), pp. 113–20. Tokyo: Ministry of Education.
Kirk, R. W. (ed.) (1977). Current veterinary therapy. *Vol. VI: Small Animal Practice*. Philadelphia, PA: W. B. Saunders.
Kleiman, D. G. (1968). Reproduction in the Canidae. *International Zoo Yearbook*, **8**, 1–7.
Kleiman, D. G. & Brady, C. A. (1978). Coyote behaviour in the context of recent canid research: problems and perspectives. In *Coyotes*, ed. M. Bekoff, pp. 163–88. New York: Academic Press.
Kleiman, D. G. & Eisenberg, J. F. (1973). Comparisons of canid and felid social systems from an evolutionary perspective. *Animal Behaviour*, **21**, 637–59.
Kleiman, D. G. & Malcolm, J. R. (1981). The evolution of male parental investment in mammals. In *Parental Care in Mammals*, ed. D. J. Gubernik & P. H. Klopfer, pp. 347–87. New York: Plenum.
Kruuk, H. (1972). *The Spotted Hyaena: a Study of Predation and Social Behaviour.* Chicago: University of Chicago Press.
Kruuk, H. & Snell, H. (1981). Prey selection by feral dogs from a population of marine iguanas (*Amblyrhynchus cristatus*). *Journal of Applied Ecology*, **18**, 197–204.
Laundre, J. W. & Keller, B. L. (1984). Home range size of coyotes: a critical review. *Journal of Wildlife Management*, **48**, 127–39.
Lindstrom, E. (1986). Territory inheritance and the evolution of group living in carnivores. *Animal Behaviour*, **34**, 1825–35.
Lowry, D. A. & MacArthur, K. L. (1978). Domestic dogs as predators on deer. *Wildlife Society Bulletin*, **6**, 38–9.
McKnight, T. (1964). *Feral livestock in Anglo–America.* Berkeley: University of California Press.
Macdonald, D. W. 1981. Dwindling resources and the

social behaviour of Capybaras, (*Hydrochoerus hydrochaeris*). *Journal of Zoology, London*, **194**, 371–91.

Macdonald, D. W. (1983). The ecology of carnivore social behaviour. *Nature*, **301**, 379–84.

Macdonald, D. W., Ball, F. G. & Hough, N. G. (1980). Evaluation of home range size and configuration. In: *A Handbook on Biotelemetry and Radio-tracking*, ed. C. J. Amlaner & D. W. Macdonald, pp. 405–24. Oxford: Pergamon Press.

Macdonald, D. W. & Carr, G. M. (1989). Food security and the rewards of tolerance. In *Comparative Socioecology: the Behavioural Ecology of Humans and Other Mammals*, ed. V. Standen & R. A. Foley, 75–99. Oxford: Blackwell Scientific Publications.

Mech, L. D. (1970). *The Wolf: the Ecology and Behaviour of an Endangered Species*. New York: Natural History Press.

Mech, L. D. (1975). Disproportionate sex ratios of wolf pups. *Journal of Wildlife Management*, **39**, 737–40.

Mech, L. D. (1983). *Handbook of Animal Radiotracking*. Minneapolis; University of Minnesota Press.

Nesbitt, W. H. (1975). Ecology of a feral dog pack on a wildlife refuge. In *The Wild Canids*, ed. M. W. Fox, pp. 391–5. New York: Van Nostrand Reinhold.

Olson, J. C. (1974). Movements of deer as influenced by dogs. *Indiana Department of Natural Resources Job Progress Report Project* W-26-R-5, Job III-b-4, pp. 1–36.

Perry, M. C. & Giles, R. H. (1971). Free running dogs. *Virginia Wildlife*, **32**, 17–19.

Progulske, D. R. & Baskett, T. S. (1958). Mobility of Missouri deer and their harassment by dogs. *Journal of Wildlife Management*, **22**, 184–92.

Samuel, M. D., Pierce, D. J. & Garton, E. O. (1985). Identifying areas of concentrated use within home ranges. *Journal of Animal Ecology*, **54**, 711–19.

Santamaria, A., Passannanti, S. & Di Franza, D. (1990). Censimento dei cani randagi in un quartiere di Napoli. *Acta Medica Veterinaria*, **36**, 201–13.

Scott, M. D. & Causey, K. (1973). Ecology of feral dogs in Alabama. *Journal of Wildlife Management*, **37**, 253–65.

Scott, J. P. & Fuller, J. L. (1965). *Genetics and the Social Behaviour of the Dog.* Chicago: University of Chicago Press.

Spencer, W. D. & Barret, R. H. (1984). An evaluation of the harmonic mean measure for defining carnivore activity areas. *Acta Zoologica Fennica*, **171**, 255–9.

Sweeney, J. R., Marchinton, R. L. & Sweeney, J. M. (1971). Responses of radiomonitored white-tailed deer chased by hunting dogs. *Journal of Wildlife Management*, **35**, 707–16.

van Ballenberghe, V. (1983). Two litters raised in one year by a wolf pack. *Journal of Mammalogy*, **64**, 171–3.

von Schantz, T. (1984). Carnivore social behaviour – does it need patches? *Nature*, **307**, 389.

World Health Organization (1988). *Report of WHO Consultation on Dog Ecology Studies Related to Rabies Control.* WHO/Rabies Research/88.25.

Zeuner, F. E., 1963. *A History of Domesticated Animals.* London: Hutchinson.

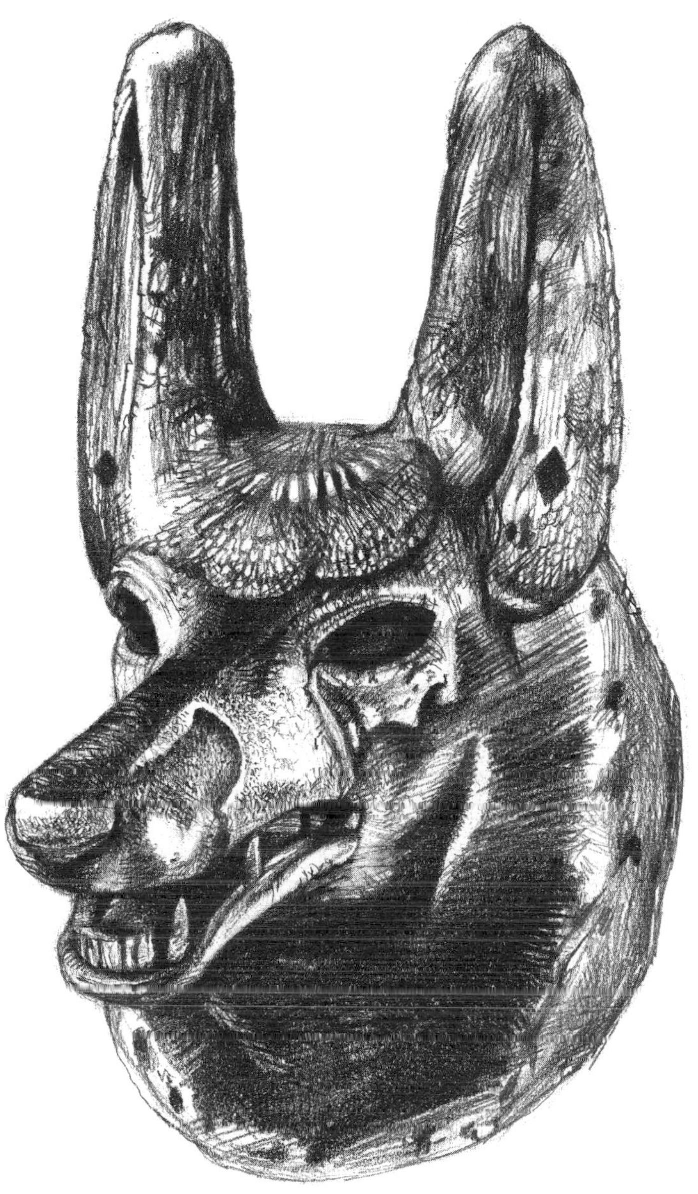

16 From paragon to pariah: some reflections on human attitudes to dogs

JAMES SERPELL

Introduction

According to traditional Judaeo–Christian philosophy, an absolute moral and conceptual barrier exists between our species and the rest of animal creation (Midgley, 1983; Thomas, 1983; Serpell, 1986). Although reinforced by an array of religious and secular attitudes and beliefs, however, this barrier is by no means impenetrable. Throughout history, human victims of political or racial oppression have found themselves summarily reclassified as 'animals' preparatory to having their legal and moral rights ignored or withdrawn (see e.g. Arluke & Sax, 1992). Society also readily assigns the same fate to anyone who is deemed to behave in a socially unacceptable way. When football hooligans are labelled 'animals' in the tabloid press, this is tantamount to saying that they should lose the right to be treated as human beings.

Under certain circumstances, the barrier between human and nonhuman can also be penetrated from the opposite direction. Just as humans can be classed as animals, so animals – for one reason or another – may be categorized and treated as persons. In the Middle Ages, dangerous animals and vermin were sometimes tried, tortured and executed for committing 'crimes', as if, like human criminals, they were guilty of malicious intent (Evans, 1906; Cohen, 1994). Conversely, animals can also acquire many of the benefits and privileges normally reserved for human beings, usually by virtue of forming strong social attachments for particular people.

In theory, any animal can attain quasi-human status in this way but, in practice, only two domestic species, the dog and cat, appear to have done so with any degree of permanence. And while cats tend to behave like temporary lodgers, retaining the ability to come and go as they please, dogs are now so thoroughly assimilated into the human domain that it is difficult to imagine them flourishing outside of this context. Indeed, as the preceding two chapters suggest, the lives of feral or free-ranging, ownerless dogs are typically short and uncomfortable (see Macdonald & Carr, Chapter 14; Boitani *et al.*, Chapter 15). Dogs and cats are also the only domestic animals not requiring physical barriers – walls, cages, fences or tethers – to enforce their association with people. But whereas even the most affectionate cats are generally more tied to places than to people, most dogs behave as if permanently attached to humans by invisible bonds. Given the opportunity, a dog will accompany its owner everywhere, and exhibit obvious signs of distress when separated involuntarily (Serpell, 1986). Moreover, this form of separation-related anxiety may become so exaggerated that the animal will howl, bark, whine, defecate, urinate, or chew up and destroy household furniture and fittings whenever it is left alone (see Mugford, Chapter 10).

Although a dog may form these strong attachments for people at any age, the process tends to occur more readily in early development during the so-called 'socialization period'. At this time, from roughly 3 to 12 weeks of age, puppies establish their primary social relationships (see Scott, 1963; Serpell & Jagoe, Chapter 6). The process of primary socialization not only determines who or what the puppy will respond to in a positive social manner, it also effectively defines the species to which it belongs. Cross-fostering experiments have shown, for instance, that if a puppy is reared exclusively with kittens during this period, it will grow up to regard cats as conspecifics rather than other dogs (Fox, 1967). It is also apparent that if a young dog is exposed to the attentions of two different species within the socialization period it will happily socialize to both. A further important aspect of early socialization is that it appears to occur independent of rewards and punishments. Scott (1963) has shown that while puppies, like all animals, react positively to rewarding stimuli and negatively to aversive ones, the process of primary socialization will proceed regardless of the nature of the accompanying stimulus. In other words, if exposed to their company at the appropriate age, a dog will develop a strong affinity for humans, whether or not it is rewarded for doing so.

This tendency to voluntarily ally itself with humans, this capacity to form strong interspecific attachments, even in the face of rejection or punishment, places the dog in a unique position relative to all other non-human animals. With the possible exception of the anthropoid apes (see Dawkins, 1993; Diamond, 1993), no other species comes as close to us as the dog in affective or symbolic terms, and, by the same token, no other species makes a stronger claim to be treated as human. Yet, far from making the dog the object of universal affection and respect, this unusual 'closeness' or affinity seems to provoke a surprising degree of ambivalence. This chapter explores the nature of this ambivalence in our relations with the domestic dog across a range of different cultural and economic contexts.

Dogs as hunting partners

Before joining forces in the Upper Palaeolithic, both humans and wolves lived in cooperative societies that specialized, at least to some extent, in hunting large game animals. Scott (1968) has suggested that, in this respect, wolves and people were already pre-adapted for life in a combined social group, and Downs (1960) has argued that the original domestication of the wolf may actually have arisen out of a kind of mutually beneficial hunting symbiosis between the two species. Whether or not this was in fact the case, it is clear from the archaeological record that the use of dogs for hunting was one of the earliest economic functions for this species (see Clutton-Brock, Chapter 2).

Given the longevity and mutual compatibility of this partnership, and the fact that it improves people's hunting success (see e.g. Lee, 1979), it is not surprising to find widespread respect and affection for hunting dogs throughout the world, especially in cultures where hunting forms a primary mode of subsistence. Among the Onges, a hunter–gatherer group from the Andaman Islands, dogs were first acquired in 1857 and are now employed extensively for hunting wild pig. Since the Onges formerly subsisted mainly on fish and shellfish, the advent of dogs has revolutionized their economy. As a result, according to one authority, the Onges have developed a 'somewhat unbalanced affection' for these animals (Cipriani, 1966, pp. 80–1). The same author goes on to state that this 'inordinate love of dogs has allowed the animals to become a pest. They already outnumber the human population; families of only three or four people may have ten or twelve dogs', and this despite the fact that the Onges suffer from constant flea infestations, are frequently bitten by their canine friends, and are kept awake at night by their continuous barking and howling. Similar intense affection for hunting dogs has also been reported among the Punan Dyaks of Malaysian Borneo (Harrison, 1965), the Vedda people of Sri Lanka (Seligmann & Seligmann, 1911), the Dorobo of Kenya (Huntingford, 1955) and the Panaré, an Amerindian group from Venezuela (Dumont, 1976).

Some of the most extreme cases of devotion to hunting dogs come from early accounts of Australian Aborigines. During a visit to Stradbroke Island off the coast of Queensland in 1828, Lockyer remarked that the 'attachment of these people to their dogs is worthy of notice'. When he attempted to purchase one from its owner in exchange for a small axe, 'his companions urged him to take it, and he was about to do so, when he looked at the dog and the animal licked his face, which settled the business. He shook his head and determined to keep him' (quoted in Bueler, 1974, pp. 102). Similarly, the Swedish explorer Carl Lumholtz (1989; p. 179) observed that the Aborigines treated these animals

> with greater care than they bestow on their own children. The dingo is an important member of the family; it sleeps in the huts and gets plenty to eat, not only of meat but also of fruit. Its master never strikes, but merely threatens it. He caresses it like a child, eats the fleas off it, and then kisses it on the snout... When hunting, sometimes it refuses to go any further, and its owner has then to carry it on his shoulders, a luxury of which it is very fond.

Although it is difficult to think of any sensible reason why anyone should harbour equivocal feelings about such a pleasant and useful companion, hunting dogs are (or were) nevertheless regarded with considerable ambivalence in a number of other societies. Among the Yurok Indians of California, dogs were valued highly, especially for hunting deer, which they were trained to drive towards awaiting hunters. Yurok folklore recounts tales of men becoming obsessionally devoted to their dogs although, in the real world, such intimacy was considered taboo. Dogs, for example, were not allowed into human habitations, and they were never named or spoken to in the belief that they might answer back, thus upsetting the natural order and provoking general catastrophe. Nevertheless, dogs that died were provided with ceremonial burial, and it was also believed that when a man died the spirit of his dog preceded him to the underworld as a guide and protector, even though the dog itself remained alive. This oddly cautious approach to dogs apparently arose, not because dogs were considered ceremonially unclean[1], but because they were too close to humans. Dogs were a potential menace because the critical psychological

[1] Some social anthropologists would argue that the dog's liminality, its closeness to the border between human and non-human, is sufficient reason in itself for regarding the species as potentially unclean or polluting (see e.g. Douglas, 1966).

line that distinguished humans from animals was constantly in danger of being effaced by their presence (Elemendorf & Kroeber, 1960).

James Jordan (1975) provided a contemporary account of similar ambivalent attitudes to dogs within a small, white, rural community in the southern American State of Georgia. Here dogs serve a variety of functions, but are valued primarily for hunting. Ownership of a good hunting dog carries with it considerable social prestige, and southern males are intensely proud of their dogs and their abilities. Yet the overall care and treatment of dogs is both callous and frequently cruel.

> On the one hand [says Jordan] the dog is highly esteemed for his various abilities in hunting, tracking, guarding, and providing companionship and trustworthy fidelity. On the other hand, the dog is used as a low-status marker within the same cultural context: to label a human a dog, to suggest that a human is the offspring of a female dog, or to liken a human as being in any way similar to a dog is to insult the human deeply . . . The dog is treated as though it is useful *and* useless; the dog is referred to as symbolically valuable *and* worthless; the dog is employed as a standard of excellence *and* of baseness.
>
> Jordan, 1975, p. 245

Jordan offers a variety of possible explanations for such apparently inconsistent attitudes. He suggest that economic constraints may be a factor and that indifference to canine suffering may simply reflect the hardships experienced by people in such communities. He also proposes that the masculine ethos of violence and toughness that enables people to endure such conditions of life is applied equally to dogs, and that the maintenance of shallow, ambivalent attachments to dogs, and seeming indifference to their welfare, may represent an unconscious defence against unpredictable losses due to frequent accidents or disease. Finally he suggests that the dog, because of its quasi-human but subordinate status, may serve as a whipping-boy; an outlet for the exercise of dominance, power and displaced anger in a community where men tend to feel dispossessed and powerless themselves (Jordan, 1975).

A somewhat similar theory has also been applied to the BaMbuti Pygmies of Zaire, who display such excessive cruelty and viscousness towards their hunting dogs, that it led one observer to remark: 'I thank God we are not pygmies. I thank Him still more that we are not pygmy women, and even still more again that we are not pygmy dogs' (quoted in Singer, 1978, p. 271). Once again, paradoxically, the BaMbuti value these dogs highly as hunting aides, and most authorities agree that it would be difficult, if not impossible, for these people to track down certain types of game successfully without canine assistance. In an attempt to explain this contradictory behaviour, the anthropologist Singer (1978) postulated that dogs serve an important function as scapegoats in these communities. The BaMbuti believe that overt demonstrations of aggression between people are extremely distasteful. The dog, however, serves as a convenient and socially acceptable outlet for repressed anger, and is a particularly suitable victim in this respect because of its closeness to humans.

In short, despite its almost universal value for hunting certain types of prey, the dog's affinity with humans seems to be able to inspire suspicion, denigration and hostility as well as devotion and respect.

Dogs for dinner

When humans slaughter and eat dogs they are, in a purely practical sense, treating them no differently from any other kind of domestic food animal. Theory would predict that, in these circumstances, people would exercise detachment and attempt to distance themselves emotionally from those animals destined for consumption (see Thomas, 1983; Serpell, 1985; 1986). In practice, however, dog-eating is often associated with more complex psychological contortions.

Dogs are, or were until recently, highly regarded as items of food in many parts of the world, including Southeast Asia and Indochina, North and Central America, parts of Africa, and the islands of the Pacific (Driver & Massey, 1957; Burkardt, 1960; Frank, 1965; Titcomb, 1969; Ishige, 1977; Olowa Ojoade, 1990). Archaeological evidence suggests that, during the Neolithic and Bronze Age, the practice of dog-eating was also widespread in Europe (Bökönyi, 1974). In present day West Africa, Korea and the Philippines, where edible dogs are still produced and reared on a commercial scale, attitudes towards them tend, superficially, to resemble our own relatively detached, Western attitudes to domestic livestock, such as pigs or chickens. But even in these circumstances, views on dog-eating appear somewhat ambivalent. Some people regard the practice as abhorrent, others only eat dog's flesh for special medicinal purposes, and it is not uncommon for people to refuse to slaughter and devour their own dogs, although evidently happy to kill and consume someone else's (Osgood, 1951, 1975; Olowo Ojoade, 1990). The true origins of dog meat are also commonly disguised through the use of euphemistic terminology[2].

In many other societies where dogs are or were traditionally eaten, however, the practice is associated with far more overt expressions of ambivalence. For example, according to George Catlin, a famous nineteenth-century explorer, the Sioux Indians were devoted to their dogs:

> The dog is more valued than in any part of the civilized world. The Indian has more time to devote to his company, and his untutored mind more nearly assimilates that of his faithful servant. He keeps his dog closer company, and draws him nearer to his heart. They hunt together and are equal sharers in the chase. Their bed is one. On rocks and on their coats of arms they carve his image as the symbol of fidelity.
>
> *in* Mooney, 1975

Yet to celebrate Catlin's arrival in their camp, the Indians promptly slaughtered a large number of their dogs and served them up as a wholesome canine stew which Catlin himself had the greatest difficulty eating, in spite of it being 'well-flavoured and palatable'. Rather than reflecting callous indifference, however, this dog-feast apparently constituted a form of ultimate sacrifice; a solemn and binding gesture of friendship. Catlin goes on to explain:

> the feast of venison or buffalo meat is due to anyone who enters an Indian's wigwam, it conveys but passive or neutral evidence of friendship and counts for nothing . . . Yet the Indian will sacrifice his faithful follower to seal a sacred pledge of friendship . . . I have seen the master take from the bowl the head of his victim and talk of its former affection and fidelity with tears in his eyes. And I have seen civilized men by my side jesting at Indian folly and stupidity. I have said in my heart they never deserved a name as honourable as that of the animal whose bones they were picking.
>
> *in* Mooney, 1975

Interestingly, the Sioux still eat dogs and, notwithstanding Catlin's somewhat overblown depiction of 'noble savagery', it is clear that the practice is still associated with considerable ritual solemnity. According to a relatively recent study, the Oglala Sioux consider the dog to be a kind of human with an individual personality of its own. Dogs are sometimes named and, significantly, named dogs are never eaten or, to put it more accurately, dogs that are destined for the stewpot are never named. Dogs are invariably slaughtered in a ritual way, and the sacrifice is performed by a medicine man and two female assistants. Before striking the death blow, the medicine man extols the virtues of the victim, calling it 'my friend' and announcing to all those assembled how difficult it is for him to sacrifice such a worthy and faithful creature. The two female assistants then put nooses round the dog's neck and draw them abruptly tight, while the medicine man strikes the dog on the head with a blunt instrument from behind. According to the Oglala, this manner of slaughter ensures that the dog's spirit will be released to travel West as a messenger to the mythical Thunder People, to whom it will plead on behalf of humanity[3]. The dog is the most suitable animal to perform this spiritual function precisely because it *is* Man's best friend (Powers & Powers, 1986).

In pre-colonial Hawaii, and other parts of Polynesia, attitudes to dogs and dog-eating were even more intricate. Roast dog cooked in the traditional earth-oven was considered a delicacy, and was sampled and apparently relished by Captain Cook and his officers during an early visit to Tahiti. At the same time,

[2] One of the more bizarre examples of this form of verbal concealment has been recently described from central Nigeria where edible dogs are referred to euphemistically as '404 station wagons' after their supposed resemblance to the Peugeot car of that name. The dog's head is known as the 'loudspeaker' or 'gearbox', its legs as '404 wheels', its tail is the 'telephone', the intestines as 'roundabout' and its ears are 'headlamps' (Olowo Ojoade, 1990).

[3] This idea of dogs acting as messengers or intermediaries between this world and the next is an extremely widespread and ancient one, exemplified by the early Egyptian canine psychopomp, Anubis, who guided and protected the souls of the dead on their journey to the Underworld. This conception of the dog also conforms with its status as a liminal creature living on the edges of the human domain. White (1991) also notes that, in folklore and mythology, the doorway or threshold of the house (*domus*) is universally associated with the dog, 'whose relationship with humans has always located it on the boundary between wildness and domesticity'.

however, the Polynesians also developed extraordinary affections for pet dogs: 'Men, women and children, of all social ranks, fondled, pampered, and talked to their pets, named them, and grieved when death or other circumstances separated them' (Luomala, 1960, p. 203). Grief over a dog might be expressed through tears and poetical eulogies, and a favourite pet was sometimes given special burial as a further token of its owner's high esteem. This curious double standard provoked George Jesse (1866, p. 299) to remark that it was 'strange that this gentle and manly race of beings should not have had their sympathies more entwined with the creatures so much partaking of their character. That mothers should suckle an animal [Polynesian women frequently suckled puppies at the breast], and yet allow that same race to become an item of food, is a singular contradiction in feeling'.

In her lengthy and detailed analysis of Polynesian attitudes to dogs, the anthropologist Katherine Luomala (1960) made no attempt to explain this apparent contradiction, although she did mention that the owners of such dogs were frequently overcome with sorrow when the animal was taken away to be slaughtered. Also, according to one of Titcomb's informants, breast-fed puppies were never eaten: 'Close physical association would rule out the animal as a creature to kill, for it had taken on some of the "being" of the nourisher' (Titcomb, 1969, p. 10). It is also clear that Polynesian dog-eating was usually, if not invariably, performed as a part of a sacrificial ritual and this may, in some way, have helped to alleviate the inevitable anxieties associated with killing and eating social companions (see Serpell, 1986), as it evidently does among the Sioux. In addition, and despite their value both as pets and as food, dogs were not necessarily always regarded in a positive light. On the contrary, according to Luomala, 'identifying a man as a dog or as being in any way like one presented the creature most dramatically as the symbol of the pariah, the degraded outcast of human society'. The comparison damaged the ego because

> it was believed that the dog originated from a human being whose social misbehaviour was punished by his human appearance being modified into that of a dog's, his power of speech removed and replaced by a howl, and his status reduced to that of a social inferior of the group which then granted him occasional favours and let him indulge, as best he could, the traits which had led to his social rejection
>
> Luomala, 1960

In ideological terms the dog was not a gregarious, wild animal which had flatteringly attached itself to human society. It was 'a transformed human being ostracized by human society and tolerated as a hanger-on of its lowest fringes' (Luomala, 1960, p. 218). Perceiving dogs in such a disparaging light perhaps made it easier to slaughter and devour them with a clear conscience.

Dogs as outcasts

Since prehistoric times, the edible waste products of human society have provided an important source of nourishment for canine scavengers. Many countries still harbour large populations of unemployed domestic dogs, some of which are truly ownerless or feral, while others maintain loose associations with people but receive little if any care or consideration in return (see e.g. Boitani *et al.*, Chapter 15). Such dogs are usually expected to fend for themselves, scavenging off carrion and human waste, and even their habit of barking vociferously at visitors or intruders is commonly regarded as something of a mixed blessing. As is often the case with animals that eat carrion, garbage or human ordure, one might reasonably expect public attitudes towards such dogs to be uniformly negative rather than ambivalent. But, once again, when we look at the relationship in detail, a more complex picture emerges. Throughout much of southern Asia, for example, religious proscriptions specifically designate these street dogs or pariahs as *unclean* or *untouchable*. Yet in these same areas, there is also a widespread reluctance to kill surplus dogs, which in some cases amounts to a religious taboo. Although there have been few detailed studies of attitudes to dogs among such communities, the few that exist describe often quite elaborate mythological reasons for exercising tolerance.

One particularly revealing and ancient example is contained in the Hindu legend of Yudhishthira. In the final scene of the Mahabharata epic, the hero Yudhishthira approaches Heaven after a lengthy mountain pilgrimage during which his queen and his four brothers have all perished. His only surviving companion at this stage is a dog that has followed him faithfully since he set out on his journey. Suddenly, Indira, the King of Heaven, appears in a blaze of light and invites Yudhishthira to complete his journey in his heavenly chariot. Yudhishthira happily accepts and stands aside to allow the dog to enter first, whereupon Indira objects strenuously on the grounds that dogs are unclean and that the animal's presence would defile Heaven itself. Yudhishthira, however, is unmoved and says that he cannot imagine happiness, even in Heaven, while haunted by the memory of casting off such a devoted, loyal and loving companion. A heated argument then ensues until Yudhishthira finally announces that he cannot conceive of a crime that would more heinous than to leave the dog behind. At this point, all is revealed. By refusing Heaven for the sake of a dog, Yudhishthira has passed his final test. The dog is suddenly transformed into Dharma, the God of Righteousness, and Yudhishthira is carried off to Heaven amidst the acclamation of radiant multitudes (Nivedita & Coomaraswamy, 1913).

The Lisu, a mountain tribe from Thailand, provide a more contemporaneous example of how tolerant attitudes to dogs are reinforced by myth. Among the Lisu, non-working dogs are tolerated because, in their mythology, the dog is regarded as a culture hero who once saved humanity from starvation by stealing rice seeds from God's paddy fields. Also, according to Lisu folklore, humans were once inconvenienced by a copulatory tie, so they petitioned God for his help. God responded by exchanging their genitals with those of dogs [perhaps in retaliation for stealing his rice], and now only dogs are inconvenienced in this way. This is not regarded as a problem, however, because dogs have no work to do (Durrenburger, 1977).

Among the Kenyah tribe of Kalimantan (Indonesian Borneo) large numbers of dogs are tolerated around the villages and houses, and they are occasionally fed with rice, although the majority serve no useful purpose. No Kenyah will ever kill a dog, but the animals are shown no respect or affection and children are discouraged from playing with them. When asked why these dogs were tolerated, one Kenyah informant merely stated that dogs 'are like children, and eat and sleep together with men in the same house' (Hose & McDougall, 1912). The authors of this study point out that the Kenyahs believe that all animals are potentially capable of thinking like humans and of understanding human speech: 'Their objection to killing their troublesome and superfluous dogs seems to be due to a somewhat similar feeling – a recognition of intelligence and emotions not unlike their own'.

Among the Beng people, a minority ethnic group from the Ivory Coast, the corpus of myths concerning dogs is both elaborate and revealing. As with the Lisu or the Kenyah, the Beng tolerate dogs but never feed them or show them any overt affection. Indeed, when the husband of the American anthropologist, Alma Gottlieb, attempted to pet one of these animals, 'both the dog and its owner looked at him in surprise – it had clearly never occurred to anyone, canine or human, that such a thing should or even could be done'. On the other hand, the Beng name their dogs (although they invariably employ foreign rather than Beng names for this purpose), and make sincere efforts to nurse them back to health when they become sick or injured. This inconsistency in the way the Beng treat their dogs is reflected in the ambiguous role that dogs play in Beng folklore. According to one legend, for instance, the dog was responsible for the origin of death. The story recounts how when death first appeared, the people sent a dog and cat as messengers to the Sky to plead for immortality on their behalf. On the way, however, the dog found some bones and became so engrossed with these that the cat, who was too stupid to understand the message correctly, reached the Sky first and relayed the wrong instructions. As a result, everyone now dies and, once dead, cannot be revived. Interestingly, the Beng do not hold cats responsible for this unfortunate outcome because they regard them as too unintelligent to have known any better. Instead, the dog is blamed for allowing his own base gluttony to bring the worst possible calamity down on the heads of humanity.

In another myth, however, the dog redeems itself by forming an alliance with humans against the other animals. In this story, the dog discovers an egg containing the primordial human couple, which it secretly hides and protects from the other animals. When the egg hatches, the first man begins killing the other animals for food, so the latter call a meeting at which they decide to launch a surprise attack on the humans to destroy them. The dog, however, betrays these plans to the man who then ambushes the animals and scatters them all over the world (Gottlieb, 1986). In other words, the dog is portrayed as both an ally of humans against animals, and a potential traitor to both humans and animals; a Jekyll and Hyde character who simultaneously embodies both craven *animal* instincts and laudable *human* virtues.

Dogs as friends

In Britain and North America, the dog's unconditional allegiance to humanity has, justifiably, earned it the title 'Man's best friend' and secured it the admiration and affection of many millions of dog owners. Like pariahs, the vast majority of dogs in the West serve no significant economic function, yet their owners cherish and pamper them to an unprecedented degree. The Western stereotype of the dog is that of the loyal and faithful companion who shares our homes, our lives and, not infrequently, our food and furniture as an equal or near-equal member of the family. Dogs are given personal names – often the same names we give to people – we stroke them, cuddle them, play with them, groom them and ensure that they receive all the exercise, social contact and medical attention they need to keep them happy and healthy (Serpell, 1986). And, in exchange for all this care and attention, pet dogs make a not insubstantial contribution to their owner's overall health and well-being (see e.g. Hart, Chapter 12).

Despite the mutually beneficial nature of this friendship between people and dogs, it would be a mistake to imagine that the relationship is entirely harmonious. In the United States alone, an estimated 5–8 million dogs are abandoned, disowned or destroyed each year by their owners (Beck & Katcher, 1983), and humane society statistics reveal that dogs are by far the most common animal victims of human negligence and abuse (see Hubrecht, Chapter 13). By moulding dogs to fit our own curious notions of canine beauty, we condemn many of them to chronic pain or ill-health through the propagation of inherited physical disorders (Wolfensohn, 1981). In various ways we also limit their ability to lead normal lives. By altering their height, the carriage of their ears and tail, and the length of their fur we prevent them from communicating effectively with each other (see Bradshaw & Nott, Chapter 8). And although we indulge them in some respects, will still deny them the freedom to express much of their natural behaviour (Beck & Katcher, 1983). The truth is that much of the dog's normal behavioural repertoire, its gluttony, sexual promiscuity, olfactory preoccupations, toilet habits, and occasional naked hostility towards strangers and visitors, can be a source of disgust or embarassment to many owners. And even the dog's characteristic loyalty, its fawning eagerness to please, appears to be the subject of mixed feelings. On the one hand, it is one of the things that makes dogs so appealing. On the other, it can also be construed as sycophantic, servile and obsequious; the sort of behaviour we associate with toadies, lovesick fools and cringeing cowards[4].

British ambivalence towards certain aspects of canine personality was brought sharply into focus in 1989, when an eleven-year-old girl was attacked and killed on a Scottish beach by two Rottweilers. Over the ensuing months, this incident, together with four or five other less serious Rottweiler attacks, ignited a wave of hysteria that swept the country. A senior government minister was obliged to issue a public statement on the subject, as if to avert a national crisis. An eminent criminologist reported that, for the first time ever, public fears concerning dogs had surpassed every other social issue, including crime, unemployment and racial harassment. And, in a bizarre incident in Kent, a young police constable was hailed as a national hero when he strangled someone's pet Rottweiler to death after it attacked and killed two rabbits. The tide of panic was further exacerbated by a veritable flood of purple prose from the tabloid press. Rottweilers were branded 'Devil Dogs' and 'Canine Terrorists' while newspaper columnists across the country vied with each other for the most arresting headlines. One report in *The Sunday Times* (4 June 1989) began with the words: 'They used to be our best friends. Now dogs, tra-

[4] These various less appealing aspects of canine behaviour are, of course, implicit in vernacular usage where the word 'dog', and its female equivalent 'bitch', has few positive connotations. When used to describe people it is almost invariably insulting or disparaging and, when applied as a prefix to things, it denotes spuriousness, baseness or inferiority, as in doggerel, dog-Latin, doghouse, dog-end, dogfish, dog rose and so on (see Paulson, 1979).

ditionally beloved by the British, have become a fang-bearing, mouth-frothing national nightmare'. The price of Rottweiler puppies plummeted, animal rescue shelters were inundated with disowned and abandoned pets, and innocent Rottweiler-owners were shouted at and abused in the street for merely taking the dog for a walk. Before long, MPs and government ministers were calling for a blanket ban on Rottweilers and, indeed, any breed of large or potentially dangerous dog. In the words of one Conservative back-bencher, 'If people are going to keep wild beasts they must be made criminally responsible for their actions'.

As it happens, drastic government action was avoided on this occasion, and the furore gradually died down. However, it re-emerged with a vengeance in the spring and summer of 1991 following two savage, and apparently unprovoked, attacks on people by pit bull terriers. This time the public and media reaction was so intense that the Government was forced to act. After the briefest possible consultation with appropriate authorities (many of whose recommendations were ignored) new legislation was swiftly drafted, and the Dangerous Dogs Act 1991 came into being. The Act imposed a ban on the breeding, sale or exchange of all 'dogs bred for fighting' which, in practice, encompassed pit bull terriers and one Japanese tosa[5] that had recently been imported as a puppy. Owners of these dogs were given the choice of having them destroyed (at taxpayer's expense) or of having them neutered, tatooed and registered with the authorities, as well as keeping them securely muzzled and leashed in public. Police were authorised to use 'reasonable force' to enter premises suspected of harbouring unneutered and unregistered animals, and Magistrates were actually *required* to order the destruction of any dog whose owner failed to comply with the regulations, regardless of whether or not it had ever shown signs of aggressiveness.

Without in any sense minimizing the potential threat posed by large and powerful dogs, such as Rottweilers or pit bull terriers, or the frightful injuries inflicted on the individual victims of their attacks, it is clear that the national spasms of horror and outrage generated by these incidents were grossly out of proportion to the actual risks. Indeed, in a nation of some 50 million people and 7.4 million dogs – perhaps half of which are large and at least potentially dangerous – it is in some respects astonishing that so *few* people are seriously injured or killed[6]. So how, then, do we account for such extreme reactions?

For whatever reason, our culture has transformed the dog into a paragon of canine virtue. He is our loyal and faithful servant and companion; a sort of amiable culture-hero whose friendship is proverbially better than that of our fellow humans. That such a devoted and trusted admirer should suddenly turn and savage one of us is both frightening and disturbing. For not only is this behaviour disloyal – a betrayal of trust – it is also grossly insubordinate. Dogs, furthermore, fulfil a childlike role in our society and, as perpetual children, we expect them to be forever innocent, playful and fun-loving (Beck & Katcher, 1983). A murderous dog, like a murderous child[7], is therefore nothing less than an abomination – a disturbance in the natural order – an unacceptable threat to the perceived security and stability of the entire community.

Discussion and Conclusions

Some of our ambivalence towards the domestic dog arises from simple conflicts of interest (see Serpell, 1985). As our voluntary companion and ally, the dog, like a faithful human employee, presses moral claims upon us that are more strident and less easily ignored than those emanating from most other domestic species. This poses relatively few ethical problems when we ask dogs to perform enjoyable tasks, such as hunting, herding livestock or providing us with company. Dogs seem to like doing these things anyway so their interests and ours, broadly-speaking,

[5] Somewhat mysteriously, two other breeds, the Fila Brasiliero and Dogo Argentino, were also banned, although neither existed in Britain at the time and neither was bred specifically for fighting. Rottweilers, significantly, were not included in the ban, probably because they were too widely employed for security purposes.

[6] To put the dangers of fatal dog attacks in perspective, people in Britain are 60 times more likely to die from accidental drowning, and about 20 times more likely to die from electrocution or from being struck by a falling object (Office of Population Censuses and Surveys, 1992; Podberscek, 1994).

[7] It may be significant in this context that a very similar wave of public and media hysteria followed the abduction and murder of a baby in Liverpool by two 10-year-old boys in 1993.

coincide. The proverbial friendliness and fidelity of dogs may, however, create a burdensome sense of guilt when we use these animals in ways which appear to betray their loyalty and affection. Where people are obliged to kill dogs either for food or to control their numbers, for example, or where they use them as beasts of burden or as experimental subjects in biomedical research, it is common to find that they either avoid affectionate contact with these animals, or employ complex psychological defence mechanisms to protect their consciences from conflict (Serpell, 1985, 1986; Arluke, 1988, 1994). Even when we make allowances for these relatively straightforward ethical dilemmas, however, it is clear that people's attitudes towards dogs remain strangely contradictory.

Constance Perin (1981) has argued that Western ambivalence towards the dog arises from this animal's peculiar symbolic role as an archetypal attachment figure; an idealized provider of love who reawakens memories of the love we once received as infants from our mothers. According to her theory, the tension that exists in our relationships with dogs denotes the re-emergence of all those unresolved love–hate tensions of infancy associated with the process of separation and individuation from our parents. Although in some ways compelling, one problem with this idea is that it is too culture-specific. Ambivalence towards dogs appears to be almost universal, but it is difficult to see how Perin's theory could be applied to societies, such as the Beng or the Kenyah, for example, who display this ambivalence in the absence of any obvious affection.

Other authors (e.g. Beck & Katcher, 1983; Burt, 1988) have resorted to a more psychoanalytic interpretation, according to which animals, such as dogs, often represent unconscious aspects of ourselves (Jung, 1959). From this perspective, our love–hate, tolerant–intolerant attitudes to dogs are simply a reflection of our own ambivalence about the *animal* within us – the Freudian 'id' – the source of all those instinctual, impulsive, forbidden feelings and desires that we tend to keep so firmly repressed. This would certainly help to explain why we are often uncomfortable with the more unbridled aspects of our dogs' behaviour, although it is possible to think of other reasons for diffidence in this respect. Dogs, after all, flatter us with their attention and devotion; we see ourselves magnified in their eyes. But the satisfaction we derive from this hero-worship is inevitably tempered by the behaviour of the worshipper. To be loved by a paragon is one thing, to be adored by a creature that eats shit, sniffs genitals and bites people is quite another.

In some respects, however, all of these theories really miss the point. As the Yurok Indians recognized, the real danger posed by the domestic dog is that its friendship threatens to dissolve or undermine the psychological barrier that distinguishes human from animal (Elmendorf & Kroeber, 1960). By adopting us and treating us as conspecifics, even in the absence of any positive encouragement to do so, the dog unwittingly represents the thin end of the wedge; a demanding and insistent reminder of the feelings, interest and moral claims not only of dogs, but of animals in general (Serpell & Paul, 1994).

In symbolic terms, the domestic dog exists precariously in the no-man's-land between the human and non-human worlds. It is an interstitial creature, neither person nor beast, forever oscillating uncomfortably between the roles of high-status animal and low-status person. As a consequence, the dog is rarely accepted and appreciated purely for what it is: a uniquely varied, carnivorous mammal adapted to a huge range of mutualistic associations with people. Instead, it has become a creature of metaphor, simultaneously embodying or representing a strange mixture of admirable and despicable traits. As a beast that voluntarily allies itself to humans, the dog often seems to lose its right to be regarded as a true animal. In many societies, it now occupies the role of a stateless refugee, tolerated and occasionally pitied as a hanger-on, but never properly assimilated or accepted. In others, it is viewed as the victim of its own depravity, a person transmogrified into a dog for engaging in immoral behaviour, a creature unworthy of humane considerations and a suitable candidate for abuse, displaced anger or the cooking pot. Elsewhere, the dog's ambiguous or intermediate status has endowed it with supernatural powers, and the ability to travel as a spiritual messenger or psychopomp between this world and the next. Such beliefs may help to engender a certain respect or even reverence for dogs, but they can also provide a convenient justification for hastening the animal on its journey and then devouring its mortal remains. In our own culture, the dog has been granted temporary personhood in return for its unfailing companionship. But, as we have seen, this privilege is swiftly withdrawn whenever the dog reveals too much of its animal nature. In other words, we love dogs and invest them with quasi-human status, but only so long as they refrain from behaving like beasts.

Whether paragon or pariah, dogs confront people with essentially the same dilemma; the problem of deciding how far, if at all, our moral responsibilities should extend beyond the taxonomic boundaries of our species. Seen in this light, our ambivalence towards the dog is ultimately an expression of the profound uncertainty we feel concerning our assumed 'right' to live at the expense of other sentient creatures.

Acknowledgements

My thanks go to Harriet Ritvo and Randall Lockwood for their valuable comments on an earlier draft of this manuscript.

References

Arluke, A. (1988). Sacrificial symbolism in animal experimentation: object or pet? *Anthrozoös*, **2**, 98–117.

Arluke, A. (1994). Managing emotions in an animal shelter. In *Animals and Human Society: Changing Perspectives*, ed. A. Manning & J. A. Serpell, pp. 145–65. London: Routledge.

Arluke, A. & Sax, B. (1992). Understanding Nazi animal protection and the Holocaust, *Anthrozoös*, **5**, 6–31.

Beck, A. M. & Katcher, A. H. (1983). *Between Pets and People*, New York: G. P. Putnam.

Bökönyi, S. (1974). *History of Domestic Mammals in Central and Eastern Europe*. Budapest: Académiai Kaidó.

Bueler, L. 1974. *Wild Dogs of The World*. London: Constable.

Burkardt, V. R. (1960). *Chinese Creeds and Customs*, vol. 3. Hong Kong: South China Morning Post Ltd.

Burt, M. R. (1988). The animal as Alter Ego; cruelty, altruism, and the work of art. In *Animals and People Sharing the World*, ed. A. N. Rowan, pp. 117–35. Hanover, NH: University Press of New England.

Cipriani, L. (1960). *The Andaman Islanders*. London: Weidenfeld & Nicholson.

Cohen, E. (1994). Animals in medieval perceptions. In *Animals and Human Society; Changing Perspectives*, ed. A. Manning, & J. A. Serpell, pp. 59–80. London: Routledge.

Dawkins, R. (1993). Gaps in the mind. In *The Great Ape Project*, ed. P. Cavalieri & P. Singer, pp. 80–7. London: Fourth Estate.

Diamond, J. M. (1993). The third chimpanzee. In *The Great Ape Project*, ed. P. Cavalieri & P. Singer, pp. 88–101. London: Fourth Estate.

Douglas, M. (1966). *Purity and Danger: an Analysis of the Concepts of Pollution and Taboo*. New York: Routledge & Kegan Paul.

Downs, J. F. (1960). Domestication: an examination of the changing social relationships between man and animals. *Kroeber Anthropological Society Papers*, **22**, 18–67.

Driver, H. E. & Massey, W. C. (1957). Comparative studies of North American Indians. *Transactions of the American Philosophical Society*, **42**, 165–456.

Dumont, J. P. (1976). *Under the Rainbow: Nature and Supernature among the Panaré Indians*. Austin: University of Texas Press.

Durrenburger, E. P. (1977). Of Lisu dogs and Lisu spirits. *Folklore*, **88**, 61–3.

Elmendorf, W. W. & Kroeber, A. L. (1960). The structure of Twana culture with comparative notes on the structure of Yurok culture. *Washington University Research Studies*, **28**(2). Pullman: Washington State University.

Evans, E. P. (1906). *The Criminal Prosecution and Capital Punishment of Animals*. London: Heinemann.

Fox, M. W. (1967). Effects of early experience on the development of inter and intra-specific social relationships in the dog. *Animal Behaviour*, **15**, 377–86.

Frank, B. (1965). *Die Rolle des Hunde im Africanischen Kulturen*. Wiesbaden; Franz Steiner Verlag.

Gottlieb, A. (1968). Dog: ally or traitor? Mythology, cosmology, and society among the Beng of Ivory Coast. *American Ethnologist*, **13**, 447–88.

Harrison, T. (1965). Three 'secret' communication systems between Borneo nomads (and their dogs). *Journal of the Malay Branch of the Royal Asiatic Society*, **38**(2), 67–86.

Hose, C. & McDougall, M. B. (1912). *The Pagan Tribes of Borneo*. London: Macmillan.

Huntingford, G. W. B. 1955. The economic life of the Dorobo. *Anthropos*, **50**, 602–34.

Ishige, N. (1977). Roasting dog (or a substitute) in an earth oven: an unusual method of preparation from Ponape. In *The Anthropologist's Cookbook*, ed. J. Kuper, pp. 204–5. New York: Universe Books.

Jesse, G. R. (1866). *Researches into the History of the British Dog*, vol. I. London: Robert Hardwicke.

Jordan, J. W. (1975). An ambivalent relationship: dog and human in the folk culture of the rural south. *Appalachian Journal*, Spring, 238–48.

Jung, C. (1959). *The Archetypes and the Collective Unconscious*. New York: Pantheon.

Lee, R. B. (1979). *The !Kung San: Men, Women and Work in a Foraging Society*. Cambridge: Cambridge University Press.

Lumholtz, C. (1989). *Among Cannibals*. London: John Murray.

Luomala, K. (1960). The native dog in the Polynesian system of values. In *Culture in History*, ed. S. Diamond, pp. 190–240. New York: Columbia University Press.

Midgley, M. (1983). *Animals and Why They Matter.* Harmondsworth, Middx.: Penguin Books.

Mooney, M. M. (1975). *George Catlin Letters and Notes on the North American Indians*. New York: Clarkson N. Potter.

Nivedita, The Sister & Coomaraswamy, A. K. (1913). *Myths of the Hindus and Buddhists*. London: Harrap & Co.

Olowo Ojoade, J. (1990). Nigerian cultural attitudes to the dog. In *Signifying Animals: Human Meaning in the Natural World*, ed. R. G. Willis, pp. 215–21. London: Unwin Hyman.

Osgood, C. (1975). *The Koreans and their Culture*. New York: Ronald Press.

Osgood, C. (1975). *The Chinese: a Study of a Hong Kong Community*. Tucson, University of Arizona Press.

Paulson, R. (1979). *Popular and Polite Art in the Age of Hogarth and Fielding*. Notre Dame, IN: University of Notre Dame Press.

Perin, C. (1981). Dogs as symbols in human development. In *Interrelations between People and Pets*, ed. B. Fogle, pp. 68–88. Springfield, IL: Charles C. Thomas.

Podberscek, A. (1994). Dog on a tightrope: the position of the dog in British society as influenced by press reports on dog attacks (1988–1992). *Anthrozoös*, **7**, 232–41.

Powers, W. K. & Powers, M. N. (1986). Putting on the dog. *Natural History*, **2**, 6–16.

Ritvo, H. (1986). Pride and pedigree: the evolution of the Victorian dog fancy. *Victorian Studies*, Winter, 227–53.

Scott, J. P. (1963). The process of primary socialization in canine and human infants. *Monographs of the Society for Research on Child Development*, **28**, 1–49.

Scott, J. P. (1968). Evolution and domestication of the dog. *Evolutionary Biology*, **2**, 243–75.

Seligmann, C. G. & Seligmann, B. A. (1911). *The Veddas*. Cambridge: Cambridge University Press.

Serpell, J. A. (1985). Best friend or worst enemy: cross-cultural variation in attitudes to the dog. In *The Human–Pet Relationship*. Vienna: IEMT – Institute for Interdisciplinary Research on the Human–Pet Relationship.

Serpell, J. A. (1986). *In the Company of Animals*. Oxford: Basil Blackwell.

Serpell, J. A. & Paul, E. S. (1994). Pets and the development of positive attitudes to animals. In *Animals and Human Society: Changing Perspectives*, ed. A. Manning & J. A. Serpell, pp. 127–44. London: Routledge.

Singer, M. (1978). Pygmies and their dogs: a note on culturally constituted defence mechanisms. *Ethos*, **6**, 270–7.

Thomas, K. (1983). *Man and the Natural World: Changing Attitudes in England 1500–1800*. London: Allen Lane.

Titcomb, M. (1969). *Dog and Man in the Ancient Pacific with Special Attention to Hawaii*. Honolulu, HI: Bernice P. Bishop Museum Special Publications 59.

White, D. G. (1991). *Myths of the Dog-Man*. Chicago: University of Chicago Press.

Wolfensohn, S. (1981). The things we do to dogs. *New Scientist*, 14 May 1981, pp. 404–7.

17 The hair of the dog

JAMES SERPELL

Domestic dogs are unusual or exceptional in so many different respects that it is difficult to conclude a book about their natural history, biology and behaviour without resorting to superlatives. To begin with, as Clutton-Brock (Chapter 2) points out, the dog – or rather its ancestor, the wolf – was the first species to be domesticated. We know that the domestication of animals and plants ushered in a radical and unprecedented shift in human subsistence patterns from hunting and gathering to farming and pastoralism. The taming of the wolf, at least 12 000 years ago, therefore heralded this revolutionary change in human affairs.

At present, we do not know whether there was a single, primary centre of domestication from where all dogs ultimately originated, or multiple centres distributed throughout Eurasia. Presumably, future archaeological discoveries, and more detailed analyses of existing subfossil material, will help to answer such questions. Similarly, we still have no clear idea why Palaeolithic hunter–gatherers chose to keep tame wolves. Whatever their point of origin, the original founder population of semi-domestic wolves expanded with great rapidity, acquiring a more-or-less worldwide distribution by about 8000 years BP. This would strongly suggest that these animals were valued highly, and were actively traded by neighbouring human groups. Precisely what they were valued for, however, remains a mystery, although the burial at Ein Mallaha (Davis & Valla, 1978) at least provides circumstantial evidence of an affectionate, rather than purely exploitative, relationship (see Serpell, 1989). Unfortunately, in the absence of supporting archaeological clues, all modern accounts of the process of wolf domestication are little better than just-so-stories. Hopefully, new discoveries will help to fill these early gaps in our knowledge.

Contrary to popular belief, our forebears did not consciously 'select' wolves for particular working roles, any more than they consciously selected them for domestication. This sort of goal-oriented selective breeding is a relatively modern invention (Ritvo, 1987; Clutton-Brock, 1994). On the contrary, as Coppinger & Schneider's (Chapter 3) thought-provoking discussion reveals, the only characteristic that was probably selected for in the early stages was temperament. Wolves can be tamed and even trained to a limited extent (see Crisler, 1958; Fentress, 1967; Woolpy & Ginsburg, 1967), but no wolf is ever as tameable or trainable as a domestic dog. Indeed, people who have attempted to live with tame, adult wolves as pets have generally found them to be over-reactive, difficult to control and at least potentially somewhat dangerous (Crisler, 1958; Fentress, 1967). A sustainable wolf–human partnership therefore required an animal that was more docile and manageable than the wild type, and this could have been achieved 'unconsciously' merely by killing or driving away temperamentally unsuitable individuals. No one, however, could have foreseen the bizarre ontogenetic ramifications of this simple process of selection for tameness. When this same process was recently applied systematically to captive foxes on a Siberian fur farm, weird things started happening. As expected, the selected lines became tamer, but they also began to look and behave increasingly like dogs, even to the extent of developing piebald coats, drooping ears and dioestrus reproductive cycles (Belaev & Trut, 1975). At least some of these effects appeared to have been due to neoteny, the retention of puppy-like characteristics into adulthood. Others, however, were more mysterious in origin and seemed to involve a partial destabilization of the entire genome. These processes are currently very poorly understood and would certainly merit further study.

According to Coppinger & Schneider, the outcome of these forces was the rapid evolution of an animal that looked and behaved very differently from a wolf. Furthermore, they argue that this creature was the true ancestor of the modern domestic dog and that we are, so to speak, barking up the wrong tree, when we look for the origins of our dogs' behaviour in the ethology of (adult) wolves. As well as being tamer, more variable and more adaptable, these 'aboriginal' dogs retained the behavioural flexibility of juveniles. They therefore provided the ideal raw material for developing new strains or breeds displaying novel and unusual patterns of behaviour. As various authors in this volume have stressed, the dog is more variable than any other single species. Indeed, variation in size and shape within the species, *Canis familiaris*, is greater than that which exists within the entire family Canidae. Yet, apart from the recent aesthetic preoccupations of dog breeders, most of this morphological variation is the indirect consequence of selection for diverse and specialized behavioural abilities.

The majority of domestic animals are products of

selection for physical or physiological traits associated with the production of food or raw materials, such as meat, eggs or fur. The dog, however, is an exception. Although it has been employed, from time to time, as an item of food, or as a source of hides or wool for clothing, in general these uses for the dog were relatively limited and localized, and most of the breeds developed for such purposes are now extinct. In contrast, virtually all of the extant breeds – or at least those originating earlier than the nineteenth century – are associated with the performance of particular tasks. At various times, and in different places, domestic dogs have served an incredible variety of different behavioural roles, including security guards, burglar alarms, beasts of burden, weapons of war, entertainers, athletes, fighters, lifeguards, shepherds, guides, garbage collectors, and instruments for detecting truffles, drugs, dry rot, explosives and oestrous pheromones in cattle. As hunting aides, they have been modified behaviourally (and physically) to assist in the capture or retrieval of everything from rodents and small birds to lions and kangaroos, and of course, since time immemorial, they have also employed their social skills to provide us with affectionate and reliable companionship. In terms of the latter ability, all dog breeds are fairly similar, although some make better companions than others. In all other respects, however, they are as dissimilar as Darwin's famous Galapagos finches, each adapted to its own narrow and specialized behavioural and ecological niche. Perhaps, like Darwin's finches, dogs can also provide us with further insights into the mechanisms of evolution, and the processes of allopatric and sympatric speciation.

The topic of breed differences in behaviour has been explored extensively in the two chapters by Willis and B. L. Hart, respectively. Willis (Chapter 4) reviews the evidence for the genetic inheritance of specialized working traits in various breeds and concludes that most behavioural traits, apart from temperament, demonstrate negligible or variable heritabilities. Several explanations for this anomaly suggest themselves. Some breeds or strains within breeds may now, as a result of intense selection and inbreeding in the past, be too genetically uniform to allow for further improvement. Alternatively, behavioural test comparisons between progeny, or between parents and progeny, may be invalidated by differences in test procedures or environmental conditions at the time of testing. In addition, of course, test performance may be affected unpredictably by the individual developmental experiences of the animals being tested, and it also seems likely that some of the traits being measured are too subjective (or anthropomorphic) to have any clear biological basis. Until we have a better scientific understanding of the underlying causation of particular dog behaviour patterns, we are unlikely to be able to measure or modify these traits effectively.

There can be little doubt that the recent overwhelming emphasis on aesthetically 'ideal' physical conformation in most purebred dogs, has had, and will continue to have, a damaging influence on behaviour and temperament. To overcome these problems, temperament and performance evaluation will need to become a prerequisite for all dogs entering the competitive show ring. Tests that can reliably predict adult behaviour early in life would also be useful, although previous puppy tests of this type have proved to be largely unsuccessful at predicting most adult behaviour patterns.

Before we can begin to iron out or eliminate some of the behavioural problems that exist in certain breeds, we need to be able to devise methods to quantify breed differences in terms of the various behavioural elements of concern to most dog owners, such as aggression, excitability, nervousness, and so on. Several potential methods are described and discussed by Hart (Chapter 5): in particular, one based on a system that allows experienced veterinarians and dog handlers to rank different breeds in terms of a range of behavioural propensities. The results of this analysis are not only useful as a guide to prospective dog owners, but may also offer a pointer to breeders concerning areas in which certain breeds appear to be in need of genetic, as well as behavioural, modification. In the future, it would also be interesting to develop survey techniques that could tap into the knowledge and experience of both the breeders and owners of particular dog breeds.

Regardless of its genetic constitution, every dog possesses the ability to adapt and change its behaviour according to circumstances, and most authorities would agree that dogs, like people, are more impressionable and adaptable during the early stages of life. You *can* teach an old dogs new tricks, but it is almost always easier to teach a young one. The two chapters by Serpell & Jagoe (Chapter 6) and Thorne

(Chapter 7) both focus on the formative effects of early experience: the former in relation to the development of so-called behaviour problems, and the latter in relation to dietary preferences. Both illustrate how little we still know about the factors influencing the early development of different behaviour patterns and preferences in dogs. The fine details of behavioural development have been studied in only a handful of domestic breeds (see e.g. Scott & Fuller, 1965), and there is a marked tendency in the literature to concentrate exclusively on the traditional 'socialization period' from 3–12 weeks, without even considering the possible impact of biologically relevant stimuli and experiences before and after this supposedly 'sensitive' phase. It is also widely assumed that behavioural development proceeds along similar lines in all the different domestic breeds, despite the fact that marked breed differences in development have already been demonstrated. Developmental studies represent a promising area for future research, particularly in the light of our improved understanding of behavioural genetics.

Differences between dog breeds, and between dogs and wolves, are re-emphasized in Bradshaw & Nott's (Chapter 8) account of social behaviour and communication in *C. familiaris*. Many of the morphological changes produced in domestic dogs, such as shaggy coats, lop-ears, squashed faces and curly tails, have reduced these animals' abilities to communicate along conventional wolfish lines. As a result, they argue, communication in the domestic dog places greater emphasis on sounds and smells that do not rely on visual perception. As every dog owner knows, barking and scent marking feature prominently in the signalling repertoires of many dogs, yet the precise functions of these and other common behaviour patterns are still surprisingly poorly understood. Further study of basic canine ethology would be valuable in this respect. Like several other authors in this volume, Bradshaw & Nott also discuss the relevance of the traditional concept of dominance hierarchies, particularly to the more 'refined' breeds of domestic dog. Compared with wolves, many of these breeds exhibit few, if any, overt expressions of dominance-related behaviour when kept in stable social groups and, where hierarchies appear to exist, they are often highly unstable. In view of the casual use of the concept of dominance in the literature on dog behaviour and behaviour problems, this issue is an important one that needs to be resolved by further research.

While the motivation to compete aggressively for social status may have been reduced or even eliminated in some breeds, it may also have been enhanced in others along with other patterns of aggressive behaviour. As Lockwood (Chapter 9) rightly points out, we have made dogs, at least to some extent, in our own image, and all too often that image has been a violent one. That some of this aggressiveness has a genetic basis is confirmed by the fact that certain breeds are over-represented in cases of dog attacks on people. Indeed, fighting and guarding breeds have been actively selected for their aggressive tendencies, although in relation to different contexts and stimuli. At the same time, it is naive to imagine that the problem of dog aggression can be solved by eliminating particular breeds. Regardless of breed, a dog's motivation to attack or defend itself aggressively is influenced by a host of internal, external and experiential factors that are unique to the individual and the situation. It is therefore impossible to generalize reliably. More research is needed to help us understand the predisposing factors leading to serious attacks, but ultimately the only way to eliminate such incidents is to penalize the owners, not their dogs.

The rational and compassionate approach to the diagnosis and treatment of dog behaviour problems is exemplified in the chapters by Mugford and O'Farrell. Surprising as it may seem, scientific principles derived from the careful study of animal and human psychology were not applied to canine behaviour problems until the late 1970s. Prior to this time, according to Mugford (Chapter 10), the problem dog, like the problem adolescent, was viewed as the product of a lack of authoritarian discipline and control on the part of its owner. Eccentric trainers, such as the notorious Barbara Woodhouse, announced on public television that there were 'no bad dogs, only bad owners!', and owners were made to feel inadequate and responsible for problems as diverse as dominance-related aggressiveness and separation-related barking. To some extent, as O'Farrell (Chapter 11) suggests, such views were endorsed by the dog-breeding community, since they helped to deflect attention from the behavioural problems inherent in some breeds. Whatever their origin, however, they resulted in domineering and often inhumane training methods, which often made problems worse rather than better. In such cases, failure of the method was invariably blamed on the owner's failure to apply the method with sufficient force. Fortunately, these archaic techniques are being gradually replaced by more enlightened and eclectic approaches. Although 'blaming' owners for their dogs' behaviour problems is generally considered to be counter-productive, it would nevertheless be helpful to know precisely how, if at all, owners' attitudes, personalities, behaviour or experience can affect their pets' behaviour.

Yet another way in which the domestic dog is unique or unusual is in the intensity of its relationships with people. No other species, domestic or wild, has ever become so inextricably involved in our lives and our affections as dogs have. Consequently, as L. A. Hart (Chapter 12) explains, no other species has played such a significant role in promoting our physical and emotional well-being. The study of human–pet relationships is another new and expanding field of scientific inquiry, although, as yet, many fundamental questions still remain unanswered. It is apparent that the companionship of animals such as dogs brings with it a number of psychosocial and medical benefits, but we are still a long way from understanding the mechanisms underlying such effects. The fact that childless adults interact more with, and become more attached to, their dogs than other people lends some support to the popular stereotype of the pet as a child surrogate, but other studies reveal that dogs mean very different things to different people and that, once again, it is a mistake to generalize.

Unfortunately, the domestic dog's extraordinary contribution to human welfare is not invariably reciprocated. In what is euphemistically referred to as the 'pet overpopulation problem', at least five million dogs are discarded and euthanized annually in the United States alone (Arkow, 1994). As Hubrecht (Chapter 13) points out, dogs are also the most common animal victims of human abuse and cruelty. They are sometimes chronically deformed by our taste for strange or comical physical features, they are regularly subjected to painful and pointless cosmetic procedures in order to fit our capricious, aesthetic preferences, and they are still one of the most widely used species in biomedical research. Yet, although the science of animal welfare has significantly improved our understanding of the biological and behavioural needs of farm animals, or captive animals in zoos or circuses, we remain surprisingly ignorant of the basic welfare needs of the domestic dog. Perhaps because dogs are so amenable and obliging by nature, we seem to take it for granted that they are happy and contented regardless of how we keep them. The evidence reviewed here by Hubrecht, however, suggests a different conclusion, and points to a need for additional research on aspects of canine welfare.

Pampered and cossetted household pets represent one extreme of the human–dog relationship continuum. At the other end of the spectrum lie the ownerless strays and feral dogs that eke out a precarious existence on the fringes of human society. The two chapters by Macdonald & Carr (Chapter 14) and Boitani *et al.* (Chapter 15) both provide us with fascinating insights into the natural history, ecology and behaviour of communities of feral and free-ranging dogs living in the mountains of central Italy. Despite thousands of years of domestication, these dogs are able to revert to an independent, scavenging existence and adjust their social organizations to suit the carrying capacity of local food supplies. Contrary to the locally accepted view, however, they are almost completely dependent on human garbage, and rarely represent a significant threat to the wild animals or livestock with which they share their habitat. Somewhat ironically, the greatest danger they pose is to their own wild relative, the wolf, whose genetic integrity and access to vital resources is seriously threatened by their existence. These studies also reveal that feral dogs in Italy are unable to maintain their numbers without periodic recruitment from the pet dog population. In other words, if people took more responsible care of their pets, stray and feral dogs would probably not exist. Such findings argue for an urgent need to develop effective educational techniques to alter public attitudes to both wolves and domestic dogs.

The existence of feral dogs, and the widespread cruelty, intolerance and indifference that dogs experience throughout the world, serves to illustrate the curious ambivalence of human attitudes towards this affectionate and supremely useful animal. The origin of this ambivalence provides the theme of the penultimate chapter (Chapter 16). By voluntarily allying itself with people some 12 000 years ago, the wolf occupied that uncertain territory which lies between the human and non-human conceptual domains.

Ever since, the domestic dog's position seems to have been more or less unstable, oscillating unpredictably between the roles of privileged and personified animal, and despised and degraded human. In a sense, our attitudes towards dogs reflect our uncertainty about our own status and moral responsibilities with respect to the rest of the animal kingdom. This uncertainty and ambivalence not only represents a barrier to knowledge and objective understanding of the dog and its role in our society, it also serves to perpetuate the often inhumane treatment of this unique and remarkable species. If this volume succeeds in undermining this barrier, even to only a small extent, it will have achieved one of its main objectives.

References

Arkow, P. (1994). A new look at pet 'over-population'. *Anthrozoös*, 7, 202–5.

Belaev, D. K. & Trut, L. N. (1975). Some genetic and endocrine effects of selection for domestication in silver foxes. In *The Wild Canids*, ed. M. W. Fox, pp. 416–26. New York: Van Nostrand Reinhold.

Clutton-Brock, J. (1994). The unnatural world: behavioural aspects of humans and animals in the process of domestication. In *Animals and Human Society: Changing Perspectives*, ed. A. Manning & J. A. Serpell, pp. 23–35. London: Routledge.

Crisler, L. (1958). *Arctic Wild*. New York: Harper.

Davis, S. J. M. & Valla, F. R. (1978). Evidence for the domestication of the dog 12 000 years ago in the Natufian of Israel. *Nature*, **276**, 608–10.

Fentress, J. C. (167). Observations on the behavioral development of a hand-reared male timber wolf. *American Zoologist*, **7**, 339–51.

Ritvo, H. (1987). *The Animal Estate*. Cambridge, MA: Harvard University Press.

Scott, J. P. & Fuller, J. L. (1965). *Genetics and the Social Behaviour of the Dog*. Chicago: University of Chicago Press.

Serpell, J. A. (1989). Pet-keeping and animal domestication: a reappraisal. In *The Walking Larder: Patterns of Domestication, Pastoralism, and Predation*, ed. J. Clutton-Brock, pp. 10–21. London: Unwin Hyman.

Woolpy, J. H. & Ginsburg, B. E. (1967). Wolf socialization: a study of temperament in a wild social species. *American Zoologist*, **7**, 357–63.

Index